The
REINDEER PEOPLE

Living with Animals and
Spirits in Siberia

Piers Vitebsky

A Mariner Book

HOUGHTON MIFFLIN COMPANY

BOSTON • NEW YORK

First Mariner Books edition 2006

Copyright © 2005 by Piers Vitebsky

Maps copyright © 2005 by Hardlines

ALL RIGHTS RESERVED

First published in Great Britain by HarperCollins Publishers, 2005

For information about permission to reproduce selections
from this book, write to Permissions, Houghton Mifflin Company,
215 Park Avenue South, New York, New York 10003.

Visit our Web site: www.houghtonmifflinbooks.com.

Library of Congress Cataloging-in-Publication Data
Vitebsky, Piers, date.
The reindeer people : living with animals and spirits in Siberia
/ Piers Vitebsky.
p. cm.
Includes bibliographical references and index.
ISBN-13: 978-0-618-21188-3
ISBN-10: 0-618-21188-8
1. Evenki (Asian people)—Domestic animals. 2. Evenki
(Asian people)—Religion. 3. Evenki (Asian people)—Social life and
customs. 4. Reindeer herding—Russia (Federation)—Verkhoiansk Range.
5. Reindeer herders—Russia (Federation)—Verkhoiansk Range. 6. Sha-
manism—Russia (Federation)—Verkhoiansk Range. 7. Verkhoiansk
Range (Russia)—History. 8. Verkhoiansk Range (Russia)—Antiquities.
9. Verkhoiansk Range (Russia)—Social life and customs. I. Title.
DK759.E83V58 2005
305.89'41—dc22   2005045994

ISBN-13: 978-0-618-77357-2 (pbk.)
ISBN-10: 0-618-77357-6 (pbk.)

Printed in the United States of America

MP 10 9 8 7 6 5 4 3 2 1

The Publisher has used its best efforts to obtain permission from the
rightsholders of all copyrighted material.

CONTENTS

PREFACE AND ACKNOWLEDGEMENTS

This book is the outcome of nearly twenty years of visits to Sebyan, a community of a few hundred Eveny people in northeastern Siberia who live in an intense partnership with a remarkable species of animal.

No-one reaches a remote place like this, let alone stays there, without a lot of help. From a base at the Scott Polar Research Institute in the University of Cambridge I have been supported on my way by many officials and scholars in Russia, especially those who took a risk in the suspicious, complicated 1980s. In Yakutsk I am grateful to Anatoly Gogolev, Vasily Ivanov, Platon Sleptsov (Plato) and Aleksandr Tomtosov, and in St Petersburg (then called Leningrad) to the late Rudolf Its and researchers at the Kunstkamera. For funding from the Russian side I thank Leningrad State University and the Academy of Sciences of the USSR (as they were called in the 1980s), and on the UK side the British Council, the British Academy, and the Economic and Social Research Council.

My wife Sally and children Patrick and Catherine have lived with my Siberian involvement for much or all of their lives. I could not have run a Siberian studies programme or accepted so much local hospitality without their support and their willingness to welcome Siberian visitors to our house in turn. Sally allowed me to travel in the early days by looking after our young children at home, and it felt like the fulfilment of a long promise when she was finally able to bring them and spend two summers nomadising with Granny and others. Sally also brought a fresh and wise eye to my view of family dynamics in the reindeer herders' tents.

In trying to understand what I have seen, I have benefited from discussions with anthropologists and reindeer specialists in several countries, and some of their insights are acknowledged in the endnotes. I have also learned much from working with my graduate students, who since 1990 have carried out long-term fieldwork of their own in remote areas across the entire Russian North and Far East. Several of them (Seona Anderson, Tanya Argounova, Joachim Otto Habeck, Vera

Skvirskaya and John Tichotsky) have also shared my life in nomadic reindeer herders' camps, and in addition Tanya's mother has provided an ever-welcoming home in her apartment in the city of Yakutsk.

The Eveny people are a branch of the Tungus family, whose languages gave the word *shaman* to the world. The Soviet regime destroyed their shamans, but on this rugged and dangerous landscape, relations with spirits and images in dreams continue to be central to the sense of each person's destiny. Like the emotions which the Eveny feel so deeply and reveal so delicately, these matters are challenging to interpret and I have been helped in seminars by members of the Magic Circle, a ritual studies group in Cambridge, and by a special session of the Psychoanalysis-Anthropology Colloquium of the American Psychoanalytic Association in New York.

Jean Briggs, Julie Cruikshank and Caroline Humphrey read a draft manuscript and gave me the vital endorsement from professional colleagues to continue to the end. I have also received useful comments from Alex King, Robert Paine and Hiroki Takakura, while Nick Tyler patiently answered my questions about reindeer biology. Anne Dunbar checked a draft for continuity. I am grateful to my editors and agents Harry Foster, Richard Johnson, Peter Robinson, and particularly Kathleen Anderson who helped me turn my first ideas into a realistic book plan. I alone am to blame for any shortcomings which remain.

My deep indebtedness to the Eveny themselves, who befriended me and kept me alive in all seasons, will leap out from every page. They are named in the book, and their characters and responses to my presence are often revealed. But in particular I could not have had these wonderful experiences without the adventurous spirit and logistic imagination of Tolya (Anatoly Alekseyev), backed up by his patient wife Varya. Vladimir Nikolayevich gave me an extraordinary immersion, which will never leave me, in the immensity and silence of his landscape: in the last few days of Vladimir Nikolayevich's life Tolya flew a translation of the chapter on our shared winter journey to his deathbed.

I have tried to explore the Eveny experience of the power of nature and the cruelty of history through the unfolding of their personal lives and the intertwining of these lives with each other – an approach which has not been attempted before in anthropological writing on Siberia – so as to give a lasting testimony to their endurance, flexibility and humour. Many people trusted me with intimate glimpses into their stories and their well-guarded feelings. Anthropologists sometimes protect their informants by disguising identities and locations in their writings. But whenever I raised this possibility, and even when I read out whole chapters to them in Russian, my Eveny friends insisted that nothing should be changed. They sensed that they were clinging to the face of the earth for a fleeting moment and wanted my book to be a record, warts and all, of who they had been and how they had lived. I have respected their openness; and in a world where it is increasingly easy for anyone to get on a plane and go anywhere, I hope the reader will do likewise.

Cambridge, January 2005

Note on names and spelling:

Russian names are not always stressed where an English-speaker would expect, eg Iván, Vladímir. Eveny today follow the general Russian pattern of having three names: a personal name, a middle name formed from their father's personal name, and a surname. Thus Vladímir Nikolyáyevich Keimetínov means Vladímir, son of Nikolái, of the Keimetínov family or clan. As an elder, he is addressed respectfully as Vladímir Nikolyáyevich: as my daughter learned to her cost (page 348), only his age mates or his own elders would address him by his first name alone, Vladímir, or by the familiar pet form Volódya. I refer to characters who are my age or younger as others do, by their first name (eg Iván, Kristína), or a familiar form of it (eg Tólya, Várya, short for Anatóly and Varvára).

Russian words in the bibliography are spelled according to the system of the US Board on Geographical Names. In the text, I have sometimes simplified these to make them easier on the English reader's eye. There is no standard orthography for the Eveny and Sakha languages, and I have spelled words as they sound.

MAPS, FIGURES AND ILLUSTRATIONS

MAPS:

PLATE SECTION:

Plates 1–15: all photographs by the author
Plate 16: Top: Author's collection, unknown photographer
Middle: Author's collection, unknown photographer
Bottom: American Museum of Natural History, New York
(Jesup collection No. 22410)

FIGURES:

Page 7 Reindeer stones from northwestern Mongolia and neighbouring areas (from Savinov 1994: 182–3)

Page 8 Mask on sacrificed horse from Barrow 1, Pazyryk, in the
form of a reindeer head (from Rudenko 1970: plate 119)

Page 9 Winged deer's head with antlers ending in heads of birds
or griffins, from Barrow 2, Pazyryk (from Rudenko 1970:
plate 142d)

DRAMATIS PERSONAE

Camp 7

Granny, a shrewd and ironic matriarch
the Old Man, her husband
their four unmarried children:
Ivan, the brigadier (head herder), somewhat morose with sudden
    flashes of dry humour
Yura, his younger brother, a lighter presence
Emmie, full of laughter, devoted to nomadic life
Masha, a teacher in the village school
also:
Gosha, a gentle man who grew up with the family
Lidia, Gosha's devoted wife, strong and passionate
Terrapin, Nikolay and the Bison, male herders
various children

Camp 8

Kesha, the brigadier, with long hair like a mythological hero
Lyuda, his courageous wife, national reindeer-racing champion
Diana, their infant daughter, with prophetic powers, perhaps a
    reincarnated shamaness
Dima, her elder brother, but still little
Dmitri Konstantinovich, Kesha's old father
Ganya, Leonid, and Boris, young male herders who live dangerously

Camp 10

Kostya, the half-Sakha brigadier, capable and steady
Arkady, his young deputy, still lacking confidence
Kristina, Arkady's thoughtful mother and camp cook
Peter, herder and winter stand-in for Kristina as cook

Ivan the Fence-Builder, a herder from another brigade enjoying a
    change of occupation
three generations of the Nikitin family, especially little Sergei, keen
    to grow up a reindeer herder
various women, children and male herders

Free spirits

Tolya, the author's first native friend, a fighter for reform and a
    trickster; at first, head of the village's civil administration
Vladimir Nikolayevich, a retired herder turned hunter; an
    extraordinarily competent person
Vitya the Wolf-Hunter, musician and son of a shamaness

The village

Petr Afanasevich, the director of the State Farm
old Efimov, one of his henchmen in the State Farm management
Afonya, gentle successor to Tolya as head of the village's civil
    administration
Baibalchan, the old headman, killed by the Soviet regime in the 1920s
Tolya's sister Anna
Tolya's wife Varya, who joins him on arduous journeys
little Sergei's mother, head of the fur-sewing workshop
Sasha the Radio Man
a retired woodcutter
teachers, doctors, vets and accountants, mostly female, some of them
    fierce

Other characters include

Motya the Music Woman and her relatives in camp 1
Valera, brigadier of camp 3
the author's family (wife Sally, children Patrick and Catherine)
radio operators, boiler attendants, and vehicle mechanics
psychics and sorcerers
living and dead shamans, some with animal doubles

Margaret Thatcher, Bill Clinton, Cleopatra, Abdullah, Sancho Panza,
    and Prime Minister Chernomyrdin, reindeer of character
reindeer doubles which die instead of their owners
Manchary, an almost wordless young hunter
daredevil Arctic aviators
wolves
bears
Communist Party bosses
presidents and governors
Korean antler mafiosi
Andrew, a US aid worker
the author's students, foreign and local
vodka bootleggers
the Sausage King of Alaska
the Energetics, employees of the state electricity corporation
Bayanay, spirit master of the forest and of its animals
a native pop group
a British film crew
a Russian exile in a tent next to a mountain of deep-frozen dogs
Nenets, Evenki, and Chukchi reindeer herders, in nomadic camps
    thousands of miles away
a leader of the Sami reindeer herders in the Norwegian Arctic
Vladimir Etylin, born a Chukchi reindeer herder, later a Member of
    Parliament in Moscow
spirits of fires and rivers
ancestors in graves
a one-eyed dog who can see into the future

PROLOGUE

Soul-flight to the Sun

In the Verkhoyansk Mountains of northeast Siberia, Eveny nomads are on the move*. Teams of reindeer pull caravans of sledges down the steep slide of a frozen mountain river. Bells tinkle on the lead reindeer while dogs on short leashes dive closely alongside through the snow like dolphins beside a boat. One man sits on the lead sledge of each caravan, his right foot stretched out in front of him and his left foot resting on the runner ready to fend off hidden rocks and snagging roots. Passengers or cargo sit on the sledges behind. The passage of each caravan is visible from afar by a cloud of frozen reindeer breath.

This is the coldest inhabited place on earth, with winter temperatures falling to −96°F (−71°C). The ice is a condition of the water for eight months of the year and by January it is 6 feet thick. Throughout the winter, warm springs continue to break through the surface of rivers, where they erupt as frozen turquoise upwellings, like igneous intrusions in rock, and freeze into jagged obstructions. Caravan after caravan jolts over the last ridge of river ice and skims across a great frozen lake in an epic sweep stretching almost from shore to shore. Deep lakes provide a more level surface and the ice that forms from their still water glows black, marbled with milky white veins snaking into the depths. The sudden speed and the spray of ice crystals flung into our faces behind the hypnotic flash of the reindeer's

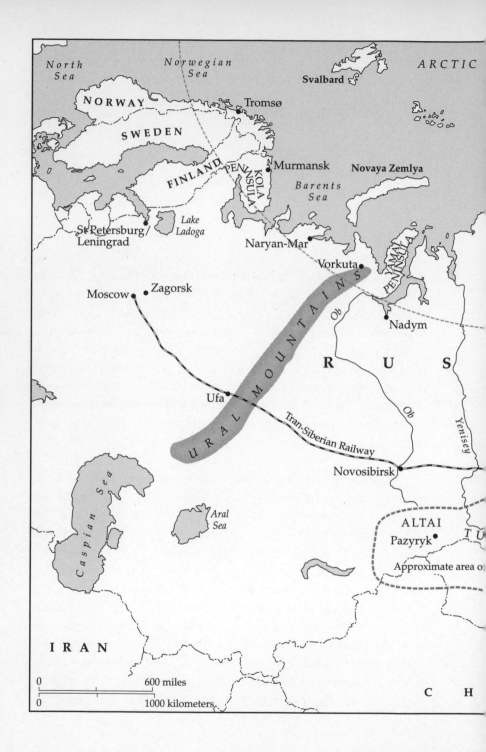

North
Sea

*Norwegian
Sea*

ARCTIC

Svalbard

NORWAY

Tromsø

SWEDEN

FINLAND

KOLA PENINSULA

Murmansk

Novaya Zemlya

*Barents
Sea*

St Petersburg
Leningrad

*Lake
Ladoga*

Naryan-Mar

Vorkuta

YAMAL PENINSULA

Moscow Zagorsk

Ob

Nadym

R   U   S

U R A L   M O U N T A I N S

Ufa

Tran-Siberian Railway

Ob

Yenisey

Novosibirsk

Caspian Sea

*Aral
Sea*

ALTAI

Pazyryk

T U

IRAN

Approximate area of

| 0 | 600 miles |
| 0 | 1000 kilometers |

C   H

OCEAN

TAIMYR

East Siberian Sea

ARCTIC CIRCLE

CHUKOTKA

Bering Sea

Anadyr

Tiksi

Kolyma

Verkhoyansk

SAKHA
REPUBLIC
(YAKUTIA)

VERKHOYANSK MTS

Sebyan

Magadan

KAMCHATKA

S I A

Yakutsk

Lena

Sakhalin

Irkutsk

Lake Baikal

Khabarovsk

SAYAN MTS

V A

MANCHURIA

ancient deer stones

Ulaan Baatar

M O N G O L I A

Vladivostok

Yellow Sea

I N A

Russia, showing main places mentioned in text.

skidding hooves make it easy to feel that we are about to take off and fly into the air.

Thousands of years before the tsarist empire taxed them and the Soviet Union relocated them into State Farms, the ancestors of today's Eveny and of their cousins the Evenki had moved out from their previous homeland in northeast China and spread for thousands of miles across forests and tundras, swamps and mountain ranges, from Mongolia to the Arctic Ocean, from the Pacific almost to the Urals, making them the most widely spread indigenous people on any landmass. Even today, elders can tell stories of journeys that make young people, tied to their villages and dependent on aircraft, smile with disbelief. The old people achieved this mobility by training reindeer to carry them on their backs and pull them on sledges. The endless succession of short migrations* from one camp site to the next, which they have shared with me, gives no more than a glimpse of the power of reindeer transport and of the way in which this creature has opened up vast swathes of the earth's surface for human habitation.

The association between reindeer and flying is very ancient – much, much older than European or American ideas about Santa Claus*. Scattered across the deserts and steppes of western Mongolia and stretching into the Altai Mountains in the west and up to the border of Manchuria in the east, stand ancient 'reindeer stones' dating from the Bronze Age* some 3,000 years ago. These upright standing stones are set above graves or surrounded by the remains of fires and sacrificed sheep and horses. They are carved with various animals, but most often with reindeer. On the earlier stones the image of the reindeer is simple, but some 500 years later it has become more ornate. On these stones, the reindeer is depicted with its neck outstretched and its legs flung out fore and aft, as if not merely galloping but leaping through the air. The antlers have grown fantastically till they reach right back to the tail, and sometimes hold the disc of

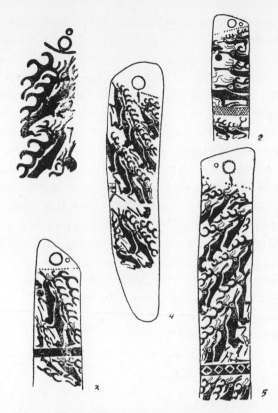

Reindeer stones from northwestern Mongolia and neighbouring areas.

the sun or a human figure with the sun as its head. The flung-out hooves seem to represent more than just a leap: it is as if the artist has caught the reindeer in the act of flying through the sky in an association with a deity of the sun.

It seems the climate of Mongolia dried out towards the end of the first millennium BC, coming closer to today's desert conditions in which reindeer can no longer live, except in one small, cool mountain region. But other evidence suggests that even where it had disappeared, the reindeer persisted in the imagination like a mythic or archetypal creature. At Pazyryk in the nearby Altai Mountains, the burial mounds of chiefs from

around 500–400 BC contain food as well as fine clothing, gold
ornaments, harps, combs, and mirrors, decorated with a range
of animals including reindeer. By the second century AD, one of
the horses sacrificed in a grave wears a face-mask made of
leather, felt, and fur and adorned with life-size antlers, clearly
dressed up to imitate a reindeer*. It seems a reindeer was still
better than a horse for riding in the afterlife. Some 1,500 years
later, in the seventeenth century, at a battle between the Oirot
Mongols and the Manchus 60 miles from Ulaan Baatar, a Mongolian chronicle tells us that the wife of the Khan Daldyn Bashig
Tu rode into battle on 'a reindeer with branching antlers'*. Since
real reindeer had been absent from this region for 2,000 years,
this probably indicates a continuation of the custom of dressing
a horse in a reindeer mask.

The reindeer appears in an even more intimate association

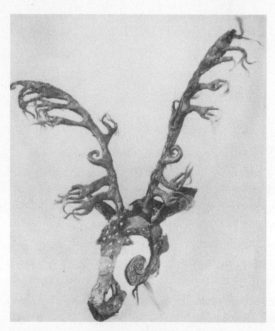

*Mask on sacrificed horse from Barrow 1, Pazyryk,
in the form of a reindeer head.*

Winged deer's head with antlers ending in heads
of birds or griffins, from Barrow 2, Pazyryk.

with the Pazyryk people – in tattoos on their bodies. After death they were eviscerated, sewn up and mummified*, as if they would be needing their flesh as well as their provisions for whatever afterlife or rebirth they were expecting. Even so, these bodies might not have survived had it not been for the water that flowed into the graves*, sometimes through the breaches left by grave robbers. This water then froze around the mummified bodies. Three of the bodies found so far bear tattoos, and have been preserved so perfectly that we can see the designs clearly. Here on the shoulders are depicted the same reindeer as on the standing stones, with their hooves flung out and their exaggerated antlers. But in the tattoos the imagery of flight is made even more explicit. The branching of the reindeers' antlers sometimes looks like the feathering of birds' wings, and on some of them each tine of the antler ends in a tiny bird's head.

When I first read about these tattoos as a child I did not

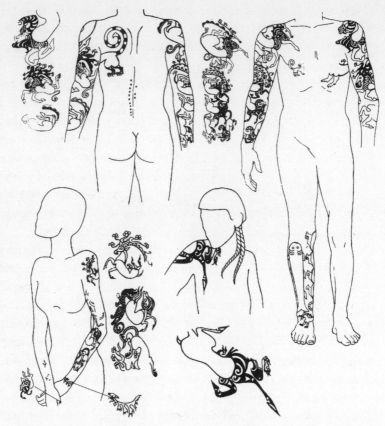

Designs of deer and birds tattooed on mummies from burials at Pazyryk.

imagine that the association of reindeer with flight had been carried by migrating populations to lands where reindeer still existed far to the north, still less that I would one day live among people who in their own childhood had taken a ritual voyage to the sun on the back of a flying reindeer. I reached this northern region in the late 1980s, and learned about this rite from my first Eveny friend, Tolya, during some of our travels together. Small but muscular, a former wrestling champion with an impish sense of humour, he was already feeling the call to abandon his role as an official in the Soviet administration and to reach back through the veils of boarding school and the Soviet Navy to rediscover the ancient traditions of his ancestors. As we rode from camp to camp, this ritual was one of Tolya's discoveries*. We crouched around darkened stoves at night, while I listened to Tolya talking intently to nomadic elders, who included his own mother, in a native language I could not yet understand. I did not know that in front of me precious words were being spoken by people who might have been the last left alive on earth capable of saying them. These words revealed a continuity of ideas, carried over thousands of miles and thousands of years, with the birds on the tips of the reindeer antlers tattooed on the shoulders of the mummies in the Altai and the carvings in Mongolia of reindeer holding the sun aloft in their antlers.

These elders told Tolya that reindeer were created by the sky god Hövki, not only to provide food and transport on earth, but also to lift the human soul up to the sun. From their childhood seventy, eighty, or more years before, they remembered a ritual that was carried out each year on Midsummer's Day, symbolizing the ascent of each person on the back of a winged reindeer. During the white night of the Arctic summer, a rope was stretched between two larch trees to represent a gateway to the sky. As the sun rose high above the horizon in the early dawn, this gateway was filled with the purifying smoke of the aromatic mountain rhododendron, which drifted over the area from two separate bonfires. Each person passed around the first fire anticlockwise, against the direction of the sun, to symbolize the death of the old year and to burn away its illnesses. They then moved around the second fire in a clockwise direction, following the sun's own motion, to symbolize the birth of the new year.

It was at this moment, while elders prayed to the sun for success in hunting, an increase in reindeer, strong sons and beautiful daughters, that each person was said to be borne aloft on the back of a reindeer which carried its human passenger towards a land of happiness and plenty near the sun. There they received a blessing, salvation, and renewal. At the highest point, the reindeer turned for a while into a crane, a bird of extreme sacredness.

I still do not understand how the old Eveny acted out the experience of flying through the air, but they would mime their return to earth by sitting on their own reindeer as if they were arriving from a long journey, expressing tiredness, unsaddling their mount, pitching a tent and lighting a fire. This rite was followed by a *hedje*, a circle dance in the direction of the sun, and a feast of plenty.

The annual soul-voyage made by the elders whom I met with Tolya was a small-scale echo of the voyages made by shamans, men and women whose souls can leave their bodies while they are in a state of trance and fly to other realms of a cosmos which is believed to have many layers. Whereas laypersons could only fly on the back of a reindeer, shamans could turn into a flying reindeer. The word *shamán* or *hamán* comes to us from the language of the Eveny and the Evenki, two closely related peoples of the Tungus language family. All Arctic peoples have comparable figures, known by various names, as do other peoples in many parts of the world. The role of the shaman is closely linked to hunting as a way of life. Before the development of agriculture around 10,000 years ago, all humans depended on hunting to survive, and it is hard to imagine that any other kind of religion could have existed. Shamans develop the ordinary hunter's skills and intuitions by flying over the landscape to monitor the movements of migratory animals and by performing rites to stimulate the vitality of animals and humans alike.

In Siberia, shamans combine a distinctive imagery of reindeer and of bird-flight. Their costumes sometimes include imitation reindeer antlers, occasionally tipped with wings or feathers, placed on the headdress or attached to the shoulders at the very point where reindeer are tattooed on the Pazyryk mummies. Like the participants in the Eveny midsummer ritual, shamans may ride to the sky on a bird or a reindeer. But their relationship with these animals goes far beyond mere riding. One shaman is suckled by a white reindeer during his initiatory vision as he incubates in a bird's nest on a branch high in the tree that links earth and sky*. Another becomes a reindeer himself by wearing its hide, while hunters with miniature bows and arrows surround him and mime the act of killing. The hide is then stretched across the broad, flat drum that the shaman will beat as accompaniment to his trance. Another shaman, seeking to consecrate his reindeer-skin drum, is guided by spirits as he combs through the forest to find the location where the reindeer was born and traces every place it has ever visited over the course of its life, right up to the point where it was killed. As he picks his way through bogs and over fallen branches, he picks up the scattered material traces of its existence – snapped twigs, dried dung – to gather together every possible part of its being, and then moulds them into a small effigy of the reindeer. When he sprinkles the effigy with a magical 'water of life', the drum comes to life. Like a reindeer itself but with enhanced power, it is now capable of bearing the shaman aloft with its throbbing beat to nine, twelve, or more levels of the heavens.

PART I

THE PARTNERSHIP OF
REINDEER AND HUMANS

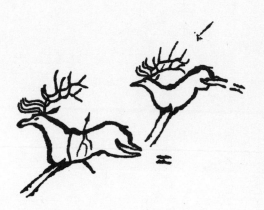

1

The prehistoric reindeer revolution

The reindeer (*Rangifer tarandus*) has been giving life to humans for hundreds of thousands of years over much of the northern hemisphere. In western Europe, where the ice sheets retreated between the late upper Paleolithic and the end of the Pleistocene era some 11,000–18,000 years ago, the sheer number of reindeer bones found in human camps has led prehistorians to call this period 'the Age of Reindeer'*. One main migration route ran from Paris to Brussels, and another from the Massif Central to the coastal plains of Bordeaux. Bones and antlers from reindeer hunts have also been found across Denmark, Germany, Poland, and the Ukraine. In North America, reindeer hunters' camps have likewise been found where the ice retreated from Michigan and Ontario around 11,000 years ago.

In the Arctic, the Age of Reindeer is not over. The world contains around 3 million wild and 2 million domesticated reindeer, and for many indigenous peoples this species remains the foundation of life today. In Canada and Alaska, Native American peoples hunt wild reindeer, which are there called caribou, using insights about herd behaviour shared with their ancestors thousands of years ago. The peoples of Siberia also hunt wild reindeer but have gone further, developing these insights to draw the reindeer into a form of domestication that is unlike that of any other animal*. Nobody knows when or how this was done. Domesticated reindeer closely resemble wild reindeer in

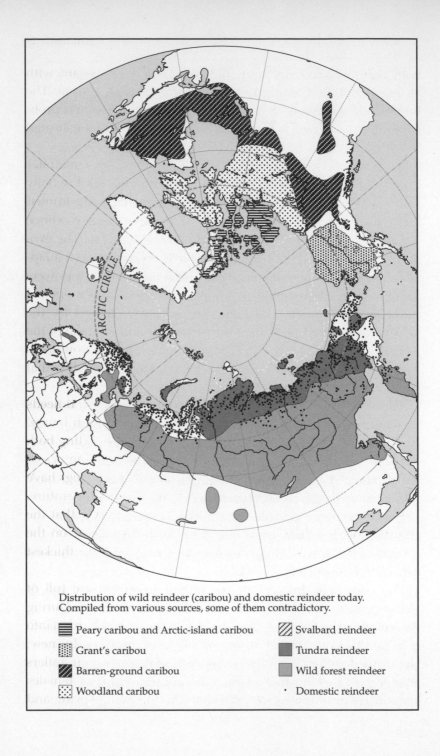

Distribution of wild reindeer (caribou) and domestic reindeer today.
Compiled from various sources, some of them contradictory.

▤	Peary caribou and Arctic-island caribou	▨	Svalbard reindeer
▦	Grant's caribou	■	Tundra reindeer
▧	Barren-ground caribou	■	Wild forest reindeer
⦂	Woodland caribou	·	Domestic reindeer

both anatomy and behaviour, and are handled by humans with an imperceptible continuity of knowledge and technique. The human relationship with reindeer in Eurasia is unique. It is probably the only animal in history that was originally domesticated for riding in order to hunt its wild cousins.

In prehistoric sites the remains of wild reindeer are sometimes found mixed with those of auroch, mammoth, and woolly rhinoceros. Of the large Pleistocene animals that were hunted for food, all except the elk and the musk-ox have become extinct, and only the reindeer has flourished in large numbers. The evolutionary success of the reindeer is due to its exceptional adaptation to cold, its gregarious nature, and its enormous powers of migration. As the glaciers retreated northward they opened up spaces which were filled by green vegetation fed by the meltwater. Massed herds of reindeer moved in the wake of the glaciers, followed by the humans and wolves who competed with each other to hunt them, in a complex relationship that continues to this day.

The reindeer's adaptation to cold seems almost total. It needs no den in winter. Above a layer of fine, inner fur, each hair of its thick outer coat is hollow and filled with air, providing both insulation against the cold and buoyancy for swimming across the icy water that often floods the Arctic landscape. Its legs have less fur than its torso and are maintained at a lower temperature. The insides of the nostrils are cleverly convoluted so that the animal loses less heat from breathing. In hot weather, on the contrary, its body is designed to keep it cool, and the thickest fur is shed in summer.

The young, fleshy antlers that sprout in spring are full of blood vessels and covered with a soft layer of velvet. During the summer, the blood vessels atrophy as the antlers harden into nothing but horn, finally dropping off to be replaced by new, blood-filled antlers the following spring. Males lose their antlers after fighting each other during the autumn rut, but females keep theirs until after they give birth the following spring and use them aggressively to defend the feeding hollows that they scrape through the snow for themselves and their calves with their powerful front hooves. Reindeer have the most flexible joints of any hoofed animal – they can scratch their ear with their back hoof and can run across rough scree where a human can barely crawl. The large surface area of their broad, forked hooves prevents them from sinking in bog or snow. A riddle among the Eveny asks: 'There is a bearded old man who lives between two cliffs. He never moves from the spot yet he runs very fast. Who is he?' The answer is the tuft of fur between the digits of a reindeer's hoof which helps it to grip the surface as it runs and skids over ice.

With the retreat of the glaciers the species moved north, and most of the world's reindeer are now found within the Arctic and sub-Arctic, across the northern forest (which in Russia is called *taigá*) and through the treeless tundra to the coast. Wild reindeer have even reached the islands of Svalbard, at 80° north, crossing hundreds of miles of frozen ocean. Beyond this, there is nowhere further to go. DNA testing of the Svalbard reindeer cannot confirm whether they came from North America via northeast Greenland or from Russia via Frantz Josef Land, but the earliest droppings so far analysed* are 5,000 years old.

In recent centuries, the impact of climate change in Europe has been accelerated by the spread of agriculture and cities. In his account of Germany written in 52 BC, Julius Caesar* wrote that the forests there contained many kinds of animal that the Romans had not seen elsewhere, including 'an ox shaped like a deer' whose horn 'sticks up higher and straighter than those of the animals we know, and branches out widely at the top like a human hand or a tree'. As late as the nineteenth century, the limit of wild reindeer in the European part of Russia ran far south of their range today, passing just east of St Petersburg on the shore of Lake Ladoga and just northeast of Moscow at Zagorsk, continuing to Ufa in Bashkortostan*. In Asia, the north was not glaciated like Europe and is not so urbanized today, and so the retreat has not been so great. Wild or domesticated reindeer still extend from north to south in an almost unbroken belt from the Arctic Ocean islands, past Lake Baikal, into Tuva and just over the border into the northwestern mountains of Mongolia. However, the numbers in the southern forests are very small and most reindeer are in the far north.

Russia today contains between 1.25 and 1.5 million wild reindeer, and of these the herd on the Taimyr Peninsula on the Arctic Ocean numbers between 800,000 and 1 million animals, double the size of the 450,000-strong Western Arctic Caribou Herd which migrates between the North Slope and the Seward Peninsula in Alaska. Reindeer populations are subject to dramatic cycles of boom and bust. When the snow cover thaws in a warm winter and refreezes into an impenetrable crust of ice, thousands may die of starvation. The Taimyr population has grown to its present size from 110,000 when the first aerial survey was made in 1959, though it is widely expected to crash either from overgrazing or from an epidemic of anthrax, in Russian called 'Siberian Ulcer', whose spores lie hidden in the soil.

The native peoples of North America never domesticated the caribou. Apart from a few Siberian reindeer introduced into Alaska in the 1890s, all domestic reindeer are in the Old World. Some half a million of these are in Scandinavian Lapland*, where over the last fifty years the Sami people have transformed their traditional reindeer techniques into a technically sophisticated modern ranching system. Distances in Scandinavia are relatively small, and infrastructure highly developed. In Russia, the situation is quite different. The grazing range of Russia's domestic reindeer, said to be 1.3 million square miles, amounts to nearly a fifth of the country's land area, with almost no roads at all. Their number has recently halved from a peak of 2.4 million in 1970 to 1.2 million in 2002*. The earlier figure was the fruit of an unsustainable Soviet system of overproduction, while the current decrease is the result of the crisis in the Russian economy since 1990.

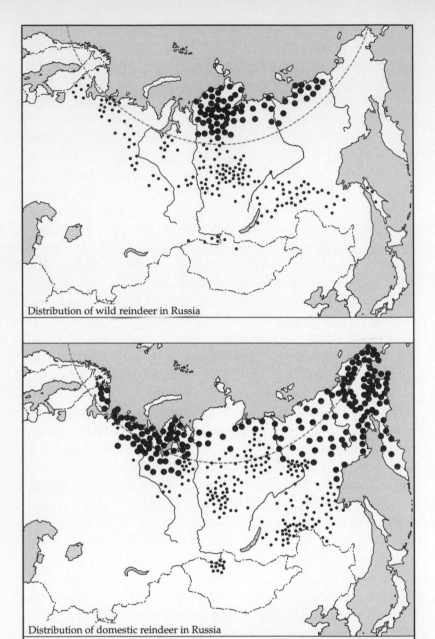

Distribution of wild reindeer in Russia

Distribution of domestic reindeer in Russia

Adapted from Syroechkovskiy 1995, pp. 7-8. These maps are based on older data and do not reflect the drastic decline during the 1990s of domestic reindeer in the far northeast and their replacement by wild reindeer (described in the Epilogue). The numbers of domestic reindeer in the west and of wild reindeer in the central area remain high today. The small dots represent 1,000 reindeer and the large dots 10,000.

Both wild and domestic reindeer show the same instinct for aggregating in herds*, and for migrating in a repetitive cycle along the same routes for years or centuries at a time. Unlike other kinds of deer, most reindeer tolerate crowding and even seek it. Strong leaders trample paths through the snow for their followers, and individuals in a packed herd are immune to wolves, which attack only isolated stragglers. They are particularly gregarious when migrating. Their spring migration follows a succession of green plants which sprout in the wake of the melting snow from May onward. The retreat of the Pleistocene ice sheets is repeated in miniature each summer in forest valleys as the reindeer migrate upstream, and on tundra plains as they move northward from the winter shelter of the tree line to the cool, insect-dispersing breezes of the coast. In autumn they migrate back downhill or inland. Green plants disappear successively through the winter, beginning with willows and ending with grasses and sedges, and the animals become increasingly dependent on lichens. While herd members may linger or disperse in summer or winter, during the spring and autumn migrations they come together and an irrepressible sense of urgency builds up like an elemental force. Reindeer can run for many hours at 20 or 30 miles an hour and in bursts at double this speed. Forest reindeer form smaller groups and their migrations are relatively short, but in a single autumn or spring migration, a wild tundra herd can travel 700–800 miles, a greater distance than any other land animal.

It is this instinct to migrate that has made human life possible, whether further south during the Pleistocene or in the Arctic today. Early humans could not keep up with a migrating herd like their rivals the wolves. Even where they invented skis*, the hunters' range was small because they could not carry heavy loads. But reindeer routes are repetitive and fairly predictable. They may vary with yearly changes in vegetation, snow cover, wind, or insect harassment, but certain bottlenecks like passes, canyons, and river crossings can remain constant over many years. During the spring migration, rivers and lakes are still frozen, but in the autumn the herd is slowed down by the need to swim across rivers. When crossing a river, reindeer will first hesitate and then blindly follow a leader, usually a mature female, and once the leader has reached the opposite bank the herd will not turn back even when under attack by hunters. Autumn is also the time when the animals are at their fattest after their summer grazing and their fur is reaching its thickest in preparation for the winter – and in the approaching winter the entire landcape becomes an open-air deepfreeze, ideal for preserving meat. Many sites have been identified where prehistoric peoples drove migrating reindeer down a narrowing funnel of cairns built in human form into snares, or speared them in the water from the river bank or from boats. Similar techniques are still employed by many northern hunters today.

Humans have also worked out numerous ways of ambushing or luring smaller groups of reindeer. The species wavers between timidity and curiosity, poised paradoxically either to flee or to explore. A reindeer sees patterns rather than detail. A man can make a reindeer run towards his companion lying in wait by putting up his parka hood so that the edging of fine fur resembles the bristling shoulder hairs of a wolf with its head down ready to attack. But if two hunters imitate the silhouette of a reindeer, one bending down like a pantomime horse and the other lifting his bow or gun as antlers, it will approach them to investigate. Siberian hunters can mimic the cry of a calf looking for its mother or the bark of a rutting male*, or roll a trumpet out of birch bark to imitate the sound of a male that has found a female, so that other males will approach to challenge it.

However, the most effective decoy is a live reindeer. This technique is many thousands of years old. A tamed reindeer is let out on the end of a long rope, which is gradually shortened when it attracts the attention of a wild reindeer, bringing it within range. Hunters may also twine a hide rope around the antlers of a rutting male in autumn, disguising their human smell by smearing the rope with birch sap, and then release the decoy to entangle a wild male in its antlers. In the 1930s the Nganasan of Taimyr Peninsula also used to fix a spear-point to the antlers, until the day when one rutting male turned around and caused carnage among its own herd.

Though humans have hunted reindeer for hundreds of thousands of years, it seems that it is only in the last 3,000 that they have begun to domesticate them. Domestication involves appropriating a wild animal's behaviour, bending it to human purposes, and continuing this relationship down through generations of animals. In the modern sense of selective breeding, the word 'domestication' can be applied to reindeer only in a very limited way until the development of large commercial herds of the twentieth century. Before then it was difficult to control their movements and their contact with wild reindeer, or even with other members of their own herd. Apart from the males trained for carrying baggage, riding, and pulling sledges, and some females when lactating, most of the animals in a herd never become very tame, in the sense of having an intimate emotional and physical contact with their human carers. Even transport reindeer may become uncooperative and recalcitrant if left unattended for a few days, and any domestic reindeer may revert to the wild if left unattended for longer. The domestication of the reindeer is a hard-earned and provisional achievement.

At the heart of domestication lies an insoluble mystery. The use of decoys in early times shows that the taming of wild individuals was once possible, but it seems almost impossible to domesticate wild reindeer today*. Though a passing wild herd can sweep away a domestic herd never to be seen again, there seems to be no reliable record of anyone successfully capturing wild reindeer to augment their domestic herd. Even if wild males impregnate domestic females, they will avoid humans and rejoin their own herd, and if a wild reindeer calf is tethered it will struggle so desperately to escape that it will often strangle itself or die of exhaustion. This behaviour has led some Russian scholars to argue that today's wild reindeer is not the ancestor of its domestic counterpart, but a cousin incapable of domestication, and that the wild strain that was domesticated no longer exists.

Like scholars, the reindeer peoples, too, see domestication as a threshold in the past, and depict this decisiveness in various ways. One legend of the Evenki, a people closely related to the Eveny, relates how a woman noticed that female reindeer and their calves would eat the moss where she had urinated, attracted by the saltiness. The woman continued to urinate nearer to her tent and the reindeer approached closer each time. At length they allowed her to reach out and touch them. When she milked them, she discovered that the milk was delicious*.

Another legend from the Eveny tells a different story. An old hunter was walking through the forest when he came across the head of a reindeer calf protruding from a vulva-shaped crack in the trunk of a larch tree. The old man stroked the calf and talked to it gently, helping it to emerge painfully, one limb at a time until it was fully born. This was the first male reindeer. When this reindeer was 3 years old it met a female (presumably born from another larch tree) and they had a family of two calves. One day the family was attacked by wolves. The parents cried helplessly to the god Hövki in the sky, while their calves fought back and impaled the wolves on their antlers.

Afterwards, the parents felt ashamed of their cowardice. When Hövki asked them why they had not fought to protect their children, they answered that they did not have the strength or the sharp antlers of the younger reindeer. Hövki asked them how they proposed to survive in the future, and the father replied, 'When I was being born from a tree, a human was kind to me. Let us go and live with humans. They will protect us from wolves and out of gratitude we will serve them.' Hövki then asked the younger reindeer how they wanted to live. They replied that however hard the life, they preferred to remain free.

Then each pair ran their separate ways to their separate destinies. For ever after, the reindeer who were born wild stayed wild; but for those who were born with human help, there was no going back. The Eveny language reflects the finality of this separation between the two kinds of reindeer by labelling them forever as different kinds of creatures. As in the languages of many reindeer peoples, there is no single word that covers both wild reindeer (*buyun*) and domestic reindeer (*oron*).

Both these stories establish domestication as an arrangement of mutual benefit to both sides, and even as a social contract between reindeer and humans. Each emphasizes a different aspect of this relationship. The first story talks of nurturing, in which a woman provides salt to reindeer in exchange for milk, and to this day Eveny women milk females after coaxing them with handfuls of salt (though not by urinating). The second is a story of danger and violence, in which the age-old triangle of reindeer, human, and wolf is realigned. Here, the bargain is protection from the wolf in exchange for all the services that the reindeer can offer to humans, even its own flesh. This is a long-term relationship of kindness in which the human is no longer a predator like the wolf, but a partner.

In the first story, the negotiation between the woman and the reindeer is tentative and wordless. In the second, the reindeer initially draws the man into a protective relationship through the greatest helplessness it is possible to imagine, the mute pleading of an unborn baby. Later, it acknowledges its continuing need when it takes the initiative by asking for domestication. One could almost wonder whether it was the reindeer who was domesticating the hunter, since domestication forced humans to adapt their own movements from intercepting herds to following their migrations, developing the cyclical form of nomadism that we see today.

Indigenous peoples' legends explore why and how reindeer were domesticated, but their narratives are set in a mythic realm. Archaeologists and biologists, however, seek to locate the origins of domestication in map space and historical time. This process would have remained forever beyond recovery if it had not been for the labours of recent generations of scholars, writing mostly in Russian. Many points remain unclear or controversial, but their work has revealed a Reindeer Revolution* which radically changed the way humans were able to move and settle throughout the Eurasian North.

The indigenous peoples of northern Asia speak languages of the Tungus-Manchu, Mongol, Turkic, and Finno-Ugrian families, as well as several isolated and unrelated languages*. All these peoples were marginal to the Chinese, Manchu, Mongol, Japanese, and Russian empires which rose and fell over the centuries. Their languages remained unwritten until the twentieth century, and their shifting identities and movements were not recorded. Nothing of their earlier experience remains in their own words, and the story of their domestication of the reindeer must be pieced together from the painstaking study of vocabularies and rituals, the genetics and behaviour of local varieties of reindeer, the designs of saddles and sledges, and the diffusion of customs such as whether or not herders milk their reindeer, use dogs to round them up, or use stirrups to ride them*.

Collars made of bone for decoy reindeer, dating from 2,000 to 2,500 years ago and similar to those used today, have been found on the lower Ob River. This is evidence for the early taming of individual animals, but no excavation in the Arctic has ever found evidence for any other kind of domestication during this period. However, dogs were already harnessed in the far North to pull sledges, and they would later inspire the invention of the reindeer sledge.

All the evidence for the first domestication of reindeer comes from further south, in the form of numerous drawings on rocks and cliffs* in the Sayan Mountains on the border of Siberia and Mongolia. This is very near the region of the carved reindeer stones and tattooed mummies. The dating of these pictures is extremely difficult, but archaeologists believe that they are more

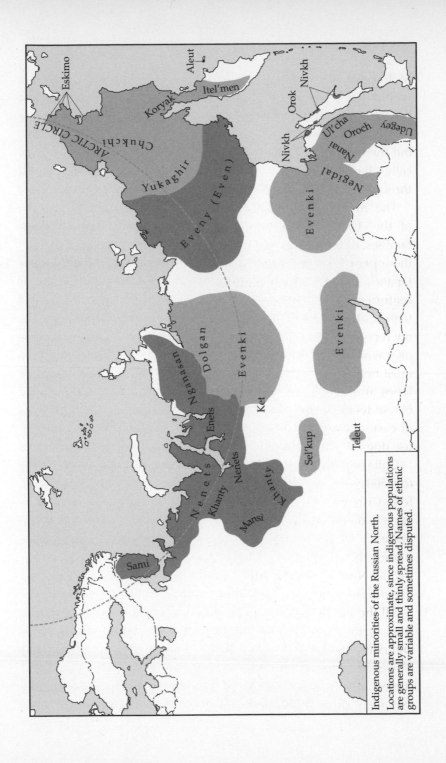

Indigenous minorities of the Russian North.

Locations are approximate, since indigenous populations are generally small and thinly spread. Names of ethnic groups are variable and sometimes disputed.

than 2,000 years old. One drawing shows humans seated on reindeer and controlling a combined herd of reindeer and camels, while another sets a reindeer rider surrounded by domestic goats and dogs among pictures of camels and (it seems) elephants. In another drawing, a human sits in the middle of the reindeer's back, in the distinctive riding position that is still used in the Sayan today*. Other evidence suggests that reindeer were used to carry baggage before they were ridden, so it is not clear how far the pictures of riding represent actual practice and how far they express a fantasy.

These drawings suggest that the domestication of the reindeer did not happen in isolation, but in relation to other kinds of domesticated animals. This region lay at the southernmost limit of the reindeer's distribution, where it met horses, yaks, camels, and cattle. These other animals were herded by Turkic peoples who rode horses and also competed for pasture, and relations were not always peaceful. With the drying of the climate of western Mongolia towards the end of the first millennium BC the forest gave way to steppe, making it more suitable for horses than for reindeer. The ancestors of the Tofalar people stayed in the more humid mountains, where Marco Polo reports them as riding* reindeer in the thirteenth century, as their descendants do today. But for other local peoples, the combination of Turkic pressure and climate change made it impossible to stay and to keep reindeer at the same time. It has been suggested that the change from plain to stylized representations of antlers on reindeer stones may correspond to the transition from a functioning reindeer economy to a world in which the reindeer was remembered only as a legendary animal that no one had ever seen, like the lion in European heraldry. For the Pazyryk people whose elite had reindeer tattooed on their bodies, it was probably this change that forced them to use a reindeer mask on a sacrificial horse. Some of them began to move northward from the dry steppe into the taiga and down the Yenisei and the Ob, two of the great Siberian rivers that drain the mountains of Inner

Asia. They eventually reached the Arctic Ocean thousands of miles away, merging with local populations to form the modern Nenets and Nganasan peoples* who live on both sides of the Urals where the mountains come down to the sea. As well as learning how to milk from the Turkic peoples who drank mare's milk, these migrants adapted the horse-riders' hard-framed saddle which gave them a far greater range.

An alternative theory* credits the independent domestication of reindeer to the Tungus peoples who at that time lived east of Lake Baikal. The ancient Tungus are the ancestors of several of today's small groups of Siberian hunters and reindeer herders, of whom the most numerous are the Evenki and my friends the Eveny. They speak languages related to the language of the Manchu who formed the last dynasty of the Chinese empire but whose language is now almost extinct. Having started at the empire's outer frontier, the Eveny and Evenki moved even further away towards the Arctic, some way east of the Pazyryk people's move towards the Urals. The argument for their independent domestication of the reindeer is based on fundamental differences in technology and technique, which it is argued could not simply have been copied from the peoples of the Sayan area. Where the Sayan rider mounts from the left and rides with stirrups on a saddle of Turkic design placed in the middle of the reindeer's back, the Tungus rider mounts from the right, has no stirrups but taps the ground with a stick for balance, and uses a Mongol kind of saddle placed on the animal's shoulders.

It is impossible to be sure how the Tungus learned to ride reindeer. But whether or not they developed the technique independently, their greatest innovation lies in their use of this new skill. Placing the saddle on the reindeer's shoulders is less of a strain on the animal and enables it to cover much greater distances without tiring. The Tungus used reindeer to colonize vast areas of the eastern taiga which had previously been impenetrable and had perhaps never before been visited by any human. To this day, short of helicopters, much of Russia's land

Contrast between Sayan (left) and Tungus (right)
methods of riding a reindeer.

surface can be reached in no other way. Moving along rivers in
winter and over passes in summer from one river system to
another, forever pushing into new territory, a Tungus family
could travel on journeys lasting for generations. Today, fewer
than 50,000 Eveny and Evenki are spread over a third of the
territory of Russia.

The reindeer sledge, central to the image of reindeer in the
outside world, was developed relatively late. Wherever southern
reindeer riders reached the tundra, they found local populations
who already had sledges pulled by dogs*, sometimes for thousands of years. The reindeer people quickly adapted this idea
to their own animals. Just as the reindeer saddle had revolutionized transport in the forest, so the reindeer sledge revolutionized life in the tundra. Reindeer have greater pulling power
than dogs, and graze by themselves on plants rather than needing to be fed on meat. Sledges harnessed to reindeer allow the
exploitation of vast areas, carrying heavier loads and travelling
enormous distances. Recent excavations have found traces of
reindeer sledges from 1,400 years ago on the Yamal Peninsula,
which juts out into the Arctic Ocean far to the north of the mouth
of the River Ob. These remains of harnesses and runners have

enabled scholars to reconstruct the way of life that they made possible. They allowed hunters to migrate in summer from the tree line to tundra lakes for fishing and to reach the northern-most coast where they could exploit walrus, seals, polar bears, and the driftwood carried downriver from the forests of the south. Toys made of wood and bone show that hunters could now take their families with them instead of leaving them behind at base camps in the forest.

Over the last two millennia the reindeer sledge spread throughout northern Eurasia, and reindeer replaced sledge-dogs almost everywhere. Around AD 870, King Alfred the Great of England received a visit from a Viking chief* called Ottar, who lived in Norway. Alfred records Ottar's account of the tribute he received from the Sami, who lived even further north, at the furthest possible point in the Eurasian North from the Sayan mountains where the domestication of reindeer probably began. This tribute included walrus tusks, bird-down, bearskins, and seal hide for making into ship's rope.

'In addition,' Alfred wrote of Ottar, 'he had six hundred of the tame animals which they call reindeer. There were six decoy reindeer, which are highly valued among the Sami, for with them they catch wild reindeer.'

Since the Sami have never ridden on the backs of reindeer, Ottar's main herd of 600 must have been used either as pack animals or to pull sledges, precursors of today's reindeer which offer sledge rides to tourists at Santa's theme park* in the Finnish part of Lapland.

Perhaps no reindeer were kept anywhere on a large scale for eating rather than transport until after 1600, when the expanding Swedish and Russian states began to encroach on the nomadic peoples on their northern and eastern frontiers. The Sami and the peoples of the European North of Russia developed large herds to supply meat to the south, while Russian adventurers moved into Siberia* and crossed the Urals in quest of furs for the tsarist State to sell on the luxury market of western Europe.

They lacked the indigenous peoples' mobility in the forest, but built wooden forts and trading posts on rivers and used their firearms to compel the natives to hunt squirrels and sables for them and to pay the notorious 'tax' called *yasak*.

As with European colonialism elsewhere, native peoples became labourers for an imperial economy, using their special skills to exploit their own natural resources for the profit of others. The Russians created new needs, supplying them with tobacco, tea, vodka, guns, and ammunition in exchange for yet more furs (and a thin veneer of Orthodox Christianity over their earlier shamanic beliefs). By the eighteenth century, much of the Siberian forest, a large portion of the earth's surface, was depleted of sable and squirrel. But a transformation had already begun in the native peoples' relationship with their reindeer*. They had to keep larger herds, not only to cover the greater distances involved in following fur-bearing animals, checking traps, and bringing the furs to distant markets or tax points, but also to supply meat to Russian settlements. Contact with Russians and access to trade goods allowed some men to become wealthy chiefs, maintaining huge herds. Some chiefs are said to have controlled up to 18,000 reindeer each, equivalent to the holdings of an entire State Farm in the Soviet heyday of the 1960s and 1970s.

These were the very people who were most violently attacked after the revolution of 1917. It took the Communists a further five years of civil war to gain complete control of northern and eastern Siberia. The principle of collectivization was extended with very little adjustment from Russian peasants to the reindeer herders of the North. Though even an ordinary family could not catch their food or travel from one camp to another without owning dozens or even hundreds of reindeer, in the 1920s and 1930s almost every reindeer in Russia was confiscated and placed collectively in large herds* run by the State as Collective (later State) Farms. Many who resisted were exiled or shot. The reindeer peoples across the country responded by abandoning their animals or eating them to prevent them from falling into State hands, and settled down to a grim test of endurance that lasted until the mid-1980s. Even with pressure to build up new herds, it took thirty years for the reindeer population to recover. In the seven years from 1927 to 1934, nearly a million reindeer disappeared in Russia as their numbers fell from 2.2 to 1.4 million.

While tsarism had exploited casually with corrupt neglect, Soviet rule overrode the reindeer peoples' self-sufficiency with a meticulous control and constraint, which made them dependent on State support. The Soviet approach was well meaning and brutal at the same time. Communist missionaries saw the Siberian natives as primitive people who needed to be rescued from backwardness.They started to 'civilize' the native peoples by building them permanent wooden villages and providing basic schooling and medical facilities, introducing State bureaucracy and teaching them Communist values. At the same time they imprisoned or killed their spiritual support, the shamans.

The Russians came for furs, but later the Soviet regime stayed for minerals. From the 1920s to the 1950s tin, gold, and finally uranium were extracted by the doomed, sickly labour of the Gulag prison camps* which depended on native caravans of reindeer to bring in their basic provisions, including reindeer meat. After Stalin's death in 1953 the Gulag prisoners were gradually replaced by well-paid free labourers. These were more cost-effective, and across many reindeer herding areas from the 1960s there began the great development of oil and gas* that today yields over half of Russia's foreign exchange earnings.

The native animal economy and the Russian mineral economy are separate universes. For Russians, Siberia is a frontier, redolent of the misery of exile, of adventure and the call of the wild, or the opportunity to make money with high bonuses. Their mining enclaves are provisioned from outside like settlements on the moon, no longer by reindeer but by air. Representatives of a peasant culture airlifted into a landscape that can support only nomads, they reassure themselves that the surface of this hostile northern planet is fit for human habitation by growing cucumbers in the short Arctic summer. They plant them on the windowsills of offices and canteens, under polythene sheeting in beds raised above the permanently frozen ground, anywhere where their pet cucumbers can multiply their cells as they race towards ripeness in the few weeks of twenty-four-hour daylight.

Before I became an anthropologist, I studied ancient languages and civilizations and spent my youth travelling between the centres and peripheries of empires. At Persepolis, the ruined capital of the ancient Persian empire, I was particularly struck by the grand staircase with bas-reliefs showing representatives of every tribe and people in the Persian empire bringing tribute to the king. I was excited by the great imperial capitals and loved the texts, inscriptions, and philosophies. But my sympathies were with the tribes at the margins, and I gradually became uneasy as I realized that ancient texts spoke only about the ideas and feelings of a small ruling class in centres of power and maintained a disdainful silence about the lives of the people in the vast hinterlands whose territory, labour, and produce sustained their civilizations. In an attempt to understand the relationship between central doctrines and life at the edge, I spent several years in the Indian jungle in the state of Orissa with the Sora, an aboriginal tribe on the margins of Hindu civilization*, where I studied their form of shamanism and became a shaman's assistant.

It was the same interest in shamanism as a marginal religion that first led me to the Tungus peoples whose languages gave the world the word 'shaman' after Peter the Great's research expeditions first encountered Siberian shamans in the eighteenth century*. At the first murmurings of *perestroika** or 'restructuring' in 1986, before the end of the Cold War, I started trying to find a way to live with one of the reindeer-herding communities of Siberia, which had been closed to westerners since the 1930s. I wanted to know how the subsistence economy and the shamanic worldview of native cultures had adapted to Soviet rule. I learned Russian and read some works of Soviet anthropology, but they were largely concerned with reconstructing the pre-revolutionary past which the Soviet regime had destroyed so effectively. The literature plotted the ancient migrations of clans and tribes, discussed the distribution of traditional designs of sledges and costumes, and transcribed chants from the last rituals of the last shamans; it contained little about what it was like to be a Siberian reindeer herder today.

But the region was almost inaccessible. It took two years of negotiations, including a three-month residence at the Museum of Anthropology and Ethnography in Leningrad (now St Petersburg), before I was able to get into northern Siberia. My main memory from that first winter in Leningrad is of loneliness and anxiety. People on the streets were frightened of each other and avoided eye contact, so unlike the overwhelming attention paid to strangers in India. Every meeting had to be formally requested and was made to feel dangerous. I spent long, dark evenings wandering alone through tall corridors lined with glass cases of mannequins dressed in exotic costumes, missing mile-stones of my baby daughter's growth at home. In the third month some of the other anthropologists, ladies with inexhaust-ible teapots in their offices, felt it safe to invite me in and I entered a quite different world of private warmth which has bound me to Russia ever since. It also turned out that Soviet anthropologists knew much more about Siberian peoples than the inhibitions of the time allowed them to commit to paper, and they taught me a great deal off the record.

In August 1988 my diplomacy bore fruit. I was put on my first flight across the 4,000-mile gulf from Leningrad to Yakutsk, six time zones away and only four time zones from the date line separating Siberia from Alaska. There, I was received by a local research institute, courteously but cautiously. The city felt grim and closed. Only a handful of foreign researchers had ever reached there, and none had been allowed outside the city for more than a few days, and even then they were always kept on a short leash.

Yakutsk is the capital of the Sakha Republic or Yakutia, a territory almost as big as the whole of India, where the Sakha or Yakut people* make up a third of the population of 1 million. Sakha scholars told me about their own culture, which was traditionally based on horse herding, but I could find very little trace of the republic's outlying indigenous minorities, the Yukaghir hunters and the reindeer-herding Eveny, Evenki, and Chukchi, fragments of the thirty or more ethnic groups that comprise the 186,000 indigenous people scattered across the Russian North and Far East. When submitting my papers a year in advance, I had made a request to visit a reindeer-herding community, using the Russian bureaucratic phrase 'in accord-ance with possibility' (*po vozmozhnosti*). I had since discovered that there existed a showcase Eveny reindeer-herding village called Topolinoye, with the most advanced facilities, which was used to impress foreign visitors and had been featured by western travellers and journalists. This would no doubt be my 'possibility'. When the Sakha director of my host institute offered me a short excursion there, I explained that this village was so well studied abroad that it would not be useful to science for me to go there. Instead, I mentioned a mountain village that I had heard about in a casual conversation with a Russian geological prospector, who told me it was the most remote and inaccessible village of all. Two days later, the director gave me an air ticket and a friendly Sakha guide called Plato.

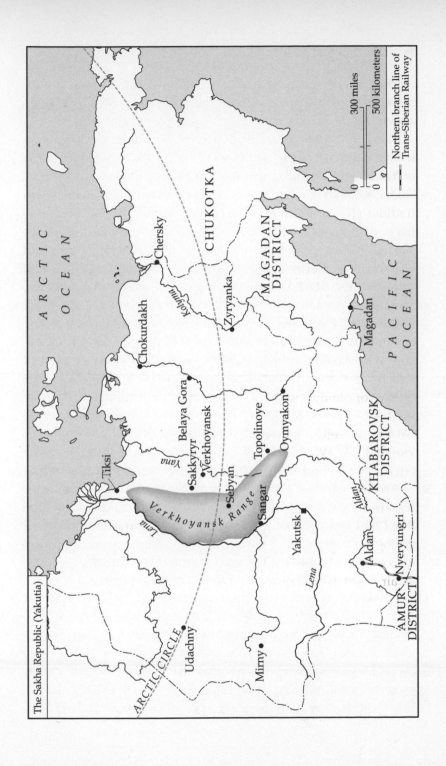

ARCTIC

OCEAN

The Sakha Republic (Yakutia)

Tiksi

Chersky

Chokurdakh

Kolyma

CHUKOTKA

Zyryanka

MAGADAN
DISTRICT

Magadan

PACIFIC
OCEAN

Belaya Gora

Sakkyryr

Verkhoyansk

Yana

Topolinoye

Oymyakon

Sebyan

Verkhoyansk Range

Sangar

Lena

Yakutsk

Aldan

KHABAROVSK
DISTRICT

Aldan

Nyeryungri

Lena

ARCTIC CIRCLE

Udachny

Mirny

AMUR
DISTRICT

Northern branch line of
Trans-Siberian Railway

300 miles

500 kilometers

0

0

2

Civilizing the nomads

The flight to Sebyan was in an Antónov-2, the little twelve-seater biplane that had become the lifeline of all northern villages since it was designed in 1947. I sat with Plato and several silent Eveny as we flew for two hours across a forested plain, snake-trailed with the River Lena like an Arctic Amazon, towards a wall of mountains 6,000 feet high. This was the Verkhoyansk Range, the outer perimeter of a crescent of peaks that ran northward in an unbroken chain for another 400 miles to the Arctic Ocean. As the still calm of winter wears on, temperatures inside this almost windless curve will sometimes plummet to an Arctic record which at −96°F* is beaten only by a site near the South Pole in Antarctica. I would later return in winter one year and feel with my skin what I knew from physics: namely that cold was the basic state of existence in the universe while heat was a limited resource.

Temperatures during the short summers could reach 86°F, so that the range from one end of the year to the other was greater than the difference between the freezing and boiling points of water, giving the biggest temperature difference in the world. Even so, the summer warming of the air did not last long enough to melt more than the top few inches of soil. Underneath, the ground remained frozen. This was the permafrost that forced even the most massive larch trees, rising slowly over centuries, to live on shallow roots. Sometimes spring erosion exposed cliffs of ice made from undergound pools of frozen water that would shear open to reveal the bodies of mammoths, which Eveny legend says scooped up sediment with their tusks at the beginning of time to form the first dry land. Trapped above the impermeable frozen soil, the melting groundwater drained badly and for a few weeks in summer the boggy countryside swarmed with mosquitoes. Like the reindeer, I would find this more unbearable than the extreme cold.

In the third hour, we pierced the mountain wall and entered a maze of peaks, sometimes flying alongside cliffs of tumbling grey shale, sometimes following streams which began as slight grooves near the bare tops of the mountains and coalesced into torrents fed by the invisible draining of huge, sloping bogs. Even in the late summer, there were permanent snowdrifts on some north-facing slopes. It was not the 200-mile distance from Yakutsk on the map, but the capriciousness of these mountains that made Sebyan so inaccessible.

The village, when it appeared, was a sudden clearing of visible human purpose. It had come into being only to control the wild space of the taiga, and had grown out of the material it was designed to tame. The solid Russian-style log cabins were built from trunks of the larch trees that covered the surrounding hills and floored with uneven larch planks. It was an absurd concentration of human presence with 750 people corralled into 1 square mile, from which a mere ninety active herders filtered out across 1.1 million hectares or 4,300 square miles, an area half the size of Wales, to look after 20,000 reindeer. Built by Russians to control not only the land but also the reindeer people, who staffed the administration themselves, the life of the village was rich and complex. In later years, as I returned from long stays in herders' camps or on hunters' trails, the village would feel like a crowded metropolis, as it did to the herders themselves. But the first time I bumped along the tiny airstrip in 1988, I felt I had reached one of the remotest places it was possible for humans to live. If the tiny red and white plane did not come back to collect me, it would take a month to reach a road that would lead me to Yakutsk, 300 miles away overland.

Yet it was not just the physical isolation of the place that made it feel so remote. Like a space traveller who visits a planet in another dimension, I appeared to be invisible. The lack of eye contact did not seem to be from fear, as in Leningrad, but because the people led a self-absorbed existence that left no room for relations with a visitor from another world. Everyone was going about their business, and this business seemed deeply unknowable. The only people who acknowledged me in any way were those who had received very careful official instructions on how to deal with me. I did not yet understand the Eveny cultural ethos of not intruding on others, though as an anthropologist I should have realized that my every move was being noticed and judged. In one house lived Tolya's sister Anna, who had an extraordinary grasp of the details of everyone's family history and was to become a most valuable informant for me. In another house one could visit old Vasily Pavlovich, son of the headman Baibalchan who was taken away by the Soviet secret police in 1928 and never seen again. Vasily Pavlovich was one of Tolya's favourite custodians of the old clan stories, an expert in Eveny poetic usage and composer of his own songs in an archaic Tungus style. Next to another house occupied by his children, an old man who had never got used to buildings lived permanently in a tent. On the outer edge of the village, the all-competent Vladimir Nikolayevich lived with his wife and grandson in a house he had built himself, sharper, fresher, and cleaner than the others.

Scattered around the houses lived the officials of the Farm, the teachers, nurses, clerks, and vets, forming a kaleidoscope of shifting alliances and jealousies around the power of the office of director. Petr Afanasevich, the director of the State Farm and his ally, Old Efimov, shared two halves of a semi-detached State Farm house like everyone else, but benefited from a separate electricity generator. Even herding families now had a house, which they regarded as their main base and storage depot, and their cyclical migration with the reindeer would be punctuated by occasional trips to the village. Herders and hunters, and their families if they visited them for the summer holiday, were almost the only people who ever journeyed into the taiga. If anyone else travelled out of the village, they went by air to Yakutsk or to the district capital of Sangar, a small coal-mining town on the Yakutsk side of the mountain range. In the 2,000 or 3,000 years since they domesticated the reindeer, villages like Sebyan had given Siberian native peoples their first experience of unloading their *uchakhs*, or transport reindeer, never to reload them again.

This fixity is just what the Soviet authorities intended*. Their goal in establishing villages from the 1920s onward was to convert northern native peoples to a 'sedentary' way of life. But this created a Catch-22, which remains the central problem of the reindeer herders' existence today. Apart from mining, there is no way that humans can make a living on this landscape except in partnership with the reindeer; and they cannot live with the reindeer except by following their perpetual migration.

The traditional response to the environment was nomadism, in which entire families lived and migrated together in an annual cycle. In common with governments around the world, the new authorities were confident that such whole-hearted nomadism was 'backward'. Their solution was to reclassify reindeer herding as nothing more than an economic activity. What they called 'nomadism as a way of life' was abolished (or 'liquidated', in the terminating language of their revolutionary fervour), and replaced by a more progressive and civilized 'industrial nomadism'.

The taiga became a giant open-air meat factory and the care of reindeer was isolated from the family and reduced to a worker's job like any other. In a stiff interview in his office, the director Petr Afanasevich explained how it worked, with a stream of statistics which I rushed to write down. The herders were organized into thirteen numbered 'brigades', a term by which the Soviet regime tried to imbue ordinary civilian work with a quasi-military discipline. A brigade usually had a core of long-term members and a number of less committed young men who came and went. A typical brigade contained six male herders and was run by a male 'brigadier' and fed by a female cook called a 'tent worker' (*chumrabotnitsa*). Each brigade looked after a herd of around 2,000 reindeer. The corresponding group-ings of humans and reindeer were referred to as 'brigade', 'herd' or 'camp x'. Over the following years I would learn how each brigade had its own collective style or tone, influenced by the landscape, the animal stock, the personality of the brigadier, personal relations among the members, and how far they were in favour or disfavour with the authorities.

The herders' lives began to oscillate between shifts on the land and periods of leave in the village, using the immigrant Russian miner or oil worker as a model. Sometimes the idea of a shift was taken so literally that the departing herders had only a few minutes under a helicopter's revolving rotor blades to pass on information about the herd's condition to the incoming team. Even the reindeer became a different creature. From being the means of transport to hunt other, wild animals, it was now meat – an end-product in its own right. The herders still kept trained reindeer to ride (males, castrated so as to be more man-ageable), but the size of the herd was increased hugely to pro-vide a constant supply of young animals for slaughter. This increase was made up mainly of breeding females*, generally amounting to 80 per cent.

Only the men who tended the herd every day were considered to constitute 'the able-bodied population directly concerned with reindeer herding, for whom nomadism is essential', and were kept on the land. The patriarch of a herding family was replaced by the brigadier, who was still surrounded in the same way by unattached young men who moved restlessly between brigades. But the women who cooked, sewed, and collected berries and herbs, as well as providing laughter, affection, and partnership, were dismissed as 'not directly involved in reindeer herding'. Being 'unutilized labour resources'*, they were removed to the village and given typical Soviet female occupations such as cook, nurse, administrator, accountant, teacher, and cleaner. 'Civilized' but loss-making forms of farming were introduced to keep them occupied, like rearing arctic foxes in cages to provide fur and looking after cows, which could not survive without hay and shelter for nine months of the year. It took two generations to drain the women off the land. By the 1980s, just one was left in each brigade as a paid tent worker to cook for the male herders.

Long before I reached Siberia I saw a socialist realist mosaic in the Moscow metro, in which representatives of the diverse peoples of the Soviet empire, each dressed in their national folk-costume, brought the fruits of their diverse ecological zones to the great capital. This was the imperial vision of Persepolis, but it was something more. Like other great empires, the Soviet Union swallowed up smaller communities, using the resources from the frontier to sustain consumption and redistribution at the centre. But this was an empire unlike any other. Whereas others would have left the natives as barbarous tribes, merely extracting tribute from them as the tsars had done, Soviet civilization undertook to incorporate these minorities into the grand plan of human evolution, giving them an education and even a path into the ruling elite.

In 1836 Pushkin, the great icon of Russian high culture, had written the lines:

> *Slukh obo mne proydet po vsey Rusi velikoy,*
> *i nazovet menya vsyak suschiy v ney yazyk,*
> *i gordiy vnuk slavyan, i finn, i nyne dikiy*
> *tungus, i drug stepey kalmyk.*

My reputation will spread through all of great Russia,
And every living being will cite me in their own tongue,
The proud descendant of the Slav, and the Finn, and the still wild
Tungus*, and the steppe-loving Kalmyk.

What Pushkin had written as hyperbole, the Soviet regime would make a reality: the Tungus peoples *would* recite him, maybe even in their own language but certainly in Russian – and they would cease to be wild. When Tolya jokes, whenever he does anything outrageous, 'I am a wild Tungus', he is making an educated person's literary joke. Between the 1920s and the 1980s, the Soviet Union turned the uneducated pre-revolutionary masses into one of the most literate populations the world had ever seen (by 'wrestling' with illiteracy and 'liquidating' it). Books on philosophy, literature, and natural history regularly sold 100,000 copies upon their first printings and even volumes of poetry or studies of the burial mounds of ancient Scythians might sell 1,000 copies. In the 1980s one would see rows of silent commuters in the crowded undergrounds of Moscow and Leningrad, their heads buried in serious books covered in newspaper to protect the binding (and, in those open but still uncertain times, perhaps to hide the title: in 1988 I was roughly arrested on the Leningrad metro by young Communist vigilantes who noticed that I was reading a book on shamanism*).

The wild northern native peoples were included in this literary programme, and many of their brightest sons and daughters were taken to the Hertzen Institute, a special native training college in Leningrad, where they were turned into teachers, administrators, and party workers before being sent back to their homes as internal missionaries for the new Soviet culture. From 1947 Antónov-2 biplanes made it all the way to the herders' villages and by the 1960s helicopters flew directly to their nomadic tents. Whole families were sent on free holidays throughout the Soviet Union and neighbouring countries of the socialist bloc. Though I knew that men had served in the Army when young, I had not expected the rugged men and women who looked as if they had never left their mountainside to tell me about their rest-cure at a health farm on the Black Sea, their tour of Genghis Khan's medieval capital in Mongolia, or their conversation about Kafka in a Prague cafe with an intellectual from Cuba.

A new official terminology came into being, which revealed how the village, surrounded by concentric circles of pasture, was itself seen as lying at the outer edge of a far grander concentric space with its centre in Moscow. Villages were designated as 'points of population' – implying that no population could exist without them or beyond them – and 'points of supply', as though supplies could come only from outside and not from the land itself. Locations were chosen for ease of access to the outside world rather than to the reindeer pasture. With a conspicuous contempt for the herders' own perspective, one official textbook explained* that if villages were sited to suit their access to pasture, this was a 'mistake' and they should be closed down or moved. The site of Sebyan has remained constant since the first huts were built in the 1920s, but many other villages across the North have been repeatedly relocated.

Every empire has a contradictory attitude towards its frontier areas, seeing them variously as a source of raw materials, a security buffer, and a social responsibility. The Soviet State expended huge efforts to seek out the remotest, tiniest communities and make them their 'own', *svoy*, a word which also implies making them less strange or alien. Soviet policy was moulded by an ideal of 'mastery' (*osvoyeniye*) of the North*, a term that literally means 'appropriation'. This mastery was to be accomplished through a combination of futuristic technological progress and tight political control. From the central point of Moscow, distance was seen as the country's greatest asset, because it encompassed infinite natural resources. Aviation and radio would harness distance and abolish remoteness. While international borders were closed to keep out spies and subversive foreign ideas, the interior of the country was turned into a homogenized space in which Soviet citizens could be moved from one end of the country to another and find almost identical conditions wherever they went.

Today, every member of a reindeer-herding community lives in a complex state of adaptation or compromise with the structures under which they have grown up. The regime standardized the management of Collective and State Farms thousands of miles apart, containing peoples with diverse cultures and reindeer skills of their own. This policy did not prevent local conditions from producing very different outcomes, however. On the Chukotka Peninsula facing Alaska, a few thousand Chukchi herded a population of reindeer that by the 1980s had reached over half a million head, an industrialization of reindeer herding that damaged their family structure and made them helplessly dependent on State Farms. The Nenets of the Yamal Peninsula*, on the other hand, managed to retain their traditional family organization as well as many of their own private reindeer and have benefited more fully from the steady increase in their huge herds. But they also live in the shadow of the same kind of massive oil and gas development that has made the lives of the neighbouring Khant and Mansi wretched* as they endure heavy construction machinery crushing and ripping up vegetation, the humming of giant gas pipelines passing ten-deep across their migration routes, and the spilling of millions of gallons of oil over their pastures.

Sebyan was a classic example of a Soviet reindeer-herding community, its inner workings clearly revealed in the almost total absence of mining or mineral exploitation* on their territory. When I first arrived in 1988, there were three kinds of authority: the Government, in the shape of the Village Council; the Communist Party; and the State Farm.

My first friend Tolya was president of the Village Council, or *soviet*. This was the tiniest level out of the many soviets – through

District Councils, Regional Councils, up to the Council of Ministers (Soviet Ministrov) – in whose name the Union of 'Soviet' Socialist Republics (USSR) had been established. When the Soviet Union was abolished after 1991, the council was renamed the 'administration', a more accurate description of its function.

The Village Council occupied a large wooden building next to the village hall, or *Klub*, which it administered. By the large double-glazed window a big, bushy fuchsia and hibiscus survived through the twenty-four-hour darkness of winter and flourished in the twenty-four-hour daylight of summer. Tolya had two desks arranged in the classic Soviet official's T-shape which I have found throughout the world, even in the Russian consulate in Kathmandu where it contrasted like a foreign visitor with the South Asian deskscapes of other local offices. The official sat behind the crossbar of the T, his desktop a background of calendars and memoranda pinned down under a heavy sheet of glass. Visitors sat ranged on either side of the stem of the T, turning their heads to face him. In the centre of the wall behind Tolya's chair hung a large oil painting of Lenin in sober shades of brown with small highlights of Communist scarlet, painted with a brush as if from life, but actually copied like thousands of others from a mass-circulated photo which itself already served as a stylized icon of the founder of the Soviet State. As visitors craned their necks towards Tolya, the painting surrounded his head like a halo and suggested an identification between the two men. As head of the civil administration, he was responsible for records of births and deaths, the electoral register, social security, and the preservation of public order. He was supported by female secretaries and accountants in the next room, and an impressive round rubber stamp.

The 'Soviet Union' was in fact controlled, not by the elected soviets, but by the unelected and self-perpetuating Communist Party. The entire structure of government throughout the country was shadowed by a parallel, and more powerful, party structure. Only members of the party were likely to be considered reliable enough to become officials in government. Tolya himself was a party member, and probably could not have stood for election to the soviet otherwise.

The Communist Party maintained its own representative (*Partkom*), a Sakha man from outside the village. He had big protruding ears, which seemed a literal expression of his political function. In the late 1980s the presence and the lingering fear of the party could be felt very strongly in big cities. But in this remote village the party agent already seemed ineffectual. The Communist Party of the Soviet Union, like the Soviet Union itself, was dissolved at midnight on 31 December 1991, but the *Partkom* had already left the village more than a year earlier.

The disappearance of the party did not strengthen Tolya's local council, as it might have done in a less isolated place. The village managed without a resident policeman or KGB officer; it would manage without a party representative, too. Because Sebyan lay at the furthest end of a thin line of supply, real power was held by the Farm's director*, an Eveny who was Tolya's relative but with whom Tolya was in conflict. It was the director who decided who could use a Farm tractor to fetch winter firewood, who had a piece of hardboard flown in so that they could cover their floor, and what could be done with every one of the Farm's 20,000 reindeer. As an indication that the director was worth keeping under surveillance more than the president of the Village Council, the party representative's office had been located next to the director's office in the building of the Farm administration, where the party man shared the same all-hearing female receptionists.

As a newly arrived anthropologist, I could not predict the long-term consequences of my first friendships. Even without being aware of it, I acquired not only my friends' friends but also their enemies. By becoming close to Tolya, I immediately found myself in a difficult position with the director's faction. The director's advice to Tolya had been to show me around (he could not avoid this since I had been cleared at a high level within the party itself), but 'whatever you do, Tolya' – I can imagine his sudden switch to an intimate, persuasive intonation – 'don't give him any *information!*'

If the director had things to hide, however, other people had things to reveal. At first, I thought I was simply witnessing a local refraction of the battle within the Kremlin between reformists and the Old Guard, the *perestroika** that was to lead within three years to the collapse of the Soviet Union. But as I came to understand the community better, I realized that this was also the latest phase in an old battle between families. Tolya's father and the director's elder brother, both long dead, had been enemies. The imposition of Communist control in the 1920s and its relaxation under *perestroika* in the late 1980s had simply provided new idioms for such conflicts, which continue to this day in the twenty-first-century idiom of the 'market economy'.

I also came to perceive a tension between the roles of the Village Council and the State Farm, two different strands in the legacy of the old headman Baibalchan. One was concerned with public services and the other with economic resources. Each was headed by a person who seemed to embody their organization's moral character, almost like two collective personalities, the one striving to provide the population with facilities, the other to withhold and sequester. The Eveny and their reindeer in Sebyan were separated by a daunting mountain range from Sangar, the capital of a low-lying district of Sakha cattle herders, Ukrainian miners, and Russian producers of natural gas. In 1988 Tolya was leading a campaign to detach Sebyan from Sangar and unite it with the Eveny village of Sakkyryr 200 miles to the north, forming a new district that would be attuned to Eveny needs*. Sakkyryr was on the same side of the mountains and would offer a new, more reliable route to the outside world.

Tolya's proposal was bitterly opposed by the director, who had much to lose by sharing his power with other Eveny leaders. His State Farm overlapped with the village almost completely, forming a total institution like a monastery, prison, or military camp. The tracks in the mud around the village ran out into the bog, with nothing but a few nomadic tents for 200 miles until the next village, and the director controlled even the allocation of seats at the airstrip. The isolation and the control reinforced each other. Almost everyone was an employee of the Farm. Even the staff of the hospital and school, though paid through the council out of separate government budgets, were equally dependent on the director for basic materials and facilities. The director's power to grant or withhold firewood, trips to a Black Sea sanatorium, or promotion in office and brigade; the awe of officialdom aroused by red banners signifying the transcendent unity of Soviet State and Communist Party: these seemed immutable, like the laws of nature that governed the annual migration of the reindeer. It was illegal in the Soviet Union to live anywhere except where you were registered, and it was difficult for anyone born in the Farm to change their registration and move to another place except by entering an urban profession or going to jail.

The State Farm's offices were much more elaborate than the office of the Village Council. They occupied two large buildings, their log skeleton fleshed over with plaster, some way from the Village Council but next to the hut where Sasha the Radio Man (*radist*) maintained the vital radio link with Yakutsk. The first building housed the office of the director and also accommodated the following officials or departments:

Chief Accountant
Accounts Office
Head of Personnel Department
Chief Engineer-Mechanic
Chief Animal Technician (a Ukrainian, much distrusted by
    the herders)
Chief Veterinary Surgeon
Senior Construction Worker
'Technicians' (i.e. cleaners)

Head of Trade Union
Chief Economist

The second building contained the offices of:

Divisional Manager (for production of animals)
Deputy Divisional Manager
Brigadier of the Village (responsible for village maintenance,
    including supply of firewood and water, and directing a
    team of mechanics, electricians, and carpenters)
Head of Animal Husbandry, Reindeer-Herding Section
Deputy Head of Animal Husbandry, Reindeer-Herding
    Section
Animal Technicians under the command of the Chief
    Animal Technician in the first building
Veterinary Surgeons under the command of the Chief
    Veterinary Surgeon in the first building
'Technicians' (i.e. cleaners)
Head of Stores
Head of Canteen

These offices employed various assistants, receptionists, secretaries, and other support staff. The various offices in the village
ran two brigades of horse herders, six brigades of woodcutters,
and several brigades of winter fur hunters (often reindeer herders redeployed during the slack season for herding); several
brigades of electricians, construction workers, and fetchers of
water from the lake; a fur-sewing workshop; an airstrip; a garage
for trucks and tractors, with mechanics; three village shops; a
day school, boarding school, nursery, hospital, and canteen; the
village hall; a fox-fur farm, later abandoned; a small farm of
cows and pigs (privatized in 1992 and the animals immediately
eaten, since no one could afford to maintain them without subsidy); a post office; a radio and communications room (Sasha
the Radio Man); a Cultural Agitation Brigade (an entertainment

troupe, run by Motya the Music Woman); a diesel generator for the light bulbs and electric sockets in every house; and several oil-fed boilers for the public buildings (houses are heated by individual wood-burning stoves). Out of a population of 750, all these activities occupied some 300–400 people, virtually the entire population of working age.

Like most members of the community, I found the goings-on in these offices compelling and seductive. The space was confined, the networks tight, and the intrigue intense. Herders, villagers and anthropologist all needed to keep abreast of every move, each for their own reasons. Out in the camps, the herders lost no opportunity to remind me that this edifice of management, accounting, and servicing existed only at the expense of ninety men who rode around the mountains on their reindeer in all weathers and came back to tiny tents where they were fed on tea and reindeer stew by thirteen women.

This was a total reindeer economy. Only reindeer herding produced a commodity that could be sold in the outside world, as the meat was laboriously flown across the peaks to the Ukrainian coal miners in Sangar and beyond. In the 1980s reindeer meat still made a substantial profit. Every other activity, even when the real cost of anything at all was masked by the Soviet tangle of cross-subsidies and phantom accounting, ran at a severe loss. Though this was to change beyond all recognition in the 1990s, reindeer herders in the 1980s were fairly well paid and well provisioned, and their exotic holidays were provided free.

Herding was highly regulated. Traditional migration routes had been rationalized to occupy exclusive, non-overlapping territories. Within each brigade's territory, the six seasonal pastures were also exclusive. These areas were prettily painted on a map in pastel wash: yellow for early autumn, dark brown for late autumn, blue for winter, pink for early spring, light brown for late spring, and green for summer. Each pastel area was marked with the precise dates when the herd should be there every year. The 'carrying capacity' of each seasonal territory had been calculated by dividing it further into small parcels and working out the number of reindeer-days' grazing that each parcel afforded. A copy of the map for each brigade's territory was supposed to be issued to each brigadier.

As with any ideal world set up by the gods who then retreated back to a distant heaven, it had been impossible to stick to this blueprint. I started out in the late 1980s by trying to visit various brigades supposedly located in specific directions from the village. But I quickly learned that entire herds had moved around in the twenty-five years since the master-map had been personally signed by its botanists and surveyors and stamped and certified by their scientific institutes in distant centres of learning. Even within the territory of a brigade, seasonal pastures were sometimes changed around or reversed, and areas that had become overgrazed (despite the minute calculations of the scientists) were left fallow to recover.

To monitor this production, the Farm maintained a painstaking system of statistics, measuring everything from weight of meat delivered at the slaughterhouse to survival rates of newborn calves, the summaries of which were collated into typed and carbon-copied tables of figures. The number of reindeer in each herd went down by several hundred after the autumn slaughter and up again after the spring calving, so that the head-count moved in two parallel bands showing a high point in June and a low point in January, when a census was taken after each of these events. At the June count in 1988 the thirteen brigades in Sebyan were officially (but unrealistically) reported as having around 2,000 reindeer each, giving a total of 26,000*.

Just as in any other Soviet factory, productivity was rewarded with radios, cash prizes, and holidays. But even more importantly, material awards were paralleled by a system of social approbation. All human qualities were submitted to the judgement of the Board of Honour in the corridor leading to the director's office, bearing photos with captions hand-painted in a neat, faded watercolour lettering that stands to this day. Here was the legendary female brigadier of camp 5, Sofia Kirillovna, Order of Labour of the Red Banner, Order of the Glory of Labour Third Grade, posing severely in a short haircut with her medals pinned on to her jacket, next to a list of her prodigious feats of meat production. Here, too, were the brigadiers of the three camps that I would come to know best over the following sixteen years, each with a list of his honours: the Old Man, patriarch of camp 7, with his tousled white hair and startled photo-studio expression; Kostya, the steady and competent younger brigadier of camp 10; and Kesha, the flamboyant brigadier of camp 8, with his eccentric flowing black tresses.

The talk in the director's office was of reindeer production, but the nearest reindeer were 30 miles away and some were six times that distance. I had spent two years obtaining permits to come as far as this village, but here was a final barrier I had not expected. There seemed to be no rules forbidding me from going to a herders' camp: my visa was as valid for one part of the State Farm's territory as another. But the only way of getting there was on the back of a horse, an animal introduced recently from the Sakha that could be brought into the village where reindeer could not. I had no idea where to go, and did not even know how to ride a horse. My enquiries were met with replies that seemed beyond challenge. All the horses were grazing far away behind the mountains, nobody was available to go looking for them, and nobody was intending to travel to a camp. I felt like an early Tungus faced with the taiga before the invention of the reindeer saddle.

I was sure the director was hoping that inertia would drive me back on to the next plane. But by then I had met Tolya, who had recently become a friend of Plato's. Tolya was thirsty for the political reform that was already in the air. He had been locked in conflict with the backward-looking director of the State Farm, a conflict that would grow into a brave and punishing struggle to improve his people's life*. Maybe he saw in me the potential ally I would later become – though he could not have foreseen that our conversations and our adventures around Siberia and abroad would lead him to become an anthropologist himself – or maybe at this stage he saw me as just a novelty. Tolya helped me to find a vet who was going out to a camp the following day. The vet agreed to take me after I assured him that I knew how to ride. The next day the horses appeared, along with several other herders returning to their camp.

The horses had not been ridden that year and still felt half-wild, rearing and kicking for an hour before they could be per-suaded to stand still and be loaded with baggage. I barely knew how to rise to the trot in a one-two rhythm, but in any case one stirrup was missing and it was hard to grip the other with my huge, floppy rubber boot, folded down in several layers around my calf. The saddle had an exposed metal ridge which bruised my coccyx at every jolt. I fell behind repeatedly and whenever we reached a dry surface, someone would slap the rump of my horse and it would break without warning into the triple rhythm of a canter. Each time we plunged down a steep slope, I seemed to be leaning horizontally over an abyss. I had no idea where we were going or when any of these horrible feelings would stop.

We camped on a stony riverbed the first night and I found time to stop being frightened and look around at the wiry figures of my companions in the flickering firelight. After my years of fieldwork in India, their weather-beaten North Asian faces seemed completely new and exotic. Some wore flat caps and some headscarves tied around the back of their heads. With their rifles and padded cotton convict jackets, they resembled brigands or mountain guerrillas. The vet asked if I had really ridden a horse before.

'Actually, no,' I said, 'but I was afraid you wouldn't take me if I said so.'

The men roared with laughter. This was my first inkling of the self-reliant and anarchic spirit that coexisted with the delicate discretion of traditional Eveny culture as well as with the nervous fear under Communism of doing anything that was not officially authorized. As I listened to their talk around the fire, I also realized that there was a life here that could not be reached through the husky, swishing sounds of the Russian language. When I had arrived in tribal India I did not know any Indian language and had to learn the tribal language in order to communicate at all. Here, however, my reliance on Russian, which the herders had learned at school to speak well, made it harder to acquire the languages they used when talking to each other*: Sakha, a language that condenses a whole sentence of meaning into a single, long word with matching vowels that race off the tongue, and Eveny, a language whose emphatic vowels lie open at the end of each shorter word or else are cut off by a clipped final consonant.

On the third day, we reached brigade 11 and I met my first Siberian reindeer. The main herd was out of sight beyond some mountains, but small groups of animals were grazing near the tents. From time to time, someone would lead an animal from one place to another and tie it up, or else saddle it and ride away, but I did not understand then about the different categories of reindeer and how they were used.

The Eveny taiga reindeer are bigger than the Sami type I had seen in Norway, and in the old days one animal could be exchanged with the Chukchi for two or three of their own smaller tundra reindeer, which the Eveny would use only for meat. As I reached out to touch one of them I felt I was face to face with a creature of considerable presence. Its fur was chestnut brown, shading whiter where it thickened around the neck. Its coat was so deep that I could run my fingers through it and hardly feel the solidity of the body beneath, but for all its depth the fur was neat rather than shaggy, with a rich, smoky smell that was to become a constant part of my life. Most compelling were the eyes, which were huge, soulful, and capable of engaging one with an intense gaze. This animal, if not eaten first by humans or wolves (or killed by a wound, virus, or parasite), would live for twelve to fifteen years. Its nose was covered with a tough, leathery skin which could be thrust into cold snow when grazing. Within this hard nose, I would later learn, lay nostrils so sensitive that they could detect not merely a human hunter upwind but the species of a plant hidden beneath the snow or the location of a mushroom growing 100 yards away. I held out some salt on the palm of my hand and the reindeer responded with a lick of its long, raspy tongue.

I stayed for a month until late September 1988, living in a bachelors' tent and taking part in the autumn tasks of sawing antlers and cutting down larch trees to build a corral*. This was also the first time I met Vladimir Nikolayevich, a jovial older man with a ruddy, round face and a singsong intonation, who had retired as brigadier of camp 9 and now travelled round other brigades joining in their work for fun. Vladimir Niko-layevich was widely admired for his competence. His knives, guns, and harnesses always worked smoothly and in years to come I would benefit from his skill and friendship when he took me on some wonderful journeys through the mountains*.

At the time, however, the Soviet Union still seemed so closed, and the chain of political contacts, official permits, and chance encounters that had brought me here so extraordinary, that I never expected to repeat the feat. Still less did I imagine that I would visit other reindeer-herding peoples across thousands of miles of the Russian North and send students to live there, or that my family would one day stay here and my children would come to have reindeer of their own.

PART II

A TALE OF TWO HERDS

3

The massacre of Granny's 2,000 reindeer, camp 7

I returned to Sebyan in the spring of 1990, and spent much of that year moving between brigades 7 and 10. At one of the great turning points of history, I saw how differently people in the same community responded to the impetus of sudden reform, just as comets that cluster together close to the gravitational field of the sun are flung out again on different trajectories, each to its own lonely destiny at the outer reaches of the solar system. In that year, brigade 10 was rewarded, and brigade 7 was destroyed.

The Soviet Union was shifting towards a 'market economy', based on the principle of 'cost-accounting' (*khozraschet*). Now every level, from the Sakha-Yakut Republic itself, through the State Farm down to each little brigade of six men and one woman, was to be 'self-financing'. The imperial duty to civilize remote northern peoples, the geopolitical concern with the security of northern frontiers – once these were held up to the harsh light of cost-accounting, the economic drain of even the most productive State Farm became exposed.

Farms across the North were instructed to convert their employees into free 'contractors' and pay them according to their productivity. Whereas the private ownership of reindeer could once be punished by imprisonment or death, now it was encouraged, and some of the Farm's reindeer were already being given to herders as prizes instead of cash or radios.

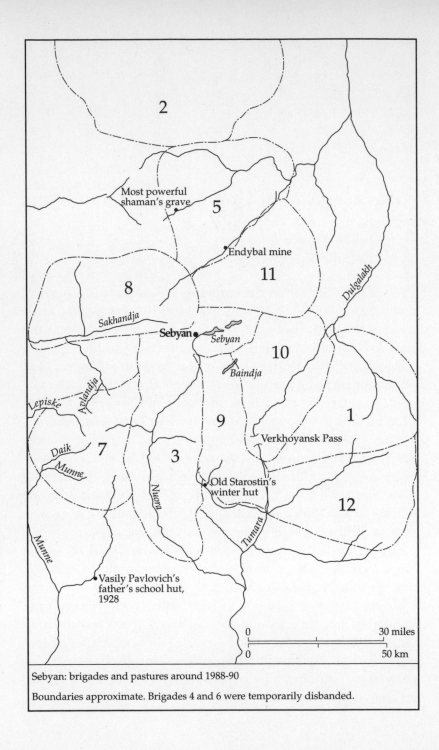

Sebyan: brigades and pastures around 1988-90

Boundaries approximate. Brigades 4 and 6 were temporarily disbanded.

The Farm managers could not all keep up with this change of policy. The director of Sebyan seemed quite confused.

'In the first place,' he 'explained' to me in his office that April, 'we need' – he paused – 'economic self-sufficiency. Through this, clearly we'll need to get ourselves technically equipped, to use modern technology. Then we'll be able to ensure more or less proper social and working conditions for the herders. But until then' – he hesitated – 'the economy is difficult, especially, um, well, how to put it?'

The director looked down intently at the papers on his glass-topped desk, as if seeking a prompt. He was wearing a grey suit and had a rather long face for a native, like an Asiatic clone of Brezhnev. After a moment his voice picked up again and his tone became more confident. 'Since the war, it was altogether difficult to put together the economy. They only rebuilt the cities, the ruined cities.'

This seemed strange, forty-five years after the war had ended and thirty years after the start of a concerted new phase of mastery of the North through industrialization and aviation. I did not see the director's point.

'Not long ago, they began to pass on to us the technology, both on land and in the air,' he continued. 'But this hasn't yet reached us, it's only the beginning. We hope that now in this period of *perestroika*, especially in the economy, a lot will be done on this question. The Supreme Soviet is studying the market economy. And if, say, a new life will begin, through this economic self-sufficiency, that means effort. You work yourself and earn your wages yourself.'

At the time, I laughed to myself at the incoherence of the director's speeches. Instead, I should have been terrified by his inability to grasp the enormity of the changes in which he was being swept up. His last point, at least, had seemed clear. But though the herders would indeed continue working through the 1990s, they would receive their wages more and more rarely. With the drying-up of subsidies from above, the Farm was already having difficulty paying them to look after its own animals and was failing in its contractual obligation to provide a marketing system so that the herders could sell their own. They would soon have almost no cash to pay for butter, sugar, tea, clothes, guns, ammunition, licences, medicines, education, or transport. The village would become a subsistence economy, in which meat and fur were plentiful, but in which almost nothing else – except for vodka – was available at all.

Brigade 10 conformed to the modern Soviet ideal, an official factory model of several male herders who worked together as a team rather than as a family, under the capable, hard-working young brigadier Kostya. Kostya, his deputy Arkady and the other members of their brigade would leave their families in the village and go out to the herd in shifts for a few weeks at a time, fed there by Arkady's mother Kristina, who was paid by the State Farm to cook for them all. A large proportion of their calves survived each year and they regularly won prizes for their high production of meat.

The structure of brigade 7 was quite different. Though they also had a house in the village, this family was the nearest anyone could get in modern times to the old nomadic way of life. They had been living in an isolated region in the mountains for eighteen years, and through selective slaughter of their reindeer stock they had built up an exceptionally strong herd. But the family's independent spirit cost them dearly. All four children were still unmarried and childless, so that there were no long-term heirs to this wonderful herd. The director of the State Farm did not share Moscow's new-found enthusiasm for the creeping privatization of the Farm's reindeer herds, and this would soon draw the family into a fateful conflict.

In the spring of 1990 I had heard about this family and wanted to include them in a television documentary*. Yet when I first made the elderly patriarch's acquaintance during one of his visits to the village, the Old Man was strangely reticent. I did not know why until I brought the crew back in September to continue filming, and the Old Man himself asked us to visit his camp and tell their story.

That hot September was the first time I saw the lovely autumn site of Tal Naldin, a wide, open platform beside a rushing stony river, fenced by high peaks and suffused with the scent of poplars. The life of the family seemed steady and calm as they worked and relaxed in the mild autumn sunshine. The Old Man had a hoarse voice and a tuft of white hair like a brush. He was slightly tubby and walked around the precipitous slopes of his landscape with a slow, weighty step, made stiff by arthritis. His wife was tiny, wrinkled, and alert. Her children called her 'Ma', and though I was slightly older than they were, I called her Granny (Upé in Eveny, or Ebé in Sakha). Granny and the Old Man would make wry little remarks about the politics of the village. Every gesture between them spoke of their lifelong partnership as joint heads of the family and of the herd.

Their daughter Emmie was in her late twenties. She appeared to be simple, but I later realized how clever she was. Whereas her sister Masha was a schoolteacher in the village, Emmie had somehow managed to avoid school altogether and had lived in the mountains all her life. She and Granny would often sit gazing at each other in silent mutual understanding. The oldest son of Granny and the Old Man, called Oleg, had recently been killed in a fight down in Sangar, so the main work with the herd was done by three young men – their sons Ivan and Yura, both in their thirties, and an orphaned relative of about thirty whom they had adopted, called Gosha. Ivan was the Old Man's successor as head of the herd – in official terms, the brigadier. Yura and Gosha breathed more easily without this responsibility. Both were gentle beings and Yura had a lightness of presence which at times seemed almost ethereal.

The Old Man had moved his family into the highest mountains in 1972 to colonize the territory. Because of its remoteness, the pasture had not been used for a long time and the winter lichen was very rich. It may also have suited them that they were several days' ride away from the Farm management. Over the years brigade 7's distinctive strain of reindeer was said to have become hardier and meatier* than any of the Farm's other animals.

But this particular brigade also provided living confirmation of the early revolutionaries' concern that a family who herded together would continue to plan their own long-term future, rather than surrendering it to the Farm. If I had known them earlier, and had understood enough of these matters, I might have predicted that the year's sudden unravelling of long-standing political structures would lead them into a clash with the Director.

'Our father spends all his time in the taiga, and he knows which reindeer have stamina and which are weak,' Ivan explained to me. 'He sends the weak ones immediately for slaughtering, and leaves the healthy ones. That's how we get this good stock.'

'You're almost the only real family left. How did you manage it?' I asked him.

'How can I explain it to you?' he replied, looking across to where the Old Man was stomping purposefully around the corral, checking reindeer. 'Our father has constantly kept us around himself, so we can live together, so it will be easy for us. He's worked, he's toiled. He wants us to be comfortable, all working together.'

Camp 7 was so far one of only two brigades in Sebyan which had switched to working on a 'family contract'. The various forms of family, brigade, or leasehold contract being offered to reindeer herders corresponded to similar contracts in all other sectors of the national economy, as the first stage in dismantling large, State-owned enterprises. The family contract seemed to match the vision that Granny and the Old Man had nurtured for so long. After sixty years of undermining the family, the State had decided that a family of herders would work together better than the groups of unrelated people who supposedly made up the brigades today. It seemed that for eighteen years camp 7 had been ahead of the times and the world had finally caught up with them.

Contracts were now drawn up between the Farm and a group of herders. What each side actually did would not be very different from before: the herders would herd and the management would manage. The brigade was given an annual working budget of 20,000 roubles (at that time US $3,000–4,000) for allowable expenses. The Farm undertook to supply the same basic services as before, such as sending peripatetic veterinary facilities to the camps and guaranteeing to buy their meat and antlers, which it would market onward to the outside world – the meat to feed local mining towns, the antlers for medicinal use in Korea*. The workers' obligation, as before, would be to deliver a specified quantity of produce. The crucial difference was that the State would not underwrite the consequences of their mistakes or ill luck, and the State Insurance Company (Gosstrakh) would cover horses, but not reindeer.

The legal status of the worker was to change from employee of the State to notionally free contractor. This was supposed to aid democratization, in the official phrase, by giving back to the herder the 'sense of being one's own master' (*chuvstvo khozyayna*). The concept of a contract was unfamiliar to Farm managers and herders alike, all of whom knew only how to work according to instructions from above – or to subvert those instructions when they did not suit. Everything depended on the spirit in which both sides discharged their obligations. Some herders saw the contracts as an economic opportunity: 'We get the money, so we're more interested in sending in first-class meat,' as one put it, adding, 'earlier, the Farm kept the profit and we made do with a certificate of honour!'

But the shrewder herders already suspected that the Farm management would find new ways to hoard profits and offload risks. Herders would now be liable for deaths in the herd. Would the Farm be liable for losses to wolves if the Farm itself failed to keep them under control? No one knew. Transport between village and camp would become the herders' own responsibility, and use of helicopters above a low basic level would be charged to the brigade's own limited budget. The homogenized space of the Soviet Union was beginning to wrinkle: no longer would one call out a helicopter because one was short of tea or sugar.

Across the North, Russia's domesticated reindeer had begun passing into private hands as it became legal to own reindeer beyond a small number of *uchakhs* for one's own transport. In 1988 the ceiling for ownership of private reindeer was raised to forty. By 1990 all limits had been removed and in Sebyan it was said that while most people still had very few, some now owned up to 100 reindeer. Rewards for fulfilling one's plan now went beyond citations on notice boards, and even cash prizes, to the transfer of the Farm's reindeer into brigade members' own possession. This policy directive from above must have pained the director, who, like many local bosses across the North, had done everything he could to impede private ownership of reindeer (except, as his opponents pointed out, among his own close relatives in camp 9). Privatized reindeer were herded alongside those of the Farm and new earmarks were invented to distinguish them. Herders looked after them more carefully, and so they survived better and multiplied faster than the animals that remained State-owned.

But Granny and the Old Man were not to be allowed to enjoy this vindication of their life's work together. During a routine veterinary visit in the spring of 1990, some of their animals were diagnosed as infected with brucellosis. The Farm decided to destroy the entire herd of 2,000 animals, sparing only a few essential *uchakhs*. The old couple were technically retired, but as a result of the slaughter, Ivan, Yura, and Gosha, the working men in their family, would have to be redeployed across other brigades hundreds of miles apart. This was a drastic reaction, and the family felt that it was also vindictive.

Under yellow leaves and a sky as blue as an Aegean summer,

10 стадо летка 9 стадо 7 стадо.

1 стадо 2 стадо. 4 стадо

6 стадо. 8 стадо 12 стадо

5 стадо 3 стадо

моя летка Гоша Kostya

Аркадий + this sign don't bother to change if given as gift – they don't live long.

All signs registered in совхоз

Earmarks for the herds in Sebyan, head facing towards the viewer. The first
four lines represent numbered camps, the last line represents the personal
earmarks of Arkady, Nikitin, and Kostya. Drawn by Arkady in camp 10.
The notes in my handwriting say 'all signs registered with sovkhoz [State
Farm]' and 'don't bother to change if given as gift – they don't live long'.

we nibbled pieces of a marmot which Yura had brought down
from a mountain while Granny and Emmie boiled endless
kettles of tea. The Old Man sat cross-legged in the sunshine, and
put down his mug on the dry, springy moss beside him.

'As part of *perestroika* we took on the family contract,' he
explained in his hoarse voice, 'but somehow it didn't work out.
And the lease contract. That also didn't work out. So now we're
in a situation of half-lease [*poluarenda*].' His weather-beaten face

creased with laughter. 'And we have a corresponding kind of life,' he wheezed. 'It's called half-living [*poluzhizn'*]!'

Ivan did not have his father's mordant humour. When I interviewed him for the camera, he spoke in a low, lugubrious voice.

'We've already agreed with the Farm, this October we'll completely liquidate this herd,' Ivan muttered. 'Completely. Except for the *uchakhs*, the riding reindeer – they'll do a blood test, to check them. As for the rest, we'll liquidate them all.'

'What will happen to the herders?' I asked.

'The herders . . . ?' He paused. 'We'll carry on working in other brigades.' He paused again. 'We'll carry on working, not in this place, but somewhere else. Somewhere where there's no disease. For two years or three years, this pasture must be left fallow.'

'It has to rest,' echoed Yura, who was standing nearby. At this first meeting, he did not display the lightness and sense of fun which I was to find so characteristic in later years.

'So that the pasture becomes' – Ivan paused – 'fresh, new. Before establishing a new herd here, the pasture needs to cleanse itself.'

'Up till now you've been one family,' I said.

'Now we're forced to live apart.' He paused for a long time, standing stock still in the dappled sunshine under a larch, his eyes filling with tears. Bells tinkled behind us* where Granny and the Old Man were working to the last with their doomed animals, tying and releasing, tying and releasing. I did not want to be the first to break Ivan's silence. At length he continued, his voice choking, 'Of course it's bad to be torn away from one's close family. We grew up in this place, we were born here, this is where we spent our childhood, everything here is connected with our childhood.'

'We're used to where we live,' added Yura. 'We can't imagine living any other way.' He, too, paused and the distant bells flowed into the silence once more. 'Our land, our mountains tug at us.'

'And your magnificent animals?'

'Ah, that's the most painful question,' responded Ivan. 'These strong reindeer will be lost. They're our personal reindeer, the best stock in the Farm. They'll disappear, it'll be the end.'

In October every year, each brigade drives 200–300 of their own reindeer, generally the weakest which will not survive the winter, to the slaughterhouse on the edge of the village where the men lasso the animals one by one and stab them at the base of the skull, causing immediate death. But in October 1990 the Old Man and his sons were forced to participate in what must have felt like a carnage rather than a harvest, and to join the Farm's vets with their own hands in wiping out their own glorious herd of 2,000 reindeer.

'The Farm slaughters 3,000 animals each autumn, so they'll just start with herd 7,' observed Tolya. 'It's all the same to them.'

I was later told that the director even received a medal for overfulfilling that year's meat procurement quota. Not having veterinary knowledge, I was puzzled as to whether this meat could be fit for human consumption. The answer was unclear.

'We couldn't eat it,' someone from another brigade told me, 'but it's all right for Russians.'

'Why?' I asked.

'Well, I suppose they cook it thoroughly,' the man replied vaguely.

But tests on the carcasses revealed that very few had been infected after all.

'Out of a sample of 300, only 10 were infected,' one senior vet told me years later, quoting a figure similar to many others I had heard, 'but the Ministry insisted. It was a pure local breed. I was against the slaughter.'

Was the decision to destroy the whole herd a technical decision, or a political one? Brucellosis invades from time to time, moving uphill from the neighbouring Verkhoyansk district. Three herds had been destroyed during an epidemic in 1970, but more recent outbreaks had been contained by culling just a few, well-chosen animals. It was strange that an entire herd could be condemned apart from the few reindeer that happened to be trained as *uchakhs*. The herd all grazed together. How could the infection engulf them with such fine discrimination? Some people smelled conspiracy.

'The director is against this family because they're economically and morally independent,' one told me.

'The family openly supported Tolya's campaign to redraw the district boundaries,' said another.

The most outrageous allegation I heard was, 'That vet was new – a relative of the director who visited the herd twice. Maybe he infected them deliberately.'

After the reindeer were liquidated, Ivan's father changed his stance on the proposal to redraw the district boundaries, and would appear at meetings in the village hall to speak against it.

'He wants to make sure his herd gets reconstituted,' someone explained to me, cynically but probably correctly.

As indeed it was. The Farm took 600 or 700 reindeer from camps 3, 5, 8, 9, 11, and 12, and within three years the family was reunited on their old pasture. But it was not the same herd. There were fewer reindeer to start with, they did not know their new territory, and the animals that other brigades gave away on command were the weaklings that they would have selected for slaughter.

'Small and scrawny,' Emmie described them, 'with no meat.'

'We had better animals to train as *uchakhs* before,' complained Gosha. 'You could ride on one reindeer all the way to Sakkyryr, but now we need two or three of the new ones just to get as far as Sebyan, with several rests and changes.'

It would take years, and skill, to train these animals and mould them by selective slaughtering into a strong herd. Moreover, the new start-up animals were State-owned. At a time when private reindeer were becoming the most important source of livelihood, this family was left with no reindeer of their own except the *uchakhs* they sat on. They were back where they had been eighteen years earlier, only older and more fatigued, and trying to build up their personal property again from scratch.

'I shall die and I can't exactly say which way my sons will turn,' the Old Man told me just before the slaughter. 'Will they work as herders? That's something I can't say, I can't say.'

He did not live to see his family's renaissance. Before they could return to their pasture, the Old Man enjoyed an extraordinarily successful hunting season but then died, it was said from a broken heart. Granny spread her fury far and wide.

'He died with a grievance in his soul [*s obidoy v dushe*],' she told me later, 'because they were killing perfectly healthy reindeer. He was all alone, he was powerless to prevent it. They brought him in from the camp on a sledge, I couldn't work out what was wrong with him. He said it was flu, something simple. The village doctor, too, she said he was O.K., nothing wrong. And then he died. I wrote a complaint, I said, "If you don't do something about it, I'll insist on an exhumation!" It was only then that they reprimanded the doctor. I'm still angry with her, because she didn't do her duty as a doctor and as a human being.'

After that, the family withdrew even more deeply into the mountains. When they did stay in the village, as Granny started doing in the wintertime, they had as little to do with the Farm management as possible.

'We just say good day and hand in our production statistics,' she told me.

I heard much sympathy expressed in the village for the family of camp 7, especially during long conversations with women over their teapots in the quiet afternoons.

'He was a respected elder, and he built up that wonderful herd,' said one.

'They're nobodies, they just live out in the mountains,' another said sarcastically.

'Such a waste of reindeer and his life's work,' lamented

another. 'It really is a nightmare. His heart couldn't stand it. It's a kind of political repression.'

The director was highly skilled at manipulating relationships in the village, the party, and the regional authorities. Everyone could work out for themselves how the Old Man had started as a frequent prize-winner, with the added advantage of being a party member. But when the tide of reform from Moscow caught up with him, it was not enough to carry him forward locally. They understood how the ruin of punishment worked in a more determined and life-changing way than the diplomas of reward, biding its time for an opportune moment which might even present itself in the form of the pathogenic micro-organism *Brucella*.

I had returned home before the Old Man died. On my next visits to Sebyan I went to other camps, and it was five years before the conditions fell into place that would enable me to visit this family again: a message from Ivan saying that they wanted to see me, a need in my research for further data about the herding practices of camp 7, and the funding to pay for several hours of flying time. By then, after quarrelling irrevocably with the director of the State Farm, Tolya had left the village and was living in Yakutsk. He, too, wanted to visit Ivan and his family. We met up in the city in 1995 to plan a major expedition, as it could be weeks before we were able to get out again and we could be caught by the beginning of winter. We collected outdoor clothing, rubber waders, and sleeping bags. We knew that Ivan's family would feed us with reindeer meat morning, noon, and night. In exchange we would take bars of chocolate, batons of luxury smoked salami, matches, torches and batteries, boxes of bullets and tins of seaweed (invitingly called 'sea cabbage', *morskáya kapústa*) from Vladivostók. Visitors were expected to bring vodka, and I bought a good supply. Though the land yields the same wild reindeer, elk, mountain sheep, fish, duck, and ptarmigan which nourished their ancestors, herders now also crave flour for heavy unleavened bread, butter, tea, and sugar. I was appalled when I realized how quickly they went through sugar, and understood their many missing teeth (Tolya calls it 'white death', the white man's revenge for the black death, which had come to Europe in the Middle Ages from Siberia). I bought a 20-pound block of butter and a sack each of flour and white death. On the way out of the shop, I added a sack of tired onions, which had travelled 8,000 miles from the hot, irrigated mountains of the Caucasus.

After the economic decay of the early 1990s, aircraft no longer flew to the most remote villages, which were cut off from the outside world almost permanently. Dismantled biplanes and helicopters littered Magan airstrip outside Yakutsk like dead dragonflies and locusts, stripped by the aviators who struggled to hold together a few whole, functioning aircraft. They would fly only to order, after complex negotiations between several interested parties over dates, routes, and finances. A biplane could only abandon us on the Sebyan village airstrip, leaving us with the task of negotiating for other people's horses to make the six-day round trip to Ivan's camp. Only a helicopter could reach camp 7 direct. We found one that was going to deliver equipment to another village. For a contribution to the relevant department's flight costs, the aviators would drop us off next to Ivan's tent.

On 1 August 1995 a minibus drove us with our baggage to a forest clearing 20 miles outside Yakutsk, where we found several known and unknown faces sitting on a jumble of sacks and waiting to hitch a ride. After an hour or two, the Mi-8 helicopter appeared, already loaded with cargo, its friendly blue and orange Aeroflot livery scarred with kerosene smoke from the engine's exhaust. The aviators, too, were old friends, Russians with wide smiles, gold-capped teeth, and smart blue uniforms. We loaded quickly and piled in. The helicopter strained for a moment on its wheels, then tugged free in a movement that you can otherwise experience only in dreams, when your own position remains still and the world suddenly drops out from under your feet.

4

Granny's herd restored:
late summer site, 1–2 August

'The fish and game are ecologically pure. The nature is wild and beautiful. It refreshes the soul,' said the co-pilot as he gestured through the window with pride.

We were clattering low over the channels and islands of the River Lena which wound several miles wide across the endless forested plain. Clipped to a board beside him was a map showing zones of geomagnetic anomaly and marked in Biro with many annotations; on his knee was a GPS, a newly introduced satellite-linked device for pinpointing one's location.

We dropped the supplies and our other passengers at a small wooden village in the foothills and pushed on into the mountains. The summer had been dry, and only thin threads of water flowed through the torrents of naked stones that marked where the spring floods had passed six weeks earlier, gnawing at the roots of larch trees which clung to the skin of unfrozen earth above the permafrost, splintering them where they toppled and nudging them towards the Arctic Ocean, where they would be found as driftwood and built into shelters and sledges by the reindeer-herding peoples of the bare tundra.

Now we were no longer flying in straight lines, but following the curves of rivers. The denser features of the land made our standard speed of 136 miles per hour seem faster, yet there still seemed no hurry as we drifted along the sides of mountains, looking above and below as if in a glass-bottomed boat from which one could see both islands and the spaces between them under the water.

With the sureness of their map knowledge the aviators found the River Sakhandjá and followed it upstream. As we nosed into the valley, their maps and instruments no longer revealed what was important to know. Tolya's body was taut, his eyes peering intently at the ground below. The aviators called him a 'human GPS' for his uncanny ability to guide them. But what Tolya registered were not coordinates in a homogenous space, nor even geophysical contours, but the threads of animal movement and purpose, and of human dependence upon these.

Every animal has its own paths. Mountain sheep migrate along the highest ridges from one end of the Verkhoyansk Range to the other, their routes scraped visibly along the sides of the cliffs. Other paths are followed by wolves as they sally out from their rocky hide-outs; the sables and squirrels whose fur brought the Russians to Siberia; the marmot, which eats mountain herbs like the sheep but whose meat is so full of fat that it is regarded as medicine rather than food and cannot safely be eaten hot; the brown bear which usually avoids humans but which is sometimes attracted to the smell of women and competes with them in autumn for berries; and the elk, which lifts its head to browse on tree branches in the densest thickets along the rivers.

Unlike the elk, the reindeer grazes with its head to the ground and seeks higher, open terrain, governed by the uphill unfolding of plants during the short summer and the quest for breezes against mosquitoes. Ivan and his family accompanied their herd of reindeer around these mountains in a never-ending cycle, camping at twenty or thirty sites every year. At the point where the Sakhandjá was joined by the smaller River Avlandjá, we looked down on a corral made of larch trunks, taken from the forest and realigned by human intention into horizontals and parallels. We had found their migration route. This was their winter pasture, which they had left at the end of April and which might not be visited again by a single reindeer or human until November. As spring ran into the short summer, the herders had quickened their pace in order to keep up with the reindeer as they flowed upstream*, following the contours of the slopes on either side of the Avlandjá.

The herders had spent May, June, and July picking their way up this stream, riding on reindeer and dismounting where slides of rock had smashed into the forest. At the junction of each tributary, wherever they could find a tiny patch of level grass, they pitched their tents, stopping for only a few days at a time before moving on. The imprint of these brief camps was usually imperceptible to my eyes, though not to Tolya's. But occasionally we saw a storage platform which they had used as a depot for supplies. Every herder's and hunter's route was staked out in depots, with caches of equipment and provisions stockpiled in huts or on platforms called *ambar*, where they were tied down under reindeer skins or canvas tarpaulins out of reach of dogs (though nothing was safe from bears). Keeping these depots stocked required long treks with reindeer or horses loaded with cargo, and much planning outside the usual migration routes and schedules. As the herders moved around, they picked up supplies from one site and dropped them off at others.

We flew over the platform built by Ivan's father shortly before he died, then another that had been wrecked by a bear and never repaired. The herders would have reached here by mid-May. The helicopter floated on steadily past an ancient child's coffin built in the old style on stilts, then over the open, grassy area where female reindeer calved around 20 May, after the long winter's gestation. This was the most stable point in the annual cycle. However much the migration route might vary from one year to the next, females preferred to give birth where they themselves were born. Having endured a drop of over 100°F on emerging from its mother's body, a newborn reindeer's body temperature was sustained by rich deposits of heat-generating brown adipose fat, especially around the heart and kidneys. It could stand after a few hours and outrun a man after a day. Most of the calves in one herd emerged within a few days of each other and the herd moved on quickly to reduce their exposure to wolves.

Beyond the lower Uchokhón (Crooked) stream, the slopes were bare except for mosses, blueberry bushes, and dwarf willows and birches standing just a few inches high. Our friends could be anywhere from here on. We rose over the last of the larch trees in the cleft of the valley, past the upper Uchokhon on the right and the stream on the opposite slope named the Lonely Tree (Khyakitachán). Tolya scanned from one side of the helicopter, I from the other, holding our heads back from the wind that sliced past the round windows which we had fastened open like portholes. Sometimes he leaned into the cockpit to share the aviators' panoramic view through the helicopter's glass nose, gesturing and shouting above the engine. The Avlandjá became no more than a trickle flowing through a bog as we traced it to its source in a small lake. If brigade 7 were not here, then they must already have crossed the long, bare Diesmin Pass, where there was no water or firewood, and nowhere to stop.

In a moment we had flown over the pass and the ground fell away once more to the upper valley of another stream. This was the Daik, named after an ancient shaman who lay buried in the forest below. Three tents stood out like little white mushrooms, pitched where the stream flowed out from a narrow gully onto a sloping grassy plateau.

We sank into the hollow between the peaks and the pilot landed 100 yards from the camp, rotor blades whipping up a shower of loose water from the bog. The altimeter read 4,000 feet, only 2,000 below the highest of the surrounding peaks. Ivan and Yura were advancing towards us, Ivan clutching a Lenin cap to his head against the wind, Yura with his head tilted sideways and what would become a familiar friendly grin. They moved with a summer gait, their thigh-length rubber waders folded down around the knee and forcing them to walk with their legs slightly apart. The inside lining of woolly socks and felt insoles insulated their feet from the permafrost, but made the boots seem huge for their small figures. Granny stayed beside her tent, eyes alert in her weather-beaten face. She was wearing a green headscarf and a faded floral frock over woollen long johns. Round her shoulders she clasped a padded black cotton jacket. The men had matured in the intervening five years, but Granny looked just the same.

As we dragged our baggage to the door and tossed it to the ground, Ivan and Yura picked it up without a word of greeting, as if we were simply in the middle of a day's work together. While we continued to move baggage over to the camp, the aviators drank tea and joked in Granny's tent, and bought a reindeer to take to the produce-starved city. As soon as Ivan had killed and butchered it, they started up their engine with a belch of orange flame and black smoke. They were reaching the limit of their rostered flying time and were anxious to begin the three-hour flight back. As the rotor blades turned, we flung ourselves across loose items of luggage that were still near the helicopter as a wave of wind and water-drops rolled across the grass. The aircraft circled once around our little plateau to gain height, and in a moment it was gone. The members of the camp returned to their own activities while Tolya and I rummaged through our baggage. It would be light all night, so there was no hurry to sort anything out.

The area beside the tents was one of the few points where the rain-starved stream came above its bed of stones. On one side of the camp, Daik's stream curved around to the right and continued down a gently sloping valley, marked by a trail of gnarled dwarf willows which the women used for kindling. On the other side, the ground rose slightly away from the stream before plunging down to the River Munne a mile away, just before it entered a gorge. Returning here another year, I would hear the gorge amplify the river's distant roar, but this year I could hear nothing; the river was so low that only rock pools remained. We would catch the trapped fish by driving them into the shallows and throwing stones at them.

Of all the species in the region, humans were the rarest. The nearest were in brigade 3 over the mountains to the east, 20 miles away on the aviators' map but 50 on the ground. To the north were Kesha, Lyuda, and their little children in brigade 8. To reach their current site one would have to go back over the pass and down the Avlandjá, across the River Sakhandjá and into the next group of mountains, 50 miles by air or 120 miles overland. I tried to work out how one would reach brigade 2, the furthest camp of all. This was run by Tolya's brother Maxim, a childhood friend of Ivan's. It lay 150 miles to the north on the map; it seemed pointless to calculate the distance on the ground. To reach it, one would have to ride for a week, crossing some twenty mountain passes. Beyond brigade 2 one would reach the outlying camps of the next village, Sakkyryr, which lay 200 miles to the north in a time zone one hour ahead of us towards the international dateline.

Within the camp, there were distinct gradations of intimacy. Granny's family occupied two tents side by side. Granny herself lived in one tent with her daughters Emmie and Masha the schoolteacher, who had come out from the village for the summer holidays. I did not think Masha found it very comfortable. Their tent also served as the restaurant and common room for the whole family. The second tent was empty during the day and was used for sleeping by the four bachelor herders: Granny's sons Ivan and Yura, their uncle Terrapin*, who was somewhat younger than Granny, and a young bachelor called Nikolay.

The third tent was occupied by Gosha, the young man who had been brought up by Granny and the Old Man. After my first visit, he had married Lidia, the widow of granny's eldest son Oleg, and taken over the upbringing of their two little girls Nadia and Lialka, who were now aged 12 and 13. Lidia and the girls had come out to the camp for the summer holiday, also bringing her nephew Stepa, around the same age. Lidia's relationship with Granny had become awkward since her re-marriage, and their tent was set some way downstream. They cooked separately, and to talk with them meant moving out of the immediate area of Ivan's family.

Gosha was out minding the reindeer and Lidia was washing pans in the stream. There would be time to talk to them later. Tolya and I headed towards Granny's tent, added our own rubber waders to the heap blocking the entrance and stepped over them in our thick socks, the loops of the weave snagging with fragments of moss and twigs. To one side of the entrance, Granny's little stove, an elongated cube made out of sheet metal, laboured to boil tea and reindeer stew for her family and her guests, its chimney-pipe protruding through a hole in the side of the tent. I ducked to avoid knocking into the boot-liners and strips of smoked reindeer meat drying in the heat that had accumulated along the ridge-pole. We edged past Terrapin and Nikolay to join Ivan at the back of the tent and sat down on the reindeer-fur mats spread over an underlay of springy larch branches.

It was cooler at floor level. Granny was sitting on an upturned log in the corner behind the stove, occupying the most com-manding position in the women's part of the tent. She had taken off her headscarf, revealing grey hair still streaked with black. Her trips outside during the day to visit other tents or collect supplies had left their trace in wisps of grass and dry lichen clinging to her carpet slippers. She was smoking a cigarette and surveying the crowded tent through shrewd, half-closed eyes. Her daughters were sitting in front of the stove, boiling kettles and pans and pressing them against the side of the stove to keep them warm. Masha filled some tin mugs with tea. Emmie darted outside and returned with a mug of reindeer milk, which was so thick and creamy that it could be obtained only in small quantities. Yura was sitting close to his mother and helping the women to hand out the tea.

'The females won't let us milk them,' he said with a laugh. 'We men look after them all day and show them nice plants to eat. But when evening comes they run straight to the women, even when we hold out a handful of salt for them to lick.'

'Granny and Emmie are very unusual,' added Ivan from the back of the tent. He was leaning back on a pile of folded khaki bedding and faded black jackets, with a floral sheet crumpled in the middle. I remembered his father leaning in the same position when he was head of the family. But now I could not work out whether Ivan was in charge, or Granny. 'They can keep the milk going far into the winter.'

'Don't they stop giving milk in the autumn, when they change from eating green leaves to lichens?' I asked.

'That's just textbook talk. It depends on how you handle them. Our women are real experts!'

As I pulled a bottle of vodka from my bag, everyone drained the last of the tea from their mugs in readiness. I opened the stove door and tossed a mugful of vodka onto the red embers. A sheet of flame shot out through the hole. There was a slight release of breath all around the tent and a murmur of approval from Granny.

Feeding the fire is the most important act upon arriving at a new site. When you do this, you are making contact with the place that is accepting you as its guest. You may accompany the offering with a silent or murmured prayer, such as 'Please draw back your feet.' It is this offering, not the pitching of a tent, which creates an inhabitable space for you in the landscape. You may maintain this contact through repeated acts of feeding the fire. I found it to be one of the easiest and most appreciated gestures, which has become part of my regular life in England. Once in quest of genealogies and clan histories I visited an old herder called Sofron where he was minding some reindeer alone at the outer reaches of brigade 1. I produced a bottle and tossed a capful of vodka into his stove before pouring a mug and offering it to the old man. My host looked at me with surprise.

'How interesting!' he mused. 'We have that custom, too!'

You could also offer the fire your first piece of meat, as the Eveny did before the coming of the Russian colonists.

'The Russians used vodka cynically to render the natives helpless,' said Tolya as I filled the mugs and handed them around. 'The Eveny didn't have drunkenness before the Russians introduced it. Or swearing,' he added.

Granny chortled. The former must be true, I thought, though the Eveny quickly perceived a life-giving quality in vodka and incorporated it into their spiritual universe. As for no swearing, could such a language exist?

'The Russians obviously filled a need,' I remarked, remembering how Eveny herders yelled Russian obscenities when the herd got out of control.

'Because we know how to feed the fire, we can live anywhere,' he declared.

'Instead of white men in rockets they should send Eveny up on flying reindeer,' I suggested. 'When Russian cosmonauts get to the moon they'll just grow cucumbers, but Eveny would light a fire and feed it with vodka!'

'We are nomads of the cosmos!' Tolya proclaimed, above Emmie's hoot of laughter.

'The fire is the foundation of life, we feed, we warm ourselves, we're nourished with its help,' Granny explained in a serious tone. 'And it protects us. Evil spirits come out at night, don't they? So before they appear you should build up the fire, you know, keep it burning, keep it burning! The fire should stay alight all night to keep the spirits away.'

In practice, of course, I knew from my own experience that this did not happen. No one stayed awake adding fresh logs all night. Each time the stove went out in winter, the temperature immediately plunged to −60°F or below until someone shivered out of their reindeer-fur sleeping bag to light it again.

The fire (*tog*) contains a spirit (Tog Muhoni). Some people say that the spirit of the fire is an old woman, but most people say it is a man. I sometimes felt that people projected an ideal of their own onto this comforting spirit. I heard one young man claim the fire spirit was a man of his own age, I thought idealizing himself. Granny saw Tog Muhoni as a venerable old man, a patriarch who nurtured and protected his family. The way she said this made me feel that she was thinking of her departed husband.

Everyone respected the domestic fire, regardless of their ideology. An old Communist Party stalwart called Pioneer Ilich in the village of Sakkyryr, who asked me to photograph him beneath a portrait of Stalin, said that he did not fear evil spirits. But this was not because he was a dialectical materialist: it was because he always took care to keep his fire well stoked with vodka. Eveny people who found themselves staying in an apartment in the city would pour their vodka onto the red-hot electric ring in the kitchen.

The domestic fire should never be contaminated by any kind of rubbish. After years in India I had become used to seeing fire used as the spiritual equivalent of a hospital incinerator, neutralizing witchcraft materials and the leavings of evil spirits by reducing everything to primal substance like the funeral pyre. But I have seen a shaman in a reindeer-herding camp doing a healing rite, and instead of casting the spiritually infected materials into the fire, he instructed his patient to go deep into the forest and hide the materials high up in a tree, as though the vastness of the forest itself had cleansing powers.

The act of feeding the fire was done discreetly, and an observer might not even notice it. But if you failed to make an offering at a new site you would have bad dreams and be unable to live well in that place. A friend called Vitya the Wolf Hunter, who rode from camp to camp dealing with epidemics of wolves, once told me how he had unthinkingly taken a swig of vodka without offering it to the fire first. He had almost reached the bottom of the glass when he started choking. Realizing what he had done and trying to put it right, he topped it up with water to its original level and fed it to the fire. But the fire was not to be hoodwinked so easily.

'That night my belly was burning,' he recalled, 'and during the night the fire said to me, "You gave me leftovers, you drank the best yourself and you gave me your diluted slops." So the following morning when I woke up I immediately fed the fire with a glass of pure spirit. You should have seen how it flared up! After that, I felt fine.'

The fire would use its knowledge to forewarn you of the future. If the fire said 'Tssss!' or a piece of ash fell out of the stove, it meant that a guest would be coming from afar. If something bad was about to happen the stove warned you by saying 'Tssss!' or 'Tsk!' in a particular way. If the fire did this persistently, it meant that someone was going to die. But the fire also said, 'Tssss!' if a hunter would be successful. How could one tell the difference between all these warnings? Some people explained it to me by making a series of noises which I thought sounded the same but which they insisted were different. Kesha, the brigadier of camp 8, once gave me a more psychological interpretation.

'It all depends on what you're thinking about,' he said. 'If a hunter is sitting sadly wondering, "Why am I not catching an animal?" then "Tssss!" is the fire's way of saying, "Go out tomorrow and an animal will be waiting for you."'

I asked Ivan's family whether the fire had foretold our arrival.

'We heard from the village that you were coming this month,' answered Granny. 'But it was the fire that told us you'd arrive today.'

'The dog knew, too,' added Ivan.

Each herder might have several dogs but this was Ivan's favourite, an old dog like a large husky with only one, blue eye. The dog had belonged to his father and was now too old to work.

'Something of my father passed into that dog,' he continued. 'I believe he's still with us. That dog can foretell the weather, and he warns me if anything bad is going to happen.'

We drank the last of the vodka from our mugs and handed them to Masha, who filled them with the juice from the cauldron of reindeer meat bubbling on the stove. The men looked with pleasure at the fat rising to the surface. In this climate, fat was as important as protein. Emmie placed a wooden board in the middle of the floor and Granny directed Masha as she ladled out large pieces of meat from one pan and a mixture of intestines and other inner organs from another. The herders and I pulled out the wooden-handled sheath knives from our belts and laid them on the table for everyone to use. We took it in turns to use the knives to reach in and stab at the meat, biting on the edge of a large hunk and slicing upward to separate the piece gripped between our teeth from the rest. I have never been comfortable with this way of eating, but do it all the same.

Suddenly Tolya asked, 'Did you know that archaeologists have discovered a tribe of long-nosed Europeans who used to live here?'

Many natives read a lot of archaeology and anthropology, and Tolya read more than most. I must have paused gratifyingly, for he went on, 'They became extinct because they kept cutting off their noses every time they tried to eat. It's obvious from Darwin or Lamarck – you've got to adapt your nose or your eating habits. Only the flat-nosed Asiatic tribes survived!'

Granny chuckled and Emmie gave a shriek.

'That's no guarantee,' said Ivan soberly. 'Mammoths had long noses. They didn't cut meat because they were vegetarian. But they died out all the same.'

Granny chortled again.

Late that night, Ivan pulled down a spare tent from a storage platform. The Eveny now use ridge tents which they sew from canvas and hang on a frame of larch trunks lashed together with reindeer-hide cords. There are no guy-ropes, and the canvas is held in place by the stability of these poles. Until the 1940s, their tents were conical like a tepee, with an open fire in the centre beneath a smoke-hole. Another year, Tolya and I would do fieldwork together and stay in this kind of tent among Nenets and Komi reindeer herders 3,000 miles to the west, on the European side of the Urals near Novaya Zemlya. In those tepees I would be struck by how the inner framework of poles resembled the ribs of an animal. Since it was summer they were covered with tarpaulin, but in winter they would be covered with over 100 reindeer hides. I came across an Eveny legend that reflected the same feeling: a reindeer rescues two little children from an evil spirit by turning its body into a living tent, with its ribs serving as a framework and its hide as a cover.

We turned the trailing edges on the canvas inward and weighted them down on the inside with rocks. The children cut larch branches for the floor and Yura laid reindeer hides over the branches around the back and sides of the tent. The pungency of the furs combined with the resin of the freshly cut larch to give a smell that could be experienced only within the first few hours after pitching a tent. When the moment came to set up the stove, Ivan and Yura discreetly disappeared. No one else should help you do this in your own tent, as this might lead to a quarrel in the future. We lit our fire and the tent crackled to life, the orange glow blending with the blue of midnight which seeped in through the weave of the canvas. The smell of burning larch wood took over from the raw needles, and we fed the fire with a drop of vodka I had discreetly saved from supper.

'What do you suppose it's saying to us?' I asked Tolya.

'It's saying it's been a long day – time to sleep!' he answered, closing his eyes and pulling a woolly bobble-cap down to his nose.

When I woke at nine the next morning, Ivan and Yura had already been out for two hours gathering the reindeer from where they had dispersed overnight. In the five years since the big slaughter, the herd had grown to 1,600 animals. I walked to the top of a small peak and watched them grazing a few miles down the Daik Valley. They undulated and creased like molten lava as they flowed over velvety bog and rugged rocks with no change of pace. Yura in his pirate headscarf floated above every ripple of the herd, as if on a magic carpet, his movements light and fluid as he rode on Sancho Panza, a reindeer he had trained to respond to the slightest signal from his body. The tiny figure of a dog darted back and forth between Yura and the herd.

The reindeer must be guided every day of every year over a vast landscape whose natural features offer no boundaries. Their natural dynamic is to spread out to graze when they feel secure and to rush together when alarmed, a pattern that resembles the slow breathing of a huge furry organism. The herder replicates this expansion and contraction but tries to enhance their feeding and protect them from harm.

During the short summer, the Arctic vegetation grows rapidly in the twenty-four-hour daylight. In these few weeks a newborn calf or an adult emaciated from the previous winter must become strong enough to survive the next winter. The reindeer's digestive system is designed for the rapid assimilation of dry matter in green plants. At midsummer the animals' appetite seems insatiable and the herd moves almost unceasingly to feed, watched over by men who stay awake around the clock in shifts. Now, at the beginning of August, the intensity of feeding had diminished*. The animals did not wander so far during the cooler half-darkness of the night and could be left unattended by the herders.

Yura's reindeer were spread out and grazing tranquilly, their heads down to the ground in a posture of security. They were facing into the breeze, which at this altitude was just enough to keep biting insects away from their faces in spite of the warm weather. He guided the herd along the stream to the far end of the Daik Valley and back again across the high slope above. Then he drove the herd downstream but along a different contour of the same slope. I could just hear the faint sound of the reindeers' constant grunting and the occasional barking of his dog.

Then I saw one animal become excited by some choice vegetation. Some of the nearest animals moved over and crowded the first reindeer out of its privileged place. The first animal moved on in the same direction and the others began to overtake it, as if competing to be the first to find the next tasty leaf*. Suddenly, the whole herd surged forward, drawn out into a long thread behind these competing leaders. This movement often dies down in a few moments, but occasionally reindeer run out of control and it can take days to gather them up again from 50 miles away. In the worst situations they simply run without feeding at all, and then they can become dangerously exhausted.

Immediately the stocky, determined figure of Ivan appeared in his Lenin cap, mounted on another reindeer and accompanied by the young dog Sharik, whose name meant 'Little Ball'. It was as if they had been waiting unseen in just the right spot for this emergency. Sharik ran to the windward side of the reindeer and Ivan followed swiftly behind.

In an echo of their ancient relationship with the wolf, the reindeer immediately bunched together and circled anticlock-wise as each animal tried to get inside the group to avoid being picked off by a predator. Much of the reindeer's herd behaviour is governed by imitation, triggered by a visual stimulus. Especially on open ground where the animals can see each other, it is enough to startle one reindeer for the reflex to ripple through an entire herd like a shock wave. Reindeer cannot see well at a distance, and where the herd is very dispersed, they may not pick up this visual signal. Ivan once showed me how to deal with this situation by tweaking a dog's ear in a special way so that it imitated the cry of a wolf. The response was instan-taneous: the reindeer came together like iron filings around a magnet. The defensive reaction of a herd to alarm is so deeply rooted that it can sometimes be precipitated by the sudden flight of a bird. Even where wolves have disappeared, as in Scandina-via, this bunching instinct provided the foundation of the herd-ers' technique*.

Ivan and Sharik moved unhurriedly around the herd till they were downwind and the reindeer no longer perceived them as

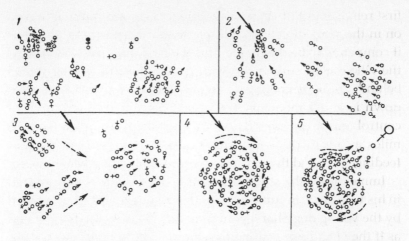

*Reaction to danger: the solid arrow marks the direction of threat. The plain
circles represent calves. (1) the dispersed reindeer are grazing quietly;
(2) they turn to face a sudden source of alarm; (3) and (4) the entire herd
converges and mills round in a circle; (5) all clear: a female leader draws
out the spiral and the herd starts to disperse.*

a threat. If they had alarmed the reindeer for too long, some
animals might have broken out of the circle and led the others
into a headlong flight. But as it was, a few reindeer started to
move out calmly. These were mostly females with calves, the
most alert to danger but also the most concerned to resume
feeding. The tightly packed spiral unwound as the rest of the
herd followed them and began to spread out again. The leaders
put their heads down to graze and the rest of the herd copied
them.

In observing the behaviour of this animal for millennia, the
reindeer people have created what is surely one of the largest
technical vocabularies in the history of human speech. One
Eveny scholar has compiled a list of 1,500 words* in his language
alone that refer to the colours and shapes of reindeer, their
body parts, harnesses, diseases, diet and moods. Though Ivan's
family, like most Eveny in this region, now spoke mostly in
Sakha, they still incorporated many of these Eveny words, which

have no parallel in Sakha or Russian. Given the thirty or more indigenous languages across the Russian North today, and the various groups of Inuit and Native American hunters of caribou in the American Arctic, there must exist tens of thousands of specialized words for talking about reindeer and their relationship with humans.

This vocabulary encourages a refinement of perception and a corresponding precision of action. One has to keep making choices between these specific words by noticing distinctions in the world and working with them. Pastoral peoples around the world have an elaborate vocabulary to describe the colour, shape, sex, and age of their cattle and other animals. But the Eveny also have words to describe their reindeers' characters and habits. A *hunyuk* is a leader and a *berne* is a reindeer that keeps getting lost and is vulnerable to wolves; a reindeer that kicks with its hooves is a *kuhitea*. A *hotamangan* is inclined to run back to the previous camp site so one always knows where to look for it; a *beymenga* is a reindeer that attacks humans and so should not be looked after by children; a *holimangan* is a salt-lover which will come running to a stream of human urine or a handful of salt held out on one's palm ('*Ma! Ma!*' one calls, meaning 'Take it!'). Many of these terms refer not to types of animals so much as to moods or phases of an animal's development. A reindeer's behaviour may change. Different animals may act as leaders on different occasions, and one that uses its antlers to be aggressive can become subdued when the antlers are cut off.

The *uchakh* reindeer that are trained for riding or harnessing to sledges are ascribed a more fully developed character than the others. *Uchakhs* live a double life between the world of reindeer and the world of humans. With a rider they become one life-form as they run around driving the other reindeer. When unladen they may move with the main herd or they may graze near the camp and even be tied up ready saddled, like parked cars with keys left in the ignition.

The *uchakh* was the key to the prehistoric Reindeer Revolution and remains the most important tool for the reindeer people today. It embodied the fullest form of partnership with humans and by comparison the rest of the herd seemed half wild. In this region an *uchakh* was always a castrated male, though I have heard of females being ridden in other areas. You selected a physically strong young adult which liked human companionship and castrated it. The initial training took about a month. At first you got him to accept the bridle and saddle and to stand motionless while you approached him to mount. In the saddle you used the bridle to train him to turn his head to the left or right. When he started moving, you could then train him to go slowly or quickly through the pressure from your heels on his flanks. A further set of exercises would train him to work in a team pulling sledges. But a mechanical description of these manoeuvres cannot convey the grace with which an accomplished *uchakh* and his rider glide up and down mountains or slip between the reindeer in the mass herd. In winter they seem even more united, clad in identical fur so that one is not sure where the human ends and the animal begins, like an antlered centaur.

My friends' *uchakhs* were workmates. As with one's human companions, they could have cooperative or obstructive moods. A reliable riding reindeer (for which there were several words) might sometimes behave like a *mechengen*, 'a sly, devious *uchakh* who throws you over his antlers'. *Uchakhs* also had individual names. Some were visually descriptive, like One-Horn, Hooknose, and Blizzard. Others might be affectionate, ironic, or whimsical. The people in brigade 7 were keen on historical figures, without distinction of gender. Their *uchakhs* included Cleopatra, Caesar, Elizabeth II, and Napoleon (who had antlers shaped like the emperor's hat).

Some names made a satisfyingly subversive political point or an autobiographical comment about the owner. I thought Margaret Thatcher was just another name plucked from world affairs, until Yura pointed out how he dominated his companions. Yura's own favourite *uchakh*, Sancho Panza, was light and graceful like himself, but I gradually came to suspect that Yura saw himself as Don Quixote and the management of the State Farm as his windmill. One of Ivan's favourite *uchakhs* was the stolid Chernomyrdin, named after the beefy Prime Minister of the Russian Federation and the only reindeer in brigade 7 strong enough to support my European frame on its back. When the human Chernomyrdin left office in the Moscow Kremlin, this was also the end for his animal namesake 4,500 miles away in the Verkhoyansk Mountains.

'Why waste resources herding an *ex*-Prime Minister?' argued Ivan in a deadpan parody of his own grave manner. 'He stopped being a useful government contact, so we ate him!'

Watching reindeer from a distance was mesmerizing. They flowed from one side of the valley to the other, approaching the brow of a hill like a wave, sometimes rolling back, sometimes straining over the rim and cascading down the other side, while men and dogs moved around them constantly, nudging back the edges when the herd moved out too far.

After three hours of solitude I turned back to the camp, which, with its dozen humans, now seemed clamorous. I approached slowly, taking in the sound of Terrapin and Nikolay on the two-man saw as they sliced tree-trunks into short pieces before splitting them with an axe. Smoke was rising out of Granny's tent and Masha was washing out the intestines of a newly slaughtered reindeer in the chilly stream. Someone had left a half-repaired saddle on the ground, with dressed reindeer hide stretched over the framework of forked antlers. Emmie was skipping down a slope in rubber boots and black ski-pants, while Lidia's girls were clambering carefully behind her in pink tracksuits. They had been collecting blueberries. In this dry year there were not many berries, but Emmie knew every plant and how to look for the places called *tarim*, where slight slopes came together to form a moist, sheltered hollow. Even from a mile away I could see that their baskets were full. She turned to say something to them, and her whoop of laughter could be heard at the camp.

I saw Lidia and Gosha near their tent. Gosha was sitting on the dry mossy ground, intent over a piece of mountain sheep's horn which he was carving for young Stepa into the silhouette of a reindeer's head. He was bent over his work and I felt that the reindeer forming in his hands was being moulded by the concentration of his gaze. I remembered Lidia from winter conversations in the village as a determined, courageous woman. She was sitting in the sunshine on a wooden log, stitching a puncture repair patch onto a rubber wader with reindeer sinew. Long strands of hennaed hair spilled from under her headscarf over her ruddy cheeks. I sat down on the ground, while she stood up and went over to a reindeer skin stretched out to dry and adjusted the slivers of wood tucked into the furls around the edge in order to expose each spot equally to the sun.

'I grew up in a herding camp over towards Sakkyryr,' she called back to me, 'but I don't really know how to do things out here. I'm learning afresh [*zanovo uchus'*].'

Tolya was passing within earshot. 'An Eveny never needs to learn afresh,' he proclaimed in ringing tones. 'The knowledge of the taiga is in their genes.'

Lidia smiled and resumed her place near me. 'I came here to be with Gosha. He lives out here all winter. I'd like to live here in winter, too, but I have to stay in the village with the girls until they're grown up. I've been working in the sewing workshop making fur coats,' she continued. 'But I don't like the village.'

I had my own prejudices about the village, and wondered whether she was making the same point. 'Because of the gossip and intrigue?' I asked provocatively.

She laughed, showing the gaps in her teeth ravaged by the white death.

The sun had sunk a little and the air now felt chilly. Emmie and the girls had reached the bottom of the slope with their blueberries. Gosha looked up from his carving, blinking into the horizontal sunlight. The reindeer would soon come right up to the camp for their evening round-up, when individuals would be lassoed for milking or for health-checks. I looked down the Daik Valley and saw the herd heading uphill towards us. They were still some 2 or 3 miles away, but we could already hear the bells of the leaders.

The speed of an approaching herd of reindeer still caught me by surprise. The animals burst upon us from the opposite bank of the stream, pouring through the willow scrub and snatching at the grey leaves with 1,600 sets of flexible lips and tugging teeth. They swarmed like bees onto their trampling ground and swirled round and round in a restless grunting mass as a defence against predators, rotating anticlockwise, as reindeer always do, for reasons that no one has ever discovered. The peat beneath them had been made bare within three evenings under their 6,400 hooves, which clicked incessantly from the tendons that snap back and forth over the bones of their lower hind legs, again for an unknown purpose. The trampled areas by the camp were the one sign of herding activity that could be seen clearly in photos taken from satellites in space. These photos allowed one to trace migrations and estimate how recently sites had been occupied. Even before a site was trampled, it could be recognized from the intense growth of grass caused by the previous year's concentration of dung.

The women and children moved around the edge of the herd, flapping their arms and shouting to drive back any reindeer that tried to leave. The men stood scrutinizing the animals beneath the forest of antlers which seethed like underwater coral. They had coiled their lassos and were stroking them, poised ready to throw. The lasso, known as the 'long arm of the herder', was the single piece of technology that made possible any kind of control over the reindeer, including their earliest use as a decoy. Made from long, thin strips of reindeer hide, or, even better, from three such strips plaited together, they were twisted to form a rope. The noose closed through a toggle, of which the best were carved from mountain sheep's horn. Toggles have been found in very early archaeological sites, but they are not necessarily evidence of domestication. The lasso was probably a continuation of a prehistoric tool for snaring, which took advantage of the branching structure of the reindeer's antler. It was coiled and thrown in such a way that the noose fell over the reindeer's antlers in a wide circle and then snapped shut at the root of the antlers without falling further down around the animal's neck. Once the noose had closed, the animal was trapped by the tension caused by its own struggle. The herder then drew himself and the reindeer together by pulling hand over hand along the rope. If the reindeer still resisted when he reached it, he twisted its antlers sideways and forced it to the ground.

Again and again, lassos spun over the heads of the herd. You could not throw a lasso directly at a running reindeer, but had to anticipate its forward movement to allow time for the throw to reach it. Sometimes a lasso would clutch at the tip of an antler and slip off, flicking back along the ground as the herder rewound it through the milling hooves. The men shouted, less to communicate than to express their satisfaction or frustration at catching an animal or missing it: *'Ai!' 'Ooh!' 'Aaaah!'*

Yura separated out the milking females, which allowed him to place a halter around their necks and hand them over to Granny and Emmie. Some of the other animals put up more of a fight. Several reindeer had hoof wounds from sharp stones and twigs, and these had become infected in the hot weather. As each injured animal was captured and pinned to the ground, the children followed Gosha into the herd and sat on its panting body as it writhed and grunted. Gosha squeezed the wound to draw out the pus, which he mopped up with sphagnum moss and gave to one of the children to burn on an open fire – too polluting for a stove, not important enough to hide in a tree.

'Look at this one,' Gosha said. 'A big rock fell on its ankle!' He poured potassium permanganate into the gaping hole and sprinkled it with antibiotic powder. 'It'll be fine in two or three days,' he assured the children. The children leaped aside and the reindeer sprang to its feet and limped away.

'There used to be vets,' Yura said to me. 'I don't understand it! They train so many in the veterinary college, but nobody comes out here.'

'Did you ever want to train as a vet?' I asked.

'You need an education for that.' He laughed. 'I was very naughty as a child. I used to skip school – I didn't care! I'd just run away, or sneak into the village hall and watch films from the back without the grown-ups noticing. The teachers were all cruel, they shouted at us all the time – how could you learn anything? They'd tell us, "You came from a reindeer camp, and that's where you'll end up – it's all you're good for." So here I am – the taiga is my teacher!'

No wonder Yura's sister Masha the teacher did not seem enthusiastic about coming back to the camp for the summer, I thought.

After an hour of activity, Ivan sent the grunting, clicking herd away with Nikolay, giving him clear instructions on where to leave them. During the night the reindeer would move uphill and into the wind. Even if they spread out over several miles, the men would know where to look for them in the morning.

When I reached Granny's tent I found Ivan already lying back on his pile of jackets and bedding, while Emmie and Masha cooked and Granny smoked and surveyed them all from her log next to the stove. I stepped past Yura and Tolya and sat in my place at the back of the tent. Ivan leaned back on his bedding and closed his eyes.

'I'm tired,' he said in Russian as I sat down beside him. 'The reindeer aren't feeding properly,' he announced to his mother and sisters, speaking in Eveny. 'This fine weather [*höke*] is too hot. If this goes on, they'll start getting hoof-rot again.'

'But surely,' I asked, 'hoof-rot comes from cold, wet, stagnant conditions?'

'Yes, sometimes,' Ivan replied, 'but right now they could get it from this dry heat.'

On questions of reindeer welfare, I could never be sure of getting it right. There was always a further nuance which seemed to contradict what I thought I had understood. I imagined Ivan reviewing the weather at the same moment the previous year, in the same tent on the same rectangle of ground.

'Was it like this last year?' I asked.

'No – last year we had a long period of *mondi*.' This was a word for weather with low clouds, rain, and mist, giving a fresh feeling as well as driving away mosquitoes. Reindeer were happy during *mondi*.

'I suppose *mondi* is always good?' I asked.

'No, you need just enough but not too much. The summer before that, it suddenly snowed during a period of *mondi* and a lot of reindeer caught pneumonia and died.'

'In the old days, shamans used to do a ritual to obtain *mondi*,' interposed Tolya. 'You won't find it in books, but controlling the weather was an important part of their work.'

'So how will you manage the weather without a shaman?' I asked.

'I could make it rain by standing outside and shaking a bear-skin. I really should,' Ivan answered. 'Though I don't believe it. I've done it several times and it's never worked.'

'Perhaps it doesn't work *because* you don't believe it?' I suggested. 'If you believed enough, it would work for you – like your father's dog.'

'That's true.' He laughed. 'I'm certain about that dog.'

'That's the worst thing the Soviet regime did,' interrupted Yura, 'killing off our shamans. Shamans were linked to nature, they were actually very progressive.'

He was surely aware of the irony of using 'progressive', that archetypal Soviet word of approval. In fact, the Soviets had called shamans 'backward'.

'If something's given by nature, it can't disappear,' Yura continued.

'So will there be shamans again in Sebyan?' I asked.

'If nature made shamans deliberately, how can they cease to exist?' he demanded vehemently. 'If only I'd been born before the revolution! I'd chase after a shaman and ask him to teach me. That's the kind of education I needed.'

I did not know at the time that Granny's father had been executed for being a shaman when she was a small child, so I did not realize how much Yura's outburst was connected to his family history. Granny said nothing, but merely murmured to Ivan in their own language.

'Someone needs to go to the forest tomorrow to cut down some more trees,' she said.

'I'll send Terrapin,' Ivan said, then switching to Russian he explained to me, 'We only stacked enough wood up here to last us a few days. Our official date for moving is 10th August, but we'll have exhausted this site before then. The grazing, too.'

I thought of all those reindeer lips stripping the willow tips in their summer hunger and converting them into winter fat. But Ivan had opened up one of my unending research questions. My notebooks were filled with sketch maps of the routes of various brigades, drawn partly by herders themselves and partly by me as I tried to capture their narratives of distances, place names, plant types, intersecting winter trails of sable hunters, ancestral hunting routes, sites of graves, and routine or extraordinary events. By now, the pursuit of migration dates had become a pedantic preoccupation.

'You were already living here when we arrived on 1st August,' I said. 'What date did you arrive?'

I was to notice many times how precise Ivan was. He was one of the few herders who kept a diary, and when he offered me an old diary of his to keep with my research material I found it contained a meticulous record of dates, weather, and reindeer movements.

'According to the State Farm plan, we were supposed to get here on 5th August and leave on the 10th,' he said. 'This was a difficult year. We got here on 27th July. That's very early.'

Granny had been talking to Emmie in a low voice, but she had been listening: 'The snow lay on the ground so long that there are fewer birds this year,' she said. 'We haven't seen any korma or chokuché.' Those were little birds which flitted from tree to tree near the tents.

'The snow in May was very deep, so we could move only slowly. The willow leaves didn't sprout till June, and the reindeer got confused,' Ivan explained. 'By then we were running late. So we didn't stop at Kuntekendjé. We lived there for a week last year, but we missed it altogether this year.'

'We stopped off for two nights on the River Sutandjá to collect supplies,' said Yura. 'But we didn't go to the ice sheet this year. We were late and it was too far. It had become hot and the insects were terrible.'

These were the sites I had imagined in their neat, unfailing sequence as we flew up the River Avlandjá two days ago. I was not even aware of the possibility of a detour to the ice sheet, which lay across another small mountain to one side. There were few insects over an ice sheet, so things must have been bad if they could not even lead the herd to that point of relief. I had experienced the midsummer swarms of mosquitoes which sometimes blotted out the surrounding landscape, sucking blood from reindeer and humans alike. Despite the beauty of the midnight sun, summer was my least favourite season. The harassment eased only slightly in the cooler nights. The insects could drive the reindeer frantic, causing them to run for fifteen hours a day and exhausting them to the point of death. There were even stories of reindeer corpses being found drained of blood. In regions where the harassment was the worst, herders would build makeshift shelters and light huge, damp fires, where the reindeer would huddle in the protection of the smoke.

Just as the reindeer had adapted to living on this land, so there were creatures adapted to living on the reindeer. Mosquitoes followed their own migratory cycle, spending ten months inside the animal's body and emerging in the few weeks of summer to reproduce. One species of gadfly lived on no other host. Each time a reindeer heard one approaching, it shook its whole body in a vain attempt to prevent it from laying eggs in its fur. The larvae then burrowed under the skin, where they grew into large grubs which could be seen clinging beneath the hide when a reindeer was skinned. The inside of a coat made of dressed reindeer fur was sprinkled with their marks like little leopard spots. I had also read in a textbook what could not be seen by the naked eye, how one kind of fly hovered by a reindeer's nostril until it inhaled and then squirted in a stream of eggs, which emerged fully grown ten months later when the reindeer sneezed in spring. Fortunately, at this altitude, the summer season and the insects were nearly over.

Yura resumed the narrative of their migration. 'We didn't stop by the child's grave, but carried on to the Lonely Tree,' he said.

'We were supposed to reach Lake Avlandjá on 16th July,' added Ivan, 'but we didn't get there until the 24th.'

We were talking without a sketch map and I was trying to follow their reasoning as well as their route. The timing did not make sense. 'If you arrived at the lake eight days late,' I asked, 'how did you get all the way here a week early, just three days after that?'

'The newborn calves were being killed by wolves. There are more than ever this year. Every brigade has noticed it. The wolves have brought us down from 1,800 reindeer last year to just 1,600. That whole side of the mountain is dangerous, so we decided to drive the herd straight on over the pass. It was 25 miles and we had a summer snowstorm at the top, but we brought 500 females and their calves to safety here.'

They had been on the move and under exceptional pressure since April. The next site, too, would be a hurried stay, and it would not be until the autumn pasture, two sites on, that they would have time to linger. Ivan's mind was also moving in the same direction.

'It's really important to move on,' he said, 'but we have to wait for it to get cooler. We can only do what suits the reindeer. I don't know when we'll be able to move.'

There was a silence, except for the crackling of the stove under our supper. I wondered whether it was saying anything, but no one else was taking any notice. Stoves spoke of arrivals, not departures.

I left the tent at one in the morning. On the ground the vegetation was turning autumnal, but in the sky it was still summer. The children were playing football and laughing in the blue light of the Arctic summer night, while Gosha and Lidia were setting out for a stroll together. I walked beyond the low voices of the camp, beyond the tinkle of our stream to the slope overlooking the big river and the gorge. A deep yellow three-quarter moon was rising behind a ridge, like a flaming bubble or gas jet. Though I was cold and knew that Tolya had lit the stove in our tent, it was some time before I could bring myself to leave. When I returned to the camp I saw that Lidia and Gosha had come back, too. She was pulling a huge polythene sheet over their tent and weighing it down with boulders. Even if Ivan had not shaken a bearskin, she seemed to be expecting rain.

The next day, every tent was shrouded in a husk of polythene, tied to the support poles with strips of reindeer hide and held down with rocks around their crumpled edges.

5

Migrating into autumn, 3–8 August

In the morning 1,000 reindeer, over half the total herd, went missing. Gosha found them later in the day down the River Munne. Towards evening he drove them headlong to meet the other part of the herd coming up the Daik Valley and they merged like two streams of syrup. Our herd was complete again for one brief day, but it was a delicate balance of control, a reminder of the provisional and incomplete nature of domestication. The reindeer were poised on their highest pasture like a roller coaster at a moment of stillness, ready for a rapid autumn descent down the River Munne as the air cooled around them. Yet so long as the weather stayed hot, they also felt a contradictory need to stay uphill for the breeze. If the herders delayed too long in moving camp, many of the reindeer would go on without them and they would lose control; but if they let the reindeer go too quickly they would lose the stragglers who were still uphill, while the leaders would reach the autumn pasture ahead of schedule and overgraze it.

A day later, the rain still had not come, yet the peaks on all sides of the Munne Valley appeared hazy. The dull lemon sun seemed out of focus and gave no direct heat. Something was going on that I did not understand.

I found Yura and Gosha standing around the camp, looking subdued.

'There's a forest fire down the Munne Valley,' said Gosha.

'This south wind's bringing it up to us. The fire's 130 miles away, but the reindeer can smell the smoke.'

'By tomorrow we'll hardly be able to see from one tent to another,' said Yura. 'A fire like this doesn't depend on trees. It can travel in the peat when it's dry. That's why there's so much smoke.' He gestured to the herd, which had gone no more than a mile away from the camp and was circling round and round in a restless whirlpool. 'It's bad for the reindeer's lungs,' he said, seeing Ivan approaching and anticipating his concern.

Where the others were despondent, Ivan was agitated and seemed almost ready to join the reindeer in their frantic circling. Ivan and his animals mirrored each other in a symbiotic ecology of mood. He was like a barometer, but he registered conditions as perceived not by a meteorologist but by a reindeer. The only respite, he explained to me, might be if a north wind came to blow the smoke away from us.

'But just now a north wind might bring snow,' he said, his agitation alternating with gloominess. 'And then the reindeer might get pneumonia from the sudden change in temperature.'

By the afternoon, we could smell the smoke, too. Though we knew the sky was cloudless, the sun had disappeared completely and the temperature had dropped by several degrees. That evening Ivan turned the conversation to dust clouds raised by giant meteorites, and their role in the extinction of the dinosaurs. Where Yura made passionate assertions of his attachment to traditional beliefs, and Granny stated them soberly as a matter of fact, Ivan made declarations about his lack of belief and then interpreted events in a typical believer's way. His tone turned apocalyptic.

'You can see what a bad year it is,' he said darkly. 'No rain, no berries, no mushrooms. Bad luck with hunting. This spring, five lads from our village died from falling through the ice. And another young herder froze to death in the snow. And now this smoke. It's all interconnected [*vse vzaymosvyazano*].'

Amid the steady routine of chopping and sawing wood,

cooking meat, and drinking tea, even the most trivial setback provoked in Ivan a fresh upsurge of anxiety.

'Terrapin's been gone for six hours cutting trees for firewood,' he complained at one point. 'Where can he have gone, how can he be taking so long to do such a simple job?'

Another time, he demanded, 'Is Vladimir Nikolayevich coming to visit us or not?'

Now a retired elder, Vladimir Nikolayevich travelled from camp to camp and was welcomed wherever he went, possessing a competence that seemed almost magical. He had friends everywhere, he was needed everywhere. Younger men like Ivan consulted him about the logistics of migration, technical problems with the herd, or the unpredictable behaviour of a rogue bear. Wherever I went, Vladimir Nikolayevich was either expected or had just left. Like Odysseus in the early chapters of Homer's *Odyssey*, he was present even through his absence.

Ivan decided to ask the village for news of his movements. Yura pushed his way up the prickly twigs of a larch tree and tied the antenna of the bush radio as high as he could reach. I tied the motor to a log, sat beside it, and began to turn the handle.

Every brigade was issued with a radio like this one. At 9 a.m., 12 noon and 6 p.m. (except Sundays and public holidays), they could make contact with Sasha the Radio Man in the village. If there had been no aurora borealis the previous night to disturb the ionosphere, conversation was usually possible. Each camp took its turn to speak, and they could even speak to each other. One speaker shouted into a telephone receiver while the others crowded around, trying to work out what was being said at the other end. Back in the village, Sasha was surrounded by people who wanted to pass on messages to their relatives in the camps. Sitting in his operations hut, he created a virtual community of people who could keep speaking to each other at fixed times, albeit in a compressed and clamorous form, even though they might not see each other's faces for years.

A camp's call sign consisted of its number preceded by the word '*snop*', as allocated by the KGB or the military authorities. The choice of term, meaning a sheaf of corn or hay, seemed arbitrary or whimsical. Reindeer people had nothing to do with hay: even their horses grazed for themselves throughout the winter, digging through the snow with their hooves like reindeer.

'The KGB officers must be Sakha from the plains who grew up on a cattle farm,' suggested Tolya. But I later found out that the word also means the sheaf-shaped jet of flame from the mouth of a cannon – perhaps another example of the Soviet militarization of everyday life.

Ivan would not allow Yura to operate the radio. 'Yura doesn't know how to do it properly. He served in the tank regiment, but I was a radio operator at the nuclear test site in Kazakhstan,' he explained, half-jokingly, reminding me of how military service moulded every man's frame of reference.

Ivan cradled the telephone receiver under his chin and manipulated the dials. When he could hear voices, he transferred the receiver to his hand.

'Snop 7,' he barged in at the first opportunity. 'This is snop 7, snop 7! Does anyone know where Vladimir Nikolayevich is? Over [*priem*].'

He was listening to other cross-cutting conversations, possibly including answers to his question. No one spoke as we huddled at the foot of the larch tree and I kept turning the handle, the motor whirring noisily. I found it tiring on the arm but if I flagged, Ivan would lose contact.

'Snop 7, snop 7!' Ivan concluded. 'Thank you, we're leaving now. Over and out. He's been in camp 10,' he said, turning to us. 'They think he was going on to camp 4, but he hasn't arrived yet and camp 4 doesn't know anything about it.'

Two days later, the weather became sunny again, but sharper and colder. The forest fire had probably not been extinguished, but a northerly wind was blowing the smoke away from us, sparing us bronchial problems for the reindeer, yet somehow not subjecting us to the snow which might have given them pneumonia. Ivan, like his reindeer, seemed calmer. He was determined to move the following day, come what may. Camp 4 was in a completely different direction, but Vladimir Nikolayevich would figure out how to find us if necessary.

Ivan left at six in the morning to ride 7 miles back towards the beginning of the pass from Avlandjá to look for any remaining reindeer which might be pursuing some purpose of their own without his protection or control. Though he knew they had probably all come downstream, he returned there twice more during the day to check, riding 42 miles on an *uchakh* over rough ground, in between the day's other tasks, to make sure the area was clear. Ivan was thorough.

'It's a law of nature that 30 per cent of new calves die each year,' he told me, 'but why give our stragglers to the wolves of Avlandjá?'

In the afternoon I visited Lidia. She was sitting outside her tent cutting up reindeer liver on a board, while Gosha sat on a log nearby gazing at a mountain. She nodded towards the liver on the ground in front of her.

'You can tell a person's character by the way they do this,' she said. 'If they cut it in an orderly way, they're a good person. If they cut roughly or irregularly, they're bad.'

Aha! A person's moral qualities could be read on a reindeer's vital organs! This was just what an anthropologist likes to hear, and I opened my notebook. In linking orderly action to goodness, I guessed that Lidia was using the liver to make a statement about her own character. Yet there seemed to be something missing. The reindeer itself was not involved: the moral message depended simply on the way a person used a knife. Could you not change your technique in order to be a good person? I was working out how to ask this, when Lidia added, 'And by the taste of the small intestine you can tell the character of the person the reindeer belonged to.'

Here was my answer. Even after it had been killed, cut up and cooked, a reindeer could carry information from one person to another. You might change the way you cut meat in order to make a good impression, but you could not control the way your reindeer tasted to somebody else. These were characteristics that could not be disguised. Even against its former owner's wishes, the reindeer could reveal that person's inner secrets.

Lidia laid the sliced liver neatly in piles and put them aside. We sat in silence for a moment, then she said quietly, 'Sometimes Ivan is very impatient.'

So she had not dropped the thread. But she had realized that I did not understand. The reindeer had belonged to Ivan. When the intestines were served in Granny's tent I did not have the skill to interpret their taste, but she was making sure that I received their message. Suddenly I understood the distance between her tent and the tents of Ivan's family. I felt torn. Eveny culture was founded on the use of animals as metaphors for relations between humans, so the topic was central to my research. Yet if I colluded with Lidia I would be disloyal to Ivan. In grappling with logistics, preparations, and delays, a brigadier was always under the pressure of people's scrutiny and expectations, from the members of his own camp to other brigadiers to the accountants in the State Farm office. It was hard for a brigadier to be liked all the time.

'He's got a lot of responsibilities,' I pointed out.

'Yes, but he keeps telling everyone what to do and complaining that nobody else works as hard as he does. My Gosha knows what to do, he's lived in the taiga since he was a child. That's why I want to live out here with him. He's had brucellosis and I'm worried about his health. Herders take a long time to adjust [*dolgo privykayut*] when they go back to the village. They fall ill immediately because their soul is open. It's another life, another world. In the taiga you know where everything is, what to do, but when you arrive in the village you have no idea.'

Lidia was speaking in Russian and an 'open soul' (*otkrytaya dusha*) was a commonplace Russian description of an honest, straightforward person. Like other Europeans as well as North Americans, Russians talked of the taiga as a pristine wilderness where a city person went to seek purification of the soul. But for Lidia the taiga was not the empty space of white man's adventure travel. It was the basic setting for one's life, the place where one knew what to do, how to be. By linking the village with illness, Lidia seemed to mean that a herder's soul was a delicate, innocent organ. Its openness was a virtue in the taiga, but in the village its open quality made the soul raw and vulnerable.

'Someone can appear to have a strong spirit when they're in the village,' Lidia continued, 'but working in the taiga reveals a person's weaknesses [*slabosti*]. Here in the camp, you see each other directly.'

From the word 'directly' (*pryamo*), I understood that Lidia believed a person had a true self or soul, which could be obscured or corrupted by social relations in the village. I thought of how I must appear to people here, so dependent, so conspicuous in my lack of skills. On this landscape I had a lot of 'weaknesses'. Once past the novelty of being a foreigner and the status (high in Russia) of being a scholar, there was nothing to fall back on but how I joined in their activities and accepted the conditions of their life. Technical inadequacy sometimes shaded into moral laziness. Lidia's remark made me remember with shame a trip I took with Vladimir Nikolayevich on which I had become so used to being looked after by him (especially since having had my appendix removed only a few weeks earlier) that I had not rushed to help him pull out his heavy snowmobile when it fell into a hole in an ice sheet. Tolya was on that trip and had scolded me: 'Go and help the old man!' he had ordered me sharply. I told Lidia about the incident.

'Tolya was speaking like a Russian,' she said.

'Wasn't he teaching me correct Eveny behaviour?' I asked.

'That's not how you learn. Vladimir Nikolayevich would

never have said anything. A person should not have to say what they're feeling – you should work it out for yourself.' Her voice became very emphatic: 'We never say straight out, "I'm in trouble" or "I need this, I need that", it's not done [*ne polozheno*]. You know by someone's face, by their behaviour, by everything. When someone says too much, our people don't like it.'

'Do you mean they communicate like telepathy?' I asked.

'In a way, yes [*V rode da*].'

I had often noticed how silence in the camp did not feel awkward. A handful of people who might see no one but each other for months on end knew how to be alone together. They were aware of the nuances in each other's moods, which they might not sense in a more verbally profligate environment. I had also learned to pick up some of the shifting feelings of warmth or irritation between tents and to tell the difference between someone going off alone to do a job and someone wanting to get away from other people.

'It's boring in the village anyway,' Gosha added, still with his back to us but following our conversation. 'There's nothing to do except sit around and watch television.' Or engage in village power politics, I thought, but that would not interest Gosha.

I could not work out how far Lidia was idealizing life in the taiga, and how far she was giving me a lucid explanation of core values in Eveny culture. Her vehemence suggested that these were important values which she was striving to assert. Was she asserting them to me, or to herself? Did my unusual presence allow her to express her true inner feelings, or to exaggerate them? Her statements were different from Tolya's, who seemed to believe in an authentic essence of Eveny-hood, which he expressed through metaphors sometimes of soul, sometimes of habit, sometimes of genetics. His idealization was overt, and we were the audience for his theatre of Eveny culture, which supported his ultimate goal to improve their life through political reform. Lidia's ideal was a tranquil one of domestic intimacy. When she spoke about learning afresh to live in the taiga, I felt that she was talking about her own life which she had refreshed by joining Gosha in the camp. Lidia had given me a gallery of portraits, painted in terms of inner organs and culminating in the one dearest to her heart. If the reindeer's liver was about her own righteousness and its intestine was about Ivan's irritability, the herder's open soul was about her husband's fragility and her determination not to be separated from him.

Gosha later gave me his old diary to take away. I do not think he checked it before he did so, for it was very different from Ivan's. The details of accounting and herd management were mixed with poetic thoughts and painstaking drafts of letters to Lidia, so that it felt indecent to read further.

It was the last evening before our migration. Ivan had retrieved his father's horses Mashina (Motor Vehicle) and Medved (the Bear), which had been grazing some miles away. Though I was not a particularly big European, I was too heavy for a reindeer to carry over a long distance and outside the sledging season I had to depend on horses.

I stood on a mountainside tracing the grey-green lines of dwarf willow that marked the rivulets all around and watching the shadows of clouds drifting across the velvet bogs. Reindeer herders could stand like this for hours and sometimes said that the beauty of the land was the great love of their life. It seemed almost unbearable that they should have so few days to drink in the loveliness of each site. Forests of spindly larches flowed up the hillsides like splashes, still green in the shelter of the deeper valleys but already yellowing where they had thinned out and faded away towards bare ridges of lichen-speckled boulders. Hidden in the trees and rocks were the droppings, snapped twigs, and tooth-marks of wild reindeer, elk, and smaller creatures, coded signs of their late summer movements and intentions, approaches and avoidances. I was not skilled enough to observe these like the herders, but I felt that I was being observed – by birds, mice, squirrels, even fish. Beyond the exposed pass behind me at Lake Avlandjá, the reindeer and their herders had been harried by wolves on their migration here a few days ago. In the forest down the river where we would migrate tomorrow, bears would be gorging on the first of the autumn cranberries.

The north wind had died down, but the smoke had not returned. Far below, I watched Gosha and Nikolay drive the reindeer into the camp for the last round-up at Daik this year, a symphony of grunting throats, clicking hooves, and clattering bells which seemed especially poignant. There were also few mosquitoes. This changed the behaviour of the herd. The milking females arrived first, slowed down to a walk and, without needing to be lassoed, headed straight to their tethering posts to wait for Granny and Emmie. As the main herd followed, it split in two and flowed around the tethered horses, which remained absolutely unmoved. The reindeer swirled around for a few minutes before gradually lying down, hardly even grunting. By the time I climbed down to the camp, the most trusted untethered *uchakhs* had moved next to the bachelors' tent, where they sat with their legs folded under them as if listening to the Russian pop music that was turned down very low on the radio inside. Gosha had promised to tell me the names of the streams we would cross the next day, but he was too exhausted and fell asleep early. Ivan, too, did not stay up for our usual late-night conversation.

On 6 August the rain started drumming on our tent very early in the morning and soon became heavy. I went to check that our baggage was properly covered and saw Yura and Gosha plodding steadily across the camp, pale grey figures adjusting stores before they set out to work with the herd. Tolya was still asleep, so I lit the stove and slid back into my sleeping bag.

Two hours later, the rain had stopped. The air was fresh, with no mosquitoes at all. Clouds hung across the tops of the mountains, while their lower slopes appeared in the hues of a watercolourist. Every tint had become an echo of its former self, almost colourless yet glowing with potential tone. The lichen's primrose yellow had become a sulphur grey, the scarlet of the miniature birch leaves a muted crimson. Each blueberry plant as far as the eye could see was picked out by pearl drops of rain on its tiny glaucous leaves, its violet tinge sprinkled with mercury.

The morning was no hastier than any other. Voices from the other tents did not start rising till nine or ten o'clock, accompanied by a plume of smoke from the chimney of each stove. Humans and reindeer had both shed their recent nerviness. The herd had gone downstream and was reported to be spread out at ease and feeding calmly. Some of them had already found their way downstream as far as our next camp site at Djus Erekit.

'Is that a problem?' I asked Ivan.

'It's all right,' he said, 'they know it's time to move there. The older reindeer know every spot at every season and the younger ones learn from them.'

'Just like us.' Yura laughed.

Ivan felt it was safe to leave the herd with the inexperienced Nikolay while he worked on packing supplies in the camp. The most vital supplies were also the most bulky: sledges and the winter clothing of gloves, thigh-length boots, sleeping bags, and massive overcoats, all sewn from reindeer fur. Another summer, when I had some unused flying time left on my chartered helicopter, Ivan asked me to move some sledges from the lower reaches of the River Avlandjá, where they had left them when the winter ended in May, to the site where the winter would begin again in October. The knobbly sledges were not designed to be contained within another space and we crouched and twisted ourselves between them as they jammed the interior of the helicopter, snagging on the cables and appendages of the aircraft's complicated interior. But the flight of forty minutes saved his men four days' trek with the sledges tied to the backs of reindeer.

I helped Tolya heave down the huge block of butter and the heavy sacks of flour and sugar that we had brought from the city, then the sprawling, floppy canvas of Ivan's spare tents. Most of our time was spent laying out the equipment for the migration itself: numerous saddles, bridles, and hide ropes for reindeer, and a few for the horses.

The saddlebags were tied to each other in pairs with reindeer-hide straps that had been repaired and knotted over many migrations. We shook them into shape and laid them open on the ground, then moved over to join the women as they pulled everything out of the tents and began to lay out the family's belongings on the ground. There were heavy Army-style sleeping bags, jackets, and floral sheets; stained and blackened cauldrons and kettles; little pouches of salt to feed the reindeer; torches, a radio, and batteries; dried medicinal herbs and an old glass bottle containing bear's fat to heal wounds; buckets, knives, and axes; and spare rifles in addition to the ones that would be kept at the ready on the journey.

The men had books: Lermontov's poetry and a Russian translation of Marguerite Yourcenar's *Memoirs of the Emperor Hadrian*; a patriotic history of the Second World War; *Algebra for Beginners*; a veterinary pamphlet on the intestinal parasites of reindeer; and a book proving that Inca gods came from another galaxy. The women were less interested in reading*. They had small bundles wrapped in cloth or hide and little bags of finely inlaid reindeer fur decorated with beadwork but worn bald and smooth from decades of use. These held women's secrets, but I believe they contained needles, sinew thread, small medicinal animal parts, and talismans. Revealed to the daylight, these intimate objects seemed tiny and defenceless in such a huge landscape.

Though place names were important, the Eveny vocabulary for relative positions focused not on the places themselves but on one's own movement through them. The word *dyu* means 'home', one's present camp site. A site is a destination for only a moment before it becomes home and the starting point for the next migration. *Amdip, dyu, erimken*: previous site, present site, next site. Today's *erimken* becomes tomorrow's *lyu*, and the previous *dyu* slips back to the status of *amdip*, yesterday's site. Granny was around seventy years old. Allowing for a few years at school, I calculated that she must have changed the location of her *dyu* some 1,500 times. Each time, she would have packed up her belongings as she was doing today and as she had done every few days for most of her life.

Men and women alike were pressing objects carefully into the saddlebags, pushing smaller bundles into the gaps so that no hard or angular surfaces would rub against an animal's flank. A pair of saddlebags must be of even weight, and occasionally a man would pick them up and test them. Ivan was very particular about this and I often overheard him directing the others to transfer stuff from one bag to its partner or from one pair of bags to another. Someone would need to remember where everything was, and there were occasional sharp differences of opinion. Each person tended to pack their own way, though the routine changed a little on every journey as the herders picked up or deposited supplies. Tolya and I had brought different baggage of our own which complicated the process. I knew that however carefully I packed my tape-recording and photographic equipment someone might undo them and repack them to suit the reindeer, so I tried to wrap each item in its own padding.

The number of *uchakhs* needed to move a camp could run into dozens. Each *uchakh* was chosen carefully to carry either a pair of saddlebags weighing around 80 pounds, or a human passenger. An average Eveny weighed not much more than 80 pounds but could also lighten the burden by responding intuitively to the animal's movement. While the reindeer stood patiently, one man waited on the far side while another picked up the saddlebags by the connecting strap, trailing the thicker strap which would be tied under the animal. He would fling the far bag over the saddle on the reindeer's back to the man standing on the other side. The bags hung quite low on the animal. To clear snags and settle the straps firmly over the saddle could take several attempts, and the men on each side might need to lift the weight of the bags off the animal several times and firm them down again. When the cargo seemed steady, the thick strap was fastened around the animal's belly.

By early afternoon the packing was complete and nothing remained except to dismantle the tents, reversing the process of their construction. The insulating mats of reindeer fur were brought out, trussed into rolls and tied across the saddles of baggage *uchakhs*. The bulky canvas tent was pulled away from under its stone weights, folded and crammed into a saddlebag, while the poles were untied and stacked neatly against a tree to use the following year. Only a ghostly shadow remained of slightly discoloured grass under the floor of tired larch branches which were already shedding their needles.

The stoves from every tent except Granny's had been taken apart and packed with their sharp corners masked by clothes or bedding. As a signal that everyone had drunk the very last of their many teas, Ivan tied the tin mugs to the outside straps of their owners' luggage. Yura took down his mother's tent, leaving her cooled stove tiny and exposed on the flat stones which were raised above the combustible peaty earth. In a gesture of finality, he shook out the ashes, the only time this ever needed to be done since local larch burned with almost no residue. He dismantled his mother's stovepipe into several sections, placed them inside the little oblong stove itself, and tied the stove onto the back of one of her reindeer.

The reindeer were roped together in single file into several separate teams, with an adult mounted on each lead animal, perhaps a child on the reindeer tied immediately behind, and the remaining reindeer loaded with cargo. True to the Tungus tradition, the Eveny rode without stirrups, using a stick instead with their legs hanging very close to the ground. To mount, you stood beside the animal's right flank, held your stick with your right hand and passed your left leg over the reindeer's back, leaning on the stick to hoist yourself onto the saddle. An experienced *uchakh* understood the difference between the ponderous mounting of a stationary caravan and the nippiness of a single herder or hunter who hardly leaned on his stick as he floated up from the ground slightly in front of the saddle, expecting the reindeer to start running even before he was in place.

The *uchakhs* stood around while their harnesses and baggage were adjusted, edged forward for a few yards and stopped for further readjustment. I sat ready on my horse and watched as each team moved off separately, with a tap from the rider's heels on the lead animal's flank. The Eveny disliked looking back and I was the only one who turned around in the saddle to take in the abandoned site as we left. In every person's life there would come a last view of each site, but one could not know when this was until it was too late.

As the movement began, there was a noticeable lightening of mood, almost a quiet elation. Team after team of reindeer, strung far apart, followed the steep uphill and downhill banks of the River Munne. There was no intentional stopping, only pauses to adjust tangled harnesses and slipping luggage as the packs settled with the movement of the animals and the animals themselves lost weight from their exertion.

Sometimes our animals' hooves flicked through groves of dwarf birch twigs on a raised plateau high above the gorge; sometimes we descended to the river and clattered over the white, rounded stones or splashed through the shallow torrents which wove channels between them. Near the cliffs which deflected the river into sharp curves we saw the sweep-mark of the spring floods in the deposited tree-trunks and swished grass high up on the bank. Weaving from one side of the river to the other, we brushed through groves of vanilla-scented poplar bushes on river islands. Sometimes we could not avoid stumbling through the slowest terrain of all: the boggy tussocks known as 'Russian heads' because of their blond tufts of dry grass, where your animal's every footstep had to be meticulously

placed in a deep oozy crevice between freestanding clumps of
fibrous roots. The journey was unhurried and strangely solitary.
We were so far apart that conversation was difficult, and singing
on the trail was forbidden by the spirits of the land who insisted
on quiet, measured behaviour.

The distance from Daik to Djus Erekit looked like 5 miles on
the map, but Ivan said that with all the crossing and weaving
the journey amounted to 25. It took about six hours. I arrived
with Tolya at nine in the evening. Granny and Emmie were
already there and Emmie was tending a kettle suspended from
three poles leaning together over the communal outdoor fire,
the ashes from the previous year overgrown with long, wet
grass. The outline of the previous year's tents was still imprinted
in the stones that had pinned them down and which had been
pushed to one side when the herders left, distorting their
rectangles.

There was a sense of completion and relaxation, baggage
flung casually on the ground and left half opened where people
had rummaged for their most immediate needs. Granny had
taken care to keep track of the tea and sugar, and I knew where
I had packed a box of chocolates labelled 'A Night at the
Ballet'. As more caravans arrived, we unloaded the tents and
pulled down the poles that had been propped against larch trees
the year before, raising the canvas, discarding some old rocks
around the edges and fetching others. Each stove had to be
balanced on flat stones and the spirit of the new site had to
be fed inside each tent with morsels from our first meal.

In Granny's tent every item of bedding, every pot and pan,
reappeared exactly where it had been before. It was only the
outside world that changed. Beyond the tent-flap, every detail
was different: which way one turned to fetch water or wood, or
where the reindeer would gather in the evening and circle
around. The open area at Djus Erekit backed uphill onto a fairy-
land of moss, some species hanging from soft-needled larch trees
and others deep-piled on the ground or submerged under little

clear streams. On the other side, a steep, difficult path ran down to the river, which at this point was fast, deep, and loud. Fetching water here would be more arduous.

Ivan was the last to arrive, joining us at midnight for his first mug of tea since midday. He was more relaxed than I had seen him that year. Everyone in the camp ate together around the outdoor fire. Much later, I noticed that Ivan was missing, and turned to Lidia who was sitting next to me.

'Has Ivan gone to sleep?' I asked.

'Probably,' she replied.

'It's just that he didn't say anything.'

'Our people do that,' she said. 'They leave without saying anything. We don't say goodbye – it's taboo*.' She used the Russian word *grekh*, an Orthodox Christian term which literally means 'sin', but I thought 'taboo' was the real meaning behind Lidia's struggle with the Russian language to express Eveny sensibilities.

'Why is it considered taboo?' I asked.

'Probably because "goodbye" would mean forever.'

How could I not have realized? I had passed through Russian cities to reach here and had stayed with Russian friends, bathed in the overt warmth of Russian embraces and farewells, each evening finalized by repeated rounds of 'Good night!' 'Good night!' 'See you tomorrow!' Even the Russian phrase for goodbye, *do svidaniya*, contains a bold confidence about the future and means literally 'till our meeting'. But I had failed to adjust. We were speaking in Russian, but the herders were not Russians. Ivan had departed from the fireside the way we had departed from our previous camp site – discreetly and without looking back. In talking sweepingly about 'her people', was Lidia idealizing now? I thought back to other departures I had seen, and realized that they fitted Lidia's explanation closely. Not only had the people departing not looked back, the others left behind had not waved or followed the travellers with their eyes, as I would do for miles, but had simply turned away and slipped back into their previous activities. The departure was made inconspicuous and gradual, so that there was no moment of sundering.

Reflecting on this later, I realized that the Eveny arrived gradually as well. When someone entered another person's tent, they parted the entrance flap and hovered there for a moment before going in, and then waited a while longer before starting to speak. I thought of the way I had seen caravans arrive unheralded in various camps and understood why we had been met from our helicopter in such an unsensational way. The astonishing teleportation of our journey, covering the distance of a year's migration in three hours, had been converted into an imperceptible blending.

From observing their arrivals and farewells, I built up a picture of Eveny sensibility in which sudden change and uneven behaviour were experienced as harsh and dangerous. All these examples seemed like different manifestations of the same concern to cushion the suddenness of a transition from one situation to another, which the Eveny felt so readily as shock. I began to see how the changes of environment in their regular migrations were softened by the constancy of the tent's interior, as well as of the fire through which they contacted the spirits of each successive place.

Information could be shocking by its very newness, and the passing on of news likewise had to be softened and made gradual. If you met someone in the taiga, you did not call out or greet each other, but came alongside and gradually started conversing. Only after a while would the other person ask, 'What news?' (in Eveny, *yav ukchenenni?* or in Sakha, *kepseen tuokh baar?*) and you would answer, 'Nothing' (*acha*, or in Sakha, *suokh*). Later, little by little, you would begin to tell the person what you had to say. Especially if you were bringing bad news that affected them personally, you could not state your message directly. You might progress to a remark about how things were not so good in the place you had just come from. Or you might say, 'Your mother is not well.' This was called 'giving someone to understand' (in Eveny, *unuvkame*). The person would then guess that their mother might have already died and would ask for further details only when they felt ready.

I had heard many warnings against the casual splashing around of words, as well as of projecting them too forcefully, from stories of people who had offended the taiga by speaking or singing loudly, to the loss of respect accorded to Russians, and later Americans, when television arrived in the village and they saw characters in dramas shouting at their families. Sharp, unguarded words could take on a force of their own and even kill, like a curse. Someone in the village once told me, 'I had a long-running battle with one of the Farm bosses, and one day I heard of a new plot he was hatching against me. I was so angry that I said out loud, "Why isn't that old bastard dead yet?" Even at the time, I had a bad feeling about those words. Well, exactly a year later that boss was at death's door and he sent for his son, who'd been my childhood friend. The son was due to go herding that day, but he stayed close to his father. It turned out that the father was too tough to die and the same night his son died instead. His son had taken on the impact of my words.'

The high value that Eveny culture put on steady, reserved behaviour seemed to contradict the passionate nature of many of my friends: Lidia as she threw herself into taiga life in order to be with her husband; Gosha as he drafted tender letters to her when they were apart; Emmie as she ran across the mountains, rejecting the constraints of school and village that bound every citizen; Yura as he dreamed of the return of shamans; even Ivan's anxiety about living up to his father's legacy. Holding together the family and the camp was the figure of Granny, who epitomized the traditional taiga values of restraint and self-control to which Lidia aspired. Granny sometimes made revealing remarks in a dry, low-key manner, and while her husband was alive I had time to experience the warmth and strength of their partnership. But her behaviour was so well moderated that I could not be sure what she 'felt', in the sense that I understood my own feelings. She told me in what seemed unemotional words about her father's execution and the deaths of four more sons, including her eldest, Oleg, in a knife fight shortly before my first visit. Perhaps the restraint and the taboos were themselves signs of emotion, like the avoidance of direct words: I had to find out Oleg's name from others, since Granny referred to him only as 'the one who died'.

For the men, the migration never stopped. The morning after we arrived, as soon as the stove had been lit and tea drunk, Ivan sent Terrapin down to the next site to start repairing the autumn fence after the spring floods. Ivan also went down there to check for wandering groups of reindeer. Some 600 or 700 had already gone beyond the broken fence.

'Do you need to chase them back?' I asked with concern, still overreacting to the solemnity of his voice.

'*Nye-e-e-t!*' he said gravely. 'They know by themselves it's time to go there.'

Six weeks had passed since the summer solstice and the fish too were migrating downstream, hurrying to reach lakes that were deep enough to stay liquid underneath as the winter ice thickened on top. I went fishing with Tolya and the children, throwing stones into a deep pool to drive the fish downstream into a net. Our fresh stones lay gleaming at the bottom and anyone passing by afterwards would be able to work out from the growth of algae just when we had been there, what kind of fish we had been chasing, and where we and the fish might have gone since.

'In the old days they believed that throwing stones into a river causes rain,' said Tolya.

'Isn't that what we're doing now?' I asked. The rain had resumed overnight and become almost continual, and the river was rising rapidly.

'Maybe Ivan shook his bearskin too hard,' he said, only half joking. I wondered whether Ivan's one-eyed dog was predicting floods.

Other camps around us were also on the move, like planets whose orbits approached each other at rare but predictable intervals without ever meeting. Moving around another river system in the next valley, Valera's brigade 3 was due to reach a site within 20 miles of our present one around 20 August. By then we would have already moved further away from them down our own river.

The next day, Ivan, Yura, and Gosha invited me to go with them to carry cargo to the next site, Tal Naldin, meaning 'Mouth of the salty-sand stream'. This was my favourite site on the entire landscape, the place where I first lived with this family when Ivan's father was alive; where I first understood how the strength of the old couple's marriage gave the family its sense of purpose; and where I saw Ivan abandon his restraint and weep at the prospect of the family's dispersal after the wholesale destruction of their herd. As we rode down to the site once more on a combination of horses and reindeer, I thought how the responsibility of running the herd had matured Ivan in the five years since he had appeared as a young man in the documentary.

The site had not been visited for ten months. Clusters of mauve pulsatilla anemones had pushed their heads through the bare stones at the edge of the main channel of the clear, rushing river. Behind them rose a late summer meadow of shimmering grass flower-heads spangled with rosebay willowherb, pale blue Jacob's ladder and ultramarine monkshood. In the branches of the trees it was already autumn. The light green larches at the foot of the surrounding peaks were turning a golden ochre and the grey-green of the scented poplars on the river islands was becoming moon yellow. The air was vibrant with a suppressed wind which only the poplars picked up as their leaves trembled on their flexible stems.

'You remember this place,' Ivan said to me quietly.

The other men, too, seemed to feel something about this site, perhaps the memory of that painful autumn, perhaps the same delight that I felt now. No one said anything for several minutes. Then Ivan pointed to two cairns of stones on the summits of two neighbouring mountains.

'Tolya's brother Maxim and I built those cairns as a contest in 1972,' he said, 'when we were in class 5 at school. We were inseparable. Children don't play on mountains like that today.'

We unloaded the animals and waded into the herbaceous stalks of grass, which met us like chest-high virgin snow. Each man dragged a pair of saddlebags still strapped together and heaved it up onto a platform. After stacking several loads, we covered everything with tent-cloth, then cut down several small larch trees, trimmed the trunks and laid them across the top. I noticed last year's tent poles stacked neatly against the side of the platform, and a few pans and a broken bottle in the grass. Once the herd reached here and the fence was closed behind them, they would be in a large area of safe grazing and the herders could live here for several weeks, sorting their supplies and equipment before moving on into the winter. It would be their first pause since the end of April.

'It gets even more beautiful further down,' said Ivan, gazing in a direction I had never been. 'Wide river, tall poplar groves, thick forest . . .' His words made me realize how sparse and exposed he considered Daik to be, with Djus Erekit just a brief halt on their way to a more settled base. 'But we don't stay on this river in winter. It's terrible for snow. By April it gets 10 feet deep. Elk get caught in it, they walk on the hard crust and fall through, then they can't get out – it's a terrible death! We found five of them this last spring.'

I reflected on this extraordinary snowscape, forming year after year with never a human witness. 'Does anyone come here in winter?' I asked.

'No one except an occasional hunter. Vladimir Nikolayevich comes this way some winters. But even sledges can't pass then.

You have to go ahead on skis pulled by reindeer to make a path, and then sledges can follow.'

After a silence, Ivan suddenly said, 'It must be hard to leave your family behind. Next time you should bring them to live with us.'

It was an invitation I would treasure, and take up.

An evening wind had started. 'You must be cold,' said Yura.

We untied the animals and put on our gloves. Our *dyu* was still one site back upstream: we would reach it by midnight. Masha and Emmie would give us reindeer stew in Granny's tent and Gosha would be fed by Lidia. In a few days, the women and children would come down and live at the site we had prepared. At the end of August, the children would be claimed by the school in Sebyan village. Lidia and Masha would go with them, Lidia to look after her girls and Masha to teach. Of the women, only Granny and Emmie would remain with the men as they guided the herd onward through the autumn pasture into the long, deep silence of winter.

We rode back upstream between mountains covered with mist, the wide, stony bed furrowed with channels made deep by the rain. It became almost dark and I lost sight of the men in front of me. The river roared all around as if I had been caught by the incoming tide on an open beach. For a while my horse's legs were under water and I could see nothing but the crests of the little waves which seemed to race up against the horse's chest but were always in the same place. I suddenly imagined that my horse was afloat and almost lost my balance as I let go of the reins and reached out for oars. For just an instant, I felt as if I might drift alone with the fallen trees down to the Arctic Ocean.

But that did not happen. Tolya had given the aviators the coordinates of Tal Naldin. Two weeks later, their helicopter found us at our new site and took us back to the city.

6

Kostya's mushroom crisis, camp 10

The long silences of herders are not empty, but represent their blending into the cyclical processes of this land, a oneness for which words seem superfluous even in the Eveny language which was fine-tuned for use here. Language was often used by Granny, Ivan, and Yura to talk of politics and to recount narratives of reversals of fortune, which they tinged in the telling with regret, irony, and humour. For this family, every step they took in their beautiful mountains was coloured by the bitter memory of the destruction of their herd five years earlier, in 1990.

The turmoil of that momentous year had worked very differently for the several men and one woman of brigade 10, whom I had met for the first time that spring. Kostya and his team did not fight the Farm but worked within its existing framework. Kostya's rise on the Board of Honour had been rapid. He had been appointed to rescue the brigade in 1985, when it was lagging last out of all the Farm's thirteen brigades. The following year he brought it up to third place, winning medals from the Farm, and in 1989 the brigade won a nationwide competition, being awarded a cash prize ('Though we haven't yet received it!' he later told me wryly).

In 1990, when I took my film crew for two periods of residence in his camp, Kostya was at the high point of his career. His friendly, open, lightly bronzed face (whose very flat nose is said to reveal his half-Sakha origin) matched his easy manner. He was highly disciplined himself, but understood that discipline came from within and that authoritarianism, like vehicle tracks, ground to a halt in the bog beyond the edge of the village.

'Of course it's not just up to the brigadier,' he explained one day in April as we stood side by side surveying the massed herd rotating near our tents. 'Everything has to be done by consensus. If members of the brigade were at loggerheads, nothing good could ever come of our work. If the brigadier tells his herders to go and have a look over there, and they refuse – who needs a brigade like that?'

The Farm had sent Kostya out from a desk job to improve brigade 10, but was reluctant to lose him to the taiga altogether. He had also been appointed to a party position in the village. In either setting, I felt that his life was a quest for orderliness. But unlike most administrators, he saw this as beginning with the work in the taiga, rather than with exhortations from on high.

'Would you choose any other life, apart from reindeer herding?' I once asked him.

'No,' he replied, 'what for? For me it would hold no interest whatever. I like work that gives results. In the village, everyone's going hither and thither, but it's not clear that they're actually working. Of course, some kinds of work are visible: I can see builders, drivers, loaders – I can see them working. But as for the rest – they go around with their tongues wagging. Is that what their work consists of?'

His deep-rooted conscientiousness was struggling with a newer cynicism, as he sought a balance between his belief in hard work and his doubts about the system that promoted it.

'They used to drag me back to the village every month for a party meeting,' he continued. 'Now with *perestroika*, I can just say phooey! I don't really care about party work, but I go in to the meetings because I do care about reindeer production. I was even appointed to the Regional Party Congress as a member of the Control and Inspection Committee [*kontrol'no-revizionnyy komitet*]. They summoned me to a Plenum, but I couldn't go. I had real work to do! Anyway, I would have just sat on the airstrip in Sangar and jostled with everyone else trying to get a seat on a plane.' The cynicism finally won out when he added, 'The committee are all white-collar bureaucrats [*chinovniki*]. They just wanted me there as a token representative of the working class!'

Kostya did not talk like this in front of his young deputy brigadier, Arkady. He was grooming Arkady as his successor and did not want to subvert the innocence that made him a good worker.

'Will Kostya eventually stay in the village, and you become brigadier of camp 10?' I once asked Arkady as we perched together on top of the fence of the spring corral.

Arkady was tall and thin, and wore a strange skullcap with ear flaps. He took a deep drag on his cigarette and smiled. 'He already wants me to take on the responsibility when he goes back and forth on party work,' he said.

'That's good for you,' I said, hoping to prompt him into a remark about his own ambitions.

Arkady laughed modestly, creasing the corners of his almond-shaped eyes and revealing the gap between his front teeth. 'This was a backward herd,' he answered. 'It's now one of the best. Kostya's done a splendid job, but he's also an outstanding party worker so they want him in the village. Anyway,' he added, 'he's got a family and house there. I'm not married yet. That would be a good reason to move to the village, but for the moment I want to stay here.'

That spring, the brigade was so all-male that even the cooking was done by Peter, one of the bachelor herders, standing in as tent worker for Arkady's mother Kristina who was away in the village looking after grandchildren.

I next visited them during the late summer, when several women and children had moved out of the village for the long school holidays and the camp had blossomed into a cluster of family tents. Kristina had returned, and was looking after those men who were still unaccompanied: Arkady, Kostya, Peter the Winter Cook, Tolya, me, the two members of my film crew, and the occasional young bachelor who always appeared around a tent worker's stove. Kostya's wife and children were not among them, but I recognized Shura, one of the radio operators from the village, who had brought her children to join her husband, a herder called Valentin. It was still light till nearly midnight, and the children fetched wood, floated model boats on the river, filled bucket after bucket with bilberries, and made toy reindeer out of twigs, then lassooed them with cast-off lengths of hide.

'I like the camp life, I prefer staying here,' Kristina said in her calm, husky voice as we sat in her large, square restaurant tent over which I had painted a board with the words U Kristiny (Chez Kristina). 'The air's fresh, everything's fresh. I like the fresh reindeer milk best of all. Even if the work's hard, it's good all the same.' But she also added, 'I live out here because I can't bear to think of my boy and his friends with no wife and no mummy to keep them fed and warm.'

The presence of a woman civilized even the remotest location, and the men were more meticulous about removing their muddy boots before entering Chez Kristina than their own bachelor tents. The herders might be out for twelve or fifteen hours at a time without food. Whenever they came back one by one, chilled and exhausted, Kristina would immediately drop her washing, sewing, or cleaning and serve them hot tea and meat.

'I milk the females, catch the calves and tie them up, boil the kettle non-stop, make the bread, cook the meals, feed everybody five or seven times a day,' she said, counting the tasks as we sat in Chez Kristina. Then she paused and searched her mind. 'That's it!' she concluded. 'I rest only when I go to sleep.'

Kristina's situation in camp 10 was quite different from Granny's in camp 7. Granny, too, had worked hard all her life at the same tasks as Kristina. But Granny had been a married partner in a family enterprise, and the air of authority that she radiated came not just from seniority, but from a lifetime's habit of making decisions about the herd jointly with her husband.

Camp 10 also contained the men of the Nikitin family. The extraordinary old grandfather had a detailed knowledge of the parentage, medical history, and moods of each one of the 2,600 animals in the herd. He was more comfortable in the company of reindeer than of humans, and always pitched his tent some way from everyone else and cooked for himself. His son worked in the herd and had been joined for the summer by his own teenage sons, Zhenya and young Sergei.

'He's like my Arkady at the same age,' Kristina said of Sergei, who was 14 and due to leave school soon. 'Arkady loved it so much, he used to come out to the herd even during the winter holiday. Sergei's mother has tried several times to get her older son, Zhenya, to study. But each time he drops out of college and comes back to the taiga.' Sergei's mother had a forceful personality. She was also manager of the village's fur workshop, where women sewed reindeer skins into heavy winter clothing. 'Sergei will be all right if his mother doesn't spoil him,' Kristina continued. 'She's got a strong influence on her boys. But if she does spoil him, then he'll never come to the taiga again.'

As her son rode out with the men on the most arduous and exhausting sorties, Sergei's little figure embodied the conflicting visions of the future between the herders in his family and their women. The other men treated Sergei like a hero, praising him as they would never do with an adult. Arkady took him around constantly, teaching and encouraging him.

Vladimir Nikolayevich had come to stay for a while, a senior reindeer expert like Old Nikitin, but more worldly and very much at ease in the company of other humans. We had talked very pleasantly at our first meeting in camp 11 two years earlier in 1988, but I did not realize how much he had taken a liking to me until one evening this time, when he announced in front of everyone, 'Why don't you come back to Sebyan this December?

I'll take you on a hunting trip. Then you'll see how we really live on the land.'

'Vladimir Nikolayevich doesn't say things like that without meaning them,' Kostya said to me afterwards. 'It's a great honour – you should definitely go.'

Along with the transient Vladimir Nikolayevich, Arkady and the Nikitins were Kostya's best reindeer herders. But he had other men working on another important job. Each day, the surrounding hills resounded with the distant felling and trimming of larch trees and the clang of the backs of axe-heads hammering the heads of long iron nails. I sometimes joined the six men, working two by two on sections of a fence, as they rammed in upright posts, supported them with leaning buttresses, and bevelled the thick ends of long trunks before hammering them in as horizontals. If a tree was in the right place, they would leave it standing and use it as a living upright. If the line of the fence passed unavoidably through a bog, they would work thigh-deep in chilly water, and when it reached a cliff they would cling on to the cliff-face until they had secured the fence to a point beyond which no reindeer could circumvent it.

Fences of larch poles seemed spindly and incongruous on this massive, jagged landscape. But even more than mounted herders with dogs, they nakedly and explicitly directed the movement of reindeer. Apart from helicopters and vaccinations, fences were the main contribution of Soviet scientific management to reindeer herding. Just as the Great Wall of China was designed to deflect invading Mongol armies, so fences rerouted swarming reindeer herds as the ultimate sign that they were no longer wild. As you stumbled over the tussocks along the route of a fence and stepped around the leaning buttresses, and as you ran your hand along the rough coniferous bark of the springy yet firm horizontals, set far enough apart for a reindeer's head to poke through but not its body, you knew you were stroking the very concept of migration control.

All brigades used fences, especially to hold back the herd from invading the autumn pasture too soon. But Kostya went even further. In order to keep his animals separated from neighbouring herds on several sides, he was building a fence that would extend for 25 miles. It had taken four summers to build and would be completed in the next few days.

'I did it to preserve pasture,' he explained. 'We've got more than 2,600 reindeer, and there's a serious risk of overgrazing.' Brigade 10 is sometimes only 30 miles away from the village and the private animals owned by brigade members and their relatives and friends had added at least 600 to the State-owned herd of 2,000*. 'They criticized me back at the Farm office,' he continued. 'They said it was too big and refused to pay for it, so I bought the nails myself. Now we're responsible for our own budget, I'm paying my own fence-builders, too.'

I watched Peter the Winter Cook and Ivan the Fence-Builder completing one more section of the fence.

'Glory to Soviet Reindeer Herding!' I exclaimed, using a well-known slogan which could be adapted to any profession or occupation.

Peter paused, supporting one end of a horizontal on his shoulder, ready to hammer it in. 'No,' he countered, screwing up his face in a grin and wagging his finger, 'herders have become too clever for that – we don't need any more glory!'

The next day, Kostya was due to go back to the village for another party meeting. Before going, he left Arkady a long list of instructions written on a page torn out of an exercise book. Their relationship was still one of admired teacher and promising but inexperienced pupil. Arkady had recently been offered the post of brigadier of camp 12, but he did not feel ready for the responsibility. When Kostya came back a few days later, he asked me in detail what everyone had been doing in his absence. He made it seem like a discussion of my research, but I was sure that he was also checking up on Arkady's management skills. Arkady, too, was insecure about his own capabilities. I noticed on one occasion that he seemed hurt when the other men consulted grandfather Nikitin, rather than him, on a point of reindeer behaviour.

'I asked Kostya to come back soon,' he confided to me during Kostya's absence. 'We can't do without him when it's time to gather up the reindeer. This is the first time I've been left alone.'

Kostya was leaving Arkady at a vulnerable moment. Autumn was the season when the village reached out from the centre to reassert its control over the outlying taiga, as the Farm audited the brigade's performance by calculating its productivity. It was no accident that the director's blow against Ivan's family had been struck during this season, through the procedure of the autumn slaughter.

Autumn was also the season when human control must reach beyond facilitating the animals' migration into directing their genetic continuity through the selective removal of antlers, the crushing of the blood supply to testicles, and the slaughter of weak animals. The goals of the Farm and of a brigade did not entirely coincide. The Farm's ultimate criterion was the demand for meat. To satisfy this, herds were kept at a ratio of fifteen to eighteen breeding females to one breeding male (so different from the ratio of six male herders to one female cook), while the meat was provided by the majority of males, slaughtered when young.

The herders had a greater sense of the continuity of the herd, and also needed a constant supply of *uchakhs*, the original focus of domestication, which were still the basis of the herders' mobility but did not show up on Farm statistics. Using their knowledge of individual animals, they retained the best males for breeding or to castrate for riding. From this perspective, meat was the by-product of a process of rejection. The animals that would be eaten were those that were not good enough to keep.

Kostya and Arkady were already taking notes, ready to weed out females with a poor calving record, uncastrated males that seemed unlikely ever to compete in mating, castrated males that would not make good *uchakhs*, loners that would be picked off by wolves, any animal with an injury, a constitutional weakness, or bad character traits, and any that had not fattened up enough in the short summer to survive without green vegetation over the six months of winter. These animals would be matched up to brigade 10's annual delivery plan and would make the October trek to the village slaughterhouse.

The breeding males had to be prepared for the rut. Since their antlers grew back each year, larger and more aggressive, the men were starting to saw them off in advance to prevent the animals from injuring or killing each other, as they sometimes do in the wild. There was a different Eveny term for uncastrated males of every age from 1 to 11 years, whereas distinctive terminology for females stopped around 5. Exploring dictionaries of Arctic languages, I have found similar patterns in other, completely unrelated languages right across northern Eurasia, reflecting the observation of reindeer people everywhere that young males are slow to reach the stage where they can compete successfully with their elders to impregnate females.

The rut would run its course from start to finish in just a few days, a natural synchrony that enabled all the females to give birth around the same moment in the spring and so shortened the time they were vulnerable to wolves. Within these few days, some males might exhaust themselves to the point of death through a combination of fierce competition and intense sexual exertion. This is when I learned that it was not only *uchakhs* that had names: one male that never tired sexually was called 'Bill Clinton'. The best females of reproductive age would stand a good chance of carrying their pregnancy to term in May. They had to have the strength to keep scraping food hollows through the snow with their hooves, as well as the antlers and the aggressive temperament to defend these hollows from each other. There were different terms for females according to their milk yield, and to whether they had successfully given birth once, twice, or five times, proved infertile, or lost a calf.

Autumn was the season that revealed domestication most starkly as an intervention in the reindeers' fundamental natural processes. Paradoxically, it was also the season when nature made this control hardest to exercise. Fish migrated downstream to find deep water; bears gorged on berries and on other animals before fasting underground for half a year; marmots, protected by squeaking guards, filled the cavities between huge boulders with pine nuts and wild currants. Reindeer would not sleep like the bear or store rich food like the marmot. But just before the sexual derangement of the rut, they would have their final feeding frenzy on the protein-rich *Boletus* mushroom.

Arkady was anticipating the reindeers' mushroom madness, like a storm on the horizon whose arrival was inevitable. Sure enough, no sooner had Kostya left for the village than a rain shower brought out a sudden burst of *Boletus* and the reindeer scattered in quest of them, driven crazy by the smell. At first I found it amusing to watch them running around frantically and chomping on mushrooms. But I soon realized that the men were becoming agitated. They had lost many reindeer and they could not find the *uchakhs* they needed to ride to look for the rest. The herders wanted their animals to be well fed, but not on these terms. Here was a test of the symbiosis between humans and reindeer: who was really in charge?

Kostya was summoned by radio, and returned from the village in haste, his party business unfinished. While Peter the Winter Cook and Ivan the Fence-Builder continued hammering poles, Kostya, Arkady, and some other men scoured the nearer mountains. I was sitting with the film crew on the damp ground taking a break with the fence-builders when Arkady came up, wearing a bobble-cap and leading his exhausted *uchakh* by the bridle. He flopped down beside us and lit a cigarette, cupping his hand against the faint breeze.

'I'm tired,' he announced, squinting around at us. 'I haven't managed to find any reindeer.' Nobody said anything. 'I haven't found them,' he repeated.

Ivan the Fence-Builder was nursing a rheumatic knee. He looked over at me with a provocative grin. 'Useless herders!' he said. 'You're all pathetic!'

'*Yolki-palki!*' retorted Arkady. This phrase peppered men's talk, but was not used by women. It started like a Russian swearword and then, before the first consonant, veered off innocuously into a word that appeared to mean 'Fir-poles!'

The mushrooms on our autumn pasture had mostly been eaten, but reports came in that our reindeer had been spotted as far away as the territories of camps 2 and 3. Kostya sent Arkady to scout. He returned the next day, having sighted some of the boldest leaders feasting on mushrooms far away.

'There are some already blackened and dried out, and a fresh lot after the rain. A huge number, as thick as the mugs on this table. If they get to Charkymbal, they'll move on towards Bugen and Hölöngchong,' Arkady predicted. He stirred his tea morosely. 'And then on to Lake Baindja.'

Kostya was startled. 'Surely there aren't any mushrooms there?' he asked, dropping his hunk of meat back on the plate a bit too quickly, knife still in front of his short, half-Sakha nose. 'Or just a few?' he added hopefully.

'I've just been there and I saw lots and lots,' insisted Arkady.

'How strange [*tuokh modatai*],' mused Kostya. 'None here, and so many over there! Which way did you go round to Ankyndja? Did you reach the first stream?'

'Yes, I got to the summer camp site,' confirmed Arkady.

'And have our reindeer reached there?'

'They have.' Kristina refilled his mug and Arkady took a long slurp of his scalding tea. 'They'd even reached the path that leads a long way beyond.'

'You mean by the second stream, the big one?'

'The path above the lake,' Arkady confirmed. 'The forest is very thick. They got there through Sutanya.'

'*Yaponsky bog!*' This was another near-miss for a Russian obscenity, and appeared to mean no more than 'Japanese god!'.

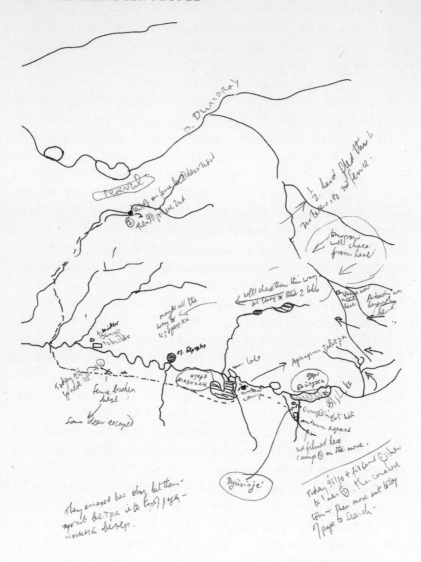

*Map of camp 10's autumn pasture, during mushroom crisis,
drawn by Kostya, wording in pencil added by author.*

By now, the men reckoned that half their herd of 2,600 was missing. This was the first time I really understood the value of the daily routine of rounding up the animals and bunching them together.

For several days reports kept coming in of small groups of reindeer being spotted where they should not be. '*YOlki-palki*, FIR-poles!' the men would respond to each other's narratives. '*YA-ponsky bog*, JA-panese god!'

The cause of the problem did not become clear until we were visited by a caravan of travellers from camp 1, who were passing through our territory. Since I had a budget for my own flights in and out of the field, I was starting to be drawn into people's lives to substitute for facilities that were failing. On one previous trip, I had happened to drop in by helicopter on the solitary tent of Valera, the brigadier of camp 3, who was living without a radio. He was in serious danger from appendicitis, and the helicopter immediately airlifted him to hospital in Sangar. This time, on the way to camp 10, a young woman had begged me to give a lift to her father, a man of 41 who was lying in Yakutsk in the last stages of liver cancer. I flew them to the village. There, he had just accepted the release of death, and the travellers were going to his funeral.

'I've had eleven children,' the mother of the dead man told me during their stopover in Kristina's tent. She had brought up these infants while working as a post courier before the days of helicopters, delivering letters to nomads on the back of a reindeer, and was now aged 80. 'Six of them were carried off in childhood by measles. One daughter and four sons grew up.' The daughter was my friend Motya the Music Woman. One son was killed by a bear, another died when a boat capsized on a lake. The third was the man to whose funeral she was travelling. 'I have one son left,' the old lady kept repeating, looking over to where he was overseeing the rearrangement of baggage on their *uchakhs*, 'just one left.'

Along their journey, her surviving son told us, they had seen some fallen horizontal poles from the section of the fence that was supposed to hold back the reindeer from the autumn pasture. The pioneers of the herd must have clambered through the hole, the others pouring after them.

'Ngggg, what a pain [*sordokh buol byt*]!' said Arkady between gulps of reindeer soup.

'What the hell are they doing, going off in that direction?' demanded Kostya. 'They're only supposed to do that if the gadflies are biting. Or when the rut begins. They'll come together and cross the boundary to the River Tumara, and then there'll be hell to pay!' No one answered. '*Yaponsky bog!*' he muttered, as an afterthought.

This situation exposed the weakness of relying on an unattended construction as a substitute for living, vigilant herders. The Great Wall of China also failed, I pointed out, and the Mongols captured Beijing in 1215.

'Kostya's ancestors!' exclaimed Tolya. Some Sakha nationalists in the city were stressing their kinship to the Mongols and claiming Genghis Khan as one of their own. 'But our Manchu cousins took the Chinese empire* from them 400 years later!'

A reindeer fence must also be built and maintained by other, less dedicated people, not only members of the brigade but men who were employed to do a job and moved on elsewhere. Now Kostya denounced the previous year's fence builders.

'It was Anastas's fence that fell down,' he exploded. 'How could that bloody Anastas have built such a crappy fence? Manchary should have noticed it, too. He went specially to check, didn't he? How could he have missed it? He was supposed to fix the fence, but he only closed the bit near Hiltiken. *Yolki-palki!* Why couldn't it have fallen down in the summer, instead of now?'

But I sensed that locating the source of the problem was the first step towards restoring control. The men now had a strategy, and their morale rose. Kristina, too, found feeding them less gloomy.

'First we've got to catch the rest of the *uchakhs*,' Kostya decided. 'Most of our *uchakhs* are over there. It's all open in that direction, so they go that way when the sun is strong, a warm day, a hot day, and they all gather together in a bunch right over there, facing into the breeze.'

Locating the reindeer was one thing, bringing them back another. At one point as we were chasing them through the forest, I heard Arkady lament under his breath, 'However much you drive them forward, there are always some left when you look behind you.'

'*Yaponsky bog!*' murmured Ivan the Fence-Builder, like an amen.

'If we don't find them all, I think we'll move camp anyway,' Arkady reckoned. 'We can go looking for the rest when we're properly settled by the corral.'

But the coming together of the herd was surprisingly swift. On the high bank across the river from the camp I saw the first 600 reindeer filtering through the forest in the dappled autumn sunshine, their brown and white mottling blending with the peeling bark of the larch trunks from which their first ancestor was born in the myth of origin. They cascaded down the slope and poured across the river almost without slowing, raising a cloud of dragonflies, translucent against the sun. All around, herders mounted on *uchakhs* were calling to one another as they massaged small knots of reindeer across obstacles, pushed them through bogs, and released them onto their trampling ground next to the camp. Now the reindeer were running to join one another like drops of water under surface tension, any drive to scatter overcome by their instinct to merge. By evening, all 2,600 reindeer were packed together beside our tents and we ate reindeer stew to the reassuring sound of their ceaseless revolving and grunting.

While the herders were out during the long summer afternoons, I would sometimes sit with Kristina by the singing kettle in her restaurant tent and listen to her thoughts as she sewed, tidied up, or tended the next pot of meat on the stove. Arkady had tried to train at veterinary school in the city, but had run back to the taiga.

'He doesn't believe in his own strength. He often asks, "How do people manage to live in the city?"' She gave a quiet laugh. 'I can't help him because I can't stand being crushed underfoot by the crowds in the village myself!'

I had formed the same impression of Arkady from his diffidence about running the herd in Kostya's absence and I was always touched by Kristina's tenderness towards him.

'I wish he would get some education first, and then come back to the camp,' she continued. 'It isn't interesting, just chasing reindeer all your life.' It was almost always the women who tried to push their children into further training, and now Kristina sounded uncannily like the mother of little Sergei, whom she accused on another occasion of spoiling her son with too much education. Sergei regarded Arkady as a role model, and Kristina's remark prompted me to wonder about Arkady at the same age.

'What was Arkady like as a boy?' I asked.

'He was devoted to herding,' answered Kristina. 'In the second and third grade he was already being paid for his work in the camp. But his character has changed a lot since childhood,' she went on, rising to her knees to lean over and prod the meat bubbling in the cauldron. 'Now he doesn't know how to talk to people, he's become very brusque [rezkiy]. This happens to reindeer herders, whatever their character. They talk abruptly to other people. He speaks at random. He doesn't think he's being sharp, but he can offend people, especially when he goes back to the village. Among herders it's different. They know each other's habits and don't pay any attention. They flare up and then it's quickly forgotten. Otherwise they can't work together.' She put down her prodding-stick and sat back again onto her reindeer-fur mat.

Just at that moment, as if to illustrate her words, Arkady led his *uchakh* into the camp, tied it up, and stomped into the tent muttering to himself. Kristina did not greet him but opened the door of the stove and adjusted the logs inside, while I sat in the corner saying nothing. Arkady picked up the kettle next to the cauldron and poured his own tea. He had not found his reindeer.

'A lot of mosquitoes,' he grumped, 'a lot of hoofprints by the big lake. If you let them run away once, they'll always do this . . .'

Kristina believed that Arkady had a girlfriend in the village, but she was not sure. I sensed that his devotion to reindeer herding was already being tested by a desire to marry, and that this was starting to make him restless with his work. Arkady himself had recently given me an earnest lecture about how men and women became estranged by spending too much time apart, so that they quarrelled when they were together. He had come to the subject suddenly, without any context, and I suspected that he was giving voice to his own preoccupation.

This was one of the basic issues of life which even Arkady's own role model, Kostya, had not got right. Kostya wanted to be a good family man as well as a good brigadier, but his wife and children had no interest in visiting the camp. Perhaps his frequent trips to the village for party meetings were also a way of bridging this gap. In camp 7, Ivan and Yura had been unable to marry at all.

The mushroom crisis had distracted Kostya from another impending problem. The camp was full of children and it would soon be 1 September, that inexorable date when every school in Russia began classes.

Long before reaching Siberia, I had seen the classic publicity photos of helicopters touring reindeer herders' camps to pick up smiling children for the start of school. These photos advertised the extraordinary resources which the Soviet State poured into remote northern communities, making actual the homogenized Soviet space in which anyone could be transported to any approved destination out of an inexhaustible fantasy budget.

But this year, under cost-accounting, helicopters were charging the equivalent of nearly US $1,000 an hour, more than a quarter of the camp's entire annual operating budget. Several of brigade 10's children were very small. How were they going to get back? Their mothers hoped that some other budget would still pay for a helicopter to go to the remote camp 12 and that it might pick up our children on the way back. Shura pulled out the bush radio and tied the motor to a log. Peter sat on the log turning the handle, while she operated the radio and spoke into the receiver.

'Snop 12,' she shouted, as camp 12 came on the line, 'I'm going to have to set out with Valentin and the children. Over.' She paused while camp 12 replied, inaudible to the rest of us.

'Snop 12, this is snop 10 calling! When will you start moving? Will you come this way? Over.' (pause)

'I hear you [*doldarom*]. So there won't be a helicopter. When do you start moving? Over.' (pause)

'Snop 12, if there's no helicopter, will you pass this way? Over.' (pause)

Kostya, practical brigadier and responsible party worker, turned aside to me. 'They'll have to transport the children on reindeer,' he said quietly. 'Camp 12 should be told, "Don't hold out for a helicopter." It costs a lot. The Farm has reindeer, lots of reindeer, and horses. If it takes four days, if it takes a week, that's all right. It's no expense, none at all.'

'The old man on the radio said most of the children had never ridden such a long way,' I pointed out. 'They're used to the helicopter. And anyway, they can't find their *uchakhs*.'

'Yes, they're used to it.' Kostya gestured to where Shura was sitting. 'Well, if I were in charge I certainly wouldn't give them a helicopter. If someone's ill, then of course you can. But not for this. The Farm has lots of problems. We could spend the money building decent houses for everyone. Why throw money into the air for helicopters?' Throwing money into the air was a Russian idiom for wasting it. It was a nice touch that Kostya meant it literally.

Camp 12 did find their *uchakhs*. They came past us a few days later, a long procession of reindeer, each group tied nose to tail behind a leader and carrying baggage, adults, and children. The children made the going slow and they had been on the trail for three days, with a further three to go. Kristina fed them in shifts while our people made final adjustments to their own reindeer. I heard Sergei promise Arkady to come back for the midwinter break. The small children were lifted onto the saddle clutching their balancing-sticks, and Shura's toddler was put in a specially designed saddle with plank walls, like an elephant's howdah, to keep him from falling out.

The augmented caravan left with a light tap of the heels to the flank of each lead reindeer. In the usual Eveny style, there was no farewell and no looking back. I followed the procession out of the camp to a point from which I could watch its progress for more than an hour across the shoulder of a huge mountain, mottled with fast-moving cloud-shadows, and imagined processions from all thirteen camps joining up as they converged on the village. This was not like a seasonal migration in which a brigade moved from one place to another around a circuit. This was a radial pattern, in which a controlling centre released its women and children for a short time before recalling them along paths that could only lead back to itself.

The village swallowed up the women and children whom it was designed to contain, but it spat out the reindeer. Shura and Valentin did not bring our *uchakhs* back but took them outside the village and released them into the taiga, where they found their own way back to the camp within five days.

After the exodus it was very quiet. No child's voice would be heard in the taiga for nine months until the following June. For thousands of miles, from the border with Finland to the border with Japan, there was not a single native child of school age left to experience the migrations and activities of late autumn, winter, spring, or early summer.

Camp 10 would not hear a woman's voice, either. Kristina had returned to her grandchildren and Peter resumed his winter role as cook and housekeeper.

INTERLUDE:

SOLITUDE AND SILENCE

Vladimir Nikolayevich's winter hunt

It was only that winter, as the land emptied of all but the most dedicated, that I began to understand how the old Eveny personality had been formed without the village's centralizing magnetism, when humans were scarce and precious.

I went home to teach classes for the autumn and returned to Sebyan in late November to take up Vladimir Nikolayevich's invitation to go hunting. We were to travel for two weeks in quest of ermine and sable, the tiny animals that had first lured the Russians into Siberia. Reindeer fur was not fashionable, or needed, among Russians and Sakha in cities like Yakutsk, whose hats, even in winter, were made from the lighter skins of otter, musk rat, fox, marmot, or wolverine. Each of these soft or fluffy furs* had its own properties. The Eveny used them only in spring and autumn, or saved them to seal joints at the wrist and neck and trim hoods of hats where the thick, heavy hairs of a reindeer would capture their breath in a crust of ice crystals. In winter, the Eveny looked and smelled like reindeer in their pungent mittens, hat, boots, and massive, all-enveloping outer coat. Without this mimicry, they would die within hours. Each hollow hair had such excellent insulating properties that the blood and organs of a dead reindeer would ferment beneath the uncut skin, while other animals just froze. These properties were appreciated 2,000 miles to the east by the Chukchi who stood for hours without moving in bitter winds on the frozen sea around the Bering Strait, waiting for a seal or walrus to haul itself onto the ice. The inland Chukchi, who herd reindeer, would keep the thickest fur from the October slaughter to trade with their cousins on the coast.

During the coldest period from November to February, one also needed double-thickness thigh-length boots for wading through snow. The inner pair was made from a reindeer slaughtered in July, when the hairs were shorter; this fur faced inward toward one's body. The outer boots came from a reindeer slaughtered in autumn, when the hairs were longer. Compared with other furs, reindeer seemed coarse, and the bristly soles of the wearer's feet gripped the ice in a padded, silent walk, with just a slight crunch of impacted snow, and sometimes a squeak as one twisted one's foot sideways. To survive overnight in a tent, one needed a double-layered sleeping bag, or *kuchu*, made in the same way. Mine was especially long, for my non-Asiatic height.

I had collected my reindeer furs a few days earlier, several weeks after sending measurements to young Sergei's mother's workshop. The designs on my clothes, held together with thread made of reindeer sinews, were quite plain. Home-made clothes were more spectacular, like the wonderful matching coat and boots that Kristina had made the previous summer for Arkady, in brown fur inlaid with a repeated pattern of white reindeer heads and antlers.

Sergei's mother enlarged every measurement to make room for underlayers. Whenever Vladimir Nikolayevich and I prepared to go out of our tent for more than a moment, we would spend twenty minutes wrapping ourselves carefully in twelve or fifteen layers, starting with Red Army long johns, flannel shirts, and heavy jumpers. Vladimir Nikolayevich wore a girdle and had given me a pair of dungarees lined with mountain sheep's fleece; I had brought polypropylene thermal underwear and silk neckscarves; he handed out strips torn from sheets to wrap around the two pairs of woollen socks we wore on each foot before inserting them into the inner boots. The thick down outer jacket I brought from my polar institute would serve as no more than the lining to my outer reindeer fur coat. The problem was not just the temperature, but of being exposed to it for hours at a time. Every joint had to be sealed with a secure overlap. If a chink opened up on the trail, it was almost impossible to seal. Movement was ponderous and I was exhausted by the sheer weight of the clothing, especially when trudging uphill in deep snow. As I stood still on a hilltop, I would feel sweat chilling on my face and would start to sneeze.

We had seventeen *uchakhs* harnessed into one caravan, alternating seven pairs of reindeer with six sledges between them, the sledges carrying our bags and boxes tied down under furs and tarpaulins. The last three animals were tied behind as spares. The distribution of reindeer acted as a brake and prevented the sledges from careering out of control when going downhill. Not all reindeer were equivalent. Some were trained to run on the right as leaders, especially when they were in front, while those harnessed on the left followed their leaders' movements. The leaders were steered by tapping them with a long, thin pole cut from a rowan tree.

Vladimir Nikolayevich sat on the front sledge, one foot stretched out in front and the other resting on the runner, while I sat on the third. The platforms were made from larch planks, the struts from willow poles, and the runners from pliable strips of willow painstakingly bent over several weeks during the summer. There were no nails or bolts: every piece was lashed together with strips of reindeer hide in loose, flexible joints so that the sledge would yield as it banged into rocks. My first lesson was to become an active passenger, fending off obstacles with my foot and swaying my fur-encumbered body to prevent my part of the caravan from capsizing. During the first few rides, Vladimir Nikolayevich stopped occasionally to check my nose and cheeks for signs of frostbite under my masks and scarves; later, he stopped worrying.

DAY 1, 22 NOVEMBER 1990

After acclimatizing to India, I was terrified of the cold and coming here was one of the biggest acts of trust in my life. Vladimir Nikolayevich had warned me to prepare for extreme physical hardship. But he was so generous and considerate that after only a few hours I felt completely safe with him.

The temperature today felt cold, but had not quite reached the threshold of −40°F*. Below −40, the school would be closed and children sent home; helicopters and biplanes were not supposed to fly; saliva solidified before it hit the ground, and if you threw hot tea up into the air, it froze and tinkled downward in a patter of tiny crystals.

It was late morning when we left the village. Vladimir Nikolayevich's wife was out in the village and their grandson had gone to school. No one noticed us from behind the double-glazed windows of the closed-off wooden houses. It was a typical inconspicuous Eveny departure, like that of Tolya's step-father Petr Stepanovich, whom I happened to notice sledging out of the village the day before.

'When will you be back?' I asked him.

'In May,' he replied, giving me a brown and white reindeer-fur hat with fox-trimmed chin-flaps which I was now wearing on top of my polypropylene balaclava. He was heading west to hunt along the River Lepiske; we were going south to the Tumara.

In five hours, we reached our first night-halt. We crossed many other travellers' tracks on the way, and the ease of winter transport made this site seem very close to the village. This was the territory of camp 10, and we flashed past sites where I had lived this last summer with Kostya and his brigade. I remembered this same journey as a long, arduous haul taking five times as long, stumbling on horseback for two fourteen-hour days through thawed bogs and over exposed boulders. Now, these impediments had been ironed out under an undulating film of ice and smoothed off with a sliding blanket of snow.

We could not plunge directly down to Lake Baindja as we would have in summer, but overshot the lake and doubled back from the east down a more gentle slope, facing the rosy afternoon glow on the southwestern horizon. Here, we planned to join up with a young hunter aged around 30, an orphan called Manchary. We found him living in a large tent, with a row of sledges outside and the floor covered with branches of bare winter larch. We released our reindeer one by one, and they set off in single file, zigzagging determinedly up a steep slope to find the warmer air above the frost-sump of the valley.

'Look at them calling, "Wait for me!"' Vladimir Nikolayevich chuckled.

We unwrapped most of our layers of clothing and entered the tent for tea. The little stove crackled welcomingly and we gradually stripped down to the shirtsleeves over our thermals, taking care to prop some furs behind our backs. The upper half of the tent was quite hot, but objects on the ground around the edge remained frozen.

Vladimir Nikolayevich had laughed yesterday as he warned me that Manchary would drive me crazy with his continual chatter. It took me a few confused minutes to get the joke: Manchary was the most silent person I had ever met. He was staying with my old friend Peter the Winter Cook of brigade 10, and this was Kristina's old restaurant tent. Chez Kristina had become Chez Peter.

Nikitin had gone to the village for supplies and Arkady had gone to sell the meat from his own reindeer. Peter was minding the herd on his own, some way from the brigade's midwinter hut. He was pleased with the vodka I had conveyed from a relative in the city, and cooked us delicious reindeer and thick soup with rice and vermicelli.

'I like Peter,' Vladimir Nikolayevich said afterwards, 'he's got a generous nature. If he wasn't there, the rest of them would die of hunger – I've never seen Kostya or Arkady cooking!'

I reviewed the performance of my clothing and equipment, and began to shed some of my fear. My feet had stayed warm throughout the journey, my hands until near the end. The silk scarves around my face made it difficult to breathe as soon as I exerted myself. High-tech goggles misted up and then froze, at times allowing no vision at all. Cameras and tape recorders were paralysed until they thawed inside the tent. My fountain pen started to write when I pulled it out from deep inside my clothing, but the ink froze in mid-word: outdoor writing would be with soft pencil only.

It was completely dark by four o'clock, as Vladimir Niko-layevich stacked two mounds of split logs by the stove. Dry wood from dead trees produced a fast flame for cooking, but damp wood burned more slowly and lasted longer at night. By seven we were asleep, fully clothed, each inside his double reindeer sleeping bag. The summer smells of fur and resin were intensified by our winter dependence on these two life-giving substances. It was as if we were curled in the soft embrace of the first reindeer, right inside the larch tree that had given birth to the species.

DAY 2, 23 NOVEMBER 1990

I woke several times during the night, but the double reindeer fur was enough to keep me warm throughout. Vladimir Niko-layevich crawled out of his sleeping bag several times to put more logs into the stove and blow on the flame. I later realized he was looking after me, as he did not bother to do this once I had become acclimatized.

For short visits outside, we wore just our inner clothes, light boots, and woolly cap. The stars were very bright indeed, and the dog – a mongrel, mostly husky – curled outside the tent looked up with big, sparkling eyes.

For breakfast, we had huge hunks of raw butter piled onto bread cut from deep-frozen loaves which split into two at one blow from a knife. Our luxury items would not last long. Vladimir Nikolayevich had a tin of Soviet 'Friendship' cheese. I had brought Italian salami from a supermarket in Helsinki. Vladimir Nikolayevich and Manchary fell on my chocolate, but were not interested in the dried apricots which were my protection against meat-induced constipation.

Our halt had been brief, and most of our baggage was still on the sledges. We wrapped the remaining items, placed them in their correct position, and tied everything down. There was no farewell with Peter and I did not even notice that Manchary had slipped away ahead of us. We would find each other later.

It would take two days to reach the headwaters of the River Tumara, which marked the beginning of our hunting ground. I had crossed this area before on horseback, but now we sought out paths of a new sort. We did not skirt laboriously around the edge of a lake, but cascaded down the embankment, each sledge bumping up against the heels of the reindeer in front, and flung ourselves onto the lake's hardened surface. The slipstream drove a blizzard into our faces. I braced myself for the jolt of rearing up the opposite bank, but instead, we swooped down onto the river that drained the lake and slid along its frozen ripples. Beneath our runners lay invisible beaches. In summer, these would be the paths of loose stones beneath our trudging hooves, but now we followed the deep central channels, floating over the sloping, glossy ice and moving from one channel to another by scraping through narrow openings in the groves of poplar that marked hidden islands and spattered snow in our faces from their springy branches. Vladimir Nikolayevich turned frequently to look back, and stopped occasionally to un-snag a harness or guide the lead pair of reindeer down a slope, running to keep up with them and leaping back onto his sledge as they picked up impetus. If a reindeer seemed tired or ran at the wrong speed, he rotated it with one of the spares at the back, like a worn tyre. The *uchakh* behind my right shoulder often banged the stump of its sawn-off antler against me as it drew level. I was sure I could be doing more as a passenger if only I had the skill.

I could stay warm indefinitely in these clothes, but suffered as soon as I removed a reindeer-fur mitten and exposed bare skin to do anything that required dexterity, like the constant tying and untying of the reindeer-hide thongs that held a herder's life together, on luggage, in the complicated skeins of harnesses, and in the erecting and dismantling of a tent. Knots were tied with a single loop so that they held firm yet pulled apart with a sharp tug at the free end, but when they were encrusted with the ice of a day's travelling, we had to pick at them with the tip of a knife. The paper of my notebook was so cold that it hurt like a burn to touch it.

By nightfall, we had left the territory of brigade 10 and reached an empty hut. This was further than I had ever been in this direction, and I never found out whose hut it was. Manchary had already arrived and we were guided by the faint glow of a candle through the absurd windowpane of torn polythene. It was −58°F and the stove and candle had difficulty catching oxygen. Inside the hut, the walls were glistening with tiny ice crystals. Though there was sharp starlight outside, the hut was full of thick fog* which took over an hour to clear.

DAY 3, 24 NOVEMBER 1990

Today was warm: −30°F. We left the hut as we had found it, with a large stack of sawn and split firewood. Our journey this morning was upstream and the pace was slower as we stopped to drag fallen larches out of our path. Vladimir Nikolayevich was concerned because the runner on one of our sledges was loose. We reached the pass that separated the inner from the outer rim of the crescent of the Verkhoyansk Range, dividing the rivers draining into the Arctic Ocean through the Yana from those that flowed into the Aldan and the Lena. Here began the Tumara, which led down to the small Sakha settlements 150 miles away on the bank of the Aldan, just before it joined the Lena. We paused to give offerings to the spirit of the pass.

'Do the Sakha downriver ever come and hunt up here?' I asked.

'They can't come this far,' replied Vladimir Nikolayevich. 'They've only got skis and snowscooters, so their range is very limited. And anyway, there's no petrol.'

The only remaining reindeer herders' camp up here was number 12, whose territory we were now touching, their winter hut two days' ride to one side of us. Our movement cut across brigade territories: herders followed animals, but we were hunters who intercepted them, as humans had once intercepted reindeer before domesticating them.

A little way down the Tumara, in a broad alley between two island ridges of tall poplars, we pulled one of our own tents off a sledge for the first time. Vladimir Nikolayevich shovelled the snow from a rectangle of ground 2 or 3 feet beneath, while Manchary chopped down medium-sized trees and I trimmed them into tent poles and dragged them over to Vladimir Nikolayevich's rectangle.

'If we were proper hunters, we'd carry our tent poles around with us. But we've got too much luggage!'

We could not find a large flat stone to place under the stove, so Vladimir Nikolayevich scraped the moss off the ground to avoid setting it alight. As in summer, there were no guy-ropes and instead of stones we shovelled snow up against the outer edge of the canvas to weigh it down.

'We're lucky there's plenty of snow,' Vladimir Nikolayevich said. 'If you can't pile it up, you've got less insulation.'

Though each site had a character quite its own, the interior of a tent stayed the same wherever one was, though Vladimir Nikolayevich was the only person I knew who had sewn little pockets all over the back wall and would unpack even the tiniest item – needle, thread, toothpaste, usefully shaped bits of twig and bone – into the same pocket each time. More of his baggage came off the sledges than at previous halts, and was opened up to reveal little packets of paper or tattered cloth. A box with a picture of an elephant and marked *chay indiyskiy* (Tea from India) contained macaroni. He saw my eye straying to an exercise book printed with the label 'Hunter's diary'. Inside, it had spaces for certifying and stamping by authorities.

'I don't always fill it in,' he said. 'I keep it just for my own use. It helps me remember next year how many traps I set and where.'

We would stop here for two nights, and there was a feeling of coming to rest. This was a site whose spirits deserved to be fed through the fire and I opened a bottle of vodka, which we also offered to Bayanay, the spirit lord of the forest who controlled its animals.

DAY 4, 25 NOVEMBER 1990

These spirits gave me my best night yet. As I woke to see Vladimir Nikolayevich washing his hands and face carefully by candlelight, I thought of my wife and children back in England, but they were far away and the tent already felt like my home.

'It's a lot colder than yesterday,' he greeted me. 'If it hisses when you breathe out, that's bad – it's below −60°F.'

As in summer, being too heavy for a reindeer's back made me very tentbound when the day's work was further afield. Vladimir Nikolayevich would ride his *uchakh* to check some traps he laid last month, while Manchary would take two reindeer and a sledge to lay some more traps further up a tributary. I felt very calm knowing we did not have to pack up again and leave, and looked forward to fixing my tape recorder, which had now stopped functioning even when warm.

Our reindeer and Manchary's had all stayed together overnight, because they were so few. But even if only three of them were worked today, they would all have to be rounded up, otherwise they would drift further and further away as they scraped craters in the snow with their front hooves to feed. Vladimir Nikolayevich drove them up at 10.30, and they scrunched their hooves and tinkled their bells around the tent.

'Don't tie them up while we're out,' he warned me, 'or they'll get hungry. They won't go far. They distinguish people by smell, sometimes by voice. They're getting used to you. One or two of them know their own name. Of course an *uchakh* knows its rider well. If several people ride it, then it gets spoiled and mutinies [*portitsya i buntuyet*].'

'Are the bells so you can find them?'

'*Nye-e-et!*' he answered. 'I can always find them. They're to warn wolves that the reindeer are protected – there are humans nearby.'

'Should I use salt?' I asked, looking at the little pouch in his hand.

'Only when you're catching them in the morning,' he replied. 'If you don't give them salt, then they'll think you're a cheat, and won't come tomorrow – watch out, they're very sly!'

The oblique late morning sun touched the highest hilltops, while everything in the valley around us remained a monochrome shade of twilight blue. Vladimir Nikolayevich harnessed two animals to a sledge and we trotted them a few hundred yards to the river, where I helped him to carve out a huge block of ice and tie it onto the sledge.

Back at the tent, I forgot the routine and asked, 'Will you go off now?'

'Of course not – we'll drink tea!' He laughed, watching Manchary hack off a portion of the ice block and stab it into fragments inside the kettle, where it was already beginning to shatter loudly over the flame.

'Drink no tea – where's your strength [*chay ne p'esh – otkuda sila?*]?' he proclaimed.

'Good thing there isn't a shortage of tea!' I remarked.

'There soon will be,' Manchary suddenly interposed. 'The balance of payments crisis with India, grubbing up plantations in Georgia . . .'

I hardly recognized Manchary's voice, and had not been aware that he had a sense of irony.

Of course, what was served at eleven was not just tea but fatty reindeer soup, with large hunks of meat. This was lunch, eaten at the time when you had gathered your reindeer for the day's task, and in anticipation of no further food until nightfall.

Vladimir Nikolayevich had been a herder all his working life and this was only the third winter that he had been hunting.

'It's the last year I'm doing this. It's too cold in a tent between November and February – I can't take it. I'm too old – nearly 60. Manchary's much more experienced,' he said, nodding towards our young companion who had fallen back into his usual silence. 'He killed a wolf last year and won a prize from the Farm. He's seen lots of hibernating bears, caught lots of bears, too. I've killed only one – I don't like doing it, but I couldn't avoid it.'

'How do you decide where to hunt?' I asked him.

'Each hunter has his own beat, it's allocated by the Farm. Two or three hunters make up a brigade, like us. Where we passed yesterday is part of our territory, then it stops again until we reach the mouth of the side stream. In between is old Starostin's territory – you'll meet him tomorrow. One mustn't trespass. Last year I was following my plan, when another hunter came and swore at me; the Farm had made a mistake and given us both the same territory!'

Vladimir Nikolayevich tied some additional bands of dog fur around his knees against the day's extra cold, and warmed the underside of the saddle over the stove as a kindness to his *uchakh*. Then he crunched off, carrying saddle, stick, bridle, and a pouch of salt for calling his mount. He and Manchary set off in different directions, while I washed up by the hum of the glowing stove.

Later, on days like this, I would have the role of cook, with instructions to start heating supper at three o'clock. But this time, I was just left alone. The snow fell silently in this windless bowl of mountains, not the big soft flakes of the moist climate of Moscow or Lapland, but dry crystalline needles with hardly enough weight to sink through the air, too tiny to make any impact when they landed.

It was the first time in my life I had been alone in an environment where I was incapable of surviving on my own. I could walk alone around the Indian jungle, where distances were shorter and the atmosphere was not like the killer surface of a hostile planet. But if my elderly friend and his taciturn companion did not come back, I would not get out alive. The remaining reindeer would not obey me and I could not chop down trees, saw them and split them fast enough to keep the stove going (which I estimated at 100 feet of larch trunk per twenty-four hours).

The midday sun was now slanting into our valley, though the light was still yellowy pink, a sunrise that would pass without a break into sunset. I waded downstream through the snow to photograph the track of a small stream falling down a cliff, where it had frozen into parallel columns like the bass pipes of a huge cathedral organ. In the few minutes while I exposed the camera to the air, the shutter speed slowed down and the light meter ceased to function. I nursed the camera back to life in the tent, where the stove had gone out.

Our lives here were an interplay between ice and fire. My bottle of ink froze solid every day on the sledge and boiled when I left it too long on the stove, though it still smelled the same and could still be used for writing. Vladimir Nikolayevich would heat an axe-head on the stove before hacking off a chunk of meat for his dog, who would eat it frozen solid. So long as we fed our little stove with a constant supply of logs, we could sit in the tent in our shirtsleeves. But when we lay down inside our double reindeer-fur sleeping bags, the warm air rushed out through the canvas within a moment to merge with the cold which stretched for thousands of miles in every direction.

Vladimir Nikolayevich arrived as I was ineptly trying to light the stove with matches and wood shavings.

'It's not burning well,' he comforted me. 'The temperature's too low.' Even when he built up the flame, I could see the contrast with yesterday's willing blaze.

'I haven't caught anything yet,' he announced. 'But I've laid three traps. It may take a few days, or two weeks. Sometimes you wait all winter, sometimes you get something within an hour. If an ermine is still alive, it's very fierce, it bites you in a second. You mustn't put your hand near it, just kill it with a stick.'

Vladimir Nikolayevich was standing in the middle of the tent, and muttered, *'Tak!'* ('So!') as he did whenever he was reflecting on a situation or wondering what to do next. The furs were to sell, but he was also longing for an interesting supper.

'Nothing fresh to eat,' he concluded, 'no ptarmigan, not many hare tracks. So it's just the same old store provisions!'

It was 3.15 and nearly dark. He went out to gather the reindeer once more, to stimulate their homing instinct. I could hear his boots scrunching into the distance, and returning a few minutes later with the reindeer. After some rummaging among the sledges, he entered the tent with a tin of stewed beef and some frozen reindeerburgers. He turned on the radio, and we heard a crackly programme of traditional Eveny songs, dedicated to the seventieth anniverary of the birth of the Eveny writer Platon Lamutsky*.

'The powdered milk is spoiled,' said Vladimir Nikolayevich, holding up a little parcel of newspaper. 'But Manchary won't complain – he's that sort of person!'

Manchary returned, the whiskers on his upper lip caked with ice, just as the radio was telling us that temperatures would drop, and that even Yakutsk city would fall to −53°F.

'That means it'll be 15° lower here,' observed Vladimir Nikolayevich.

'Colder than yesterday,' Manchary added laconically, his first words since his remark about tea.

Manchary's taciturnity was good-natured. He was open to other people: it was just that he did not need to talk. He was once hired by an expedition of geological prospectors from Sangar to manage a herd of 600 reindeer for their transport and food, but gave up because he could not stand the responsibility and the commotion. I wondered whether Vladimir Nikolayevich was drawn to him because he was another orphan. They did talk occasionally, always in Eveny, and always about the immediate job at hand. It was then that Manchary might become briefly vehement about the behaviour of a reindeer or a wild animal.

Manchary had already been this way a month earlier. Checking his earlier traps along the river, he had now found a white ermine, worth 16 roubles, wedged in a trap and had shot a glossy black squirrel, worth 5. His best find was a sable in another trap. This would fetch 150 roubles. He sat down on the floor and neatly peeled the skin inside out off the sable's tiny body onto a wooden framework, gently drawing the five bony little fingers of each paw as if through a sleeve.

'If you make a mistake and cut through the wrong place, it's worthless – you've lost everything,' Vladimir Nikolayevich commented.

The economic viability of fur hunting was highly variable, especially if you caught only low-value squirrel and ermine. Last year, Vladimir Nikolayevich earned almost nothing but the year before, along the River Lepiske, he and a companion caught about seventy sables between them, earning nearly 5,000 roubles each. For half a year's work, even this was only just a living; it was now 1990, and the Farm had already begun to default on paying herders' wages, a situation that no one yet realized would lead to total breakdown. Even though an average herder's pension had been raised this January from 120 roubles a month to 215, pensions were already lagging behind the cost of living.

Soon, hunting furs would be almost the only hope of a cash income.

The Farm set each hunter a 'plan' of so many ermines, squirrels, and sables, which it would sell to the State fur procurement agency in Yakutsk. The best furs might ultimately reach the international auctions in Leningrad and Copenhagen. But the Farm's hunting coordinator, a Sakha from outside the village, had left and his position was vacant. Hunters were forbidden to sell their furs for a much higher price on the black market; but soon, this might be the only market.

Even when it was functioning well, the Farm's plans might not suit workers. During supper, Vladimir Nikolayevich criticized the Farm management for setting the wrong ratio of males and females in the herd of reindeer.

'They don't allow enough males for selecting and training as *uchakhs*,' he said, and then added sarcastically, 'how are the herders supposed to go looking for the reindeer – on foot?'

I saw an opportunity to draw him out on village politics, but at this early stage of our acquaintance, I sensed his caution. It seemed to me that he was a little quick to deny that Ivan's family in camp 7 was being persecuted in the brucellosis incident.

But then he pointed out, very reasonably, 'The vets did their analysis, and they gave the instruction that the whole lot had to be slaughtered. If the Farm went against the vets' instructions, people would say, "What are you doing? Are you trying to spread the disease, or what?" If they didn't kill them, there would be a big scandal. People would say they weren't following the proper regime of hygiene and disease control. And how could the Old Man know whether his reindeer had the disease or not? He wasn't the one who did the blood test.'

Perhaps Vladimir Nikolayevich was simply not a conspiracy theorist, since he did not support the management either. In a review of other brigades, he identified weak herders and brigadiers, and criticized the Farm for deploying them.

'Camp 11 was shut down because they'd appointed a useless

brigadier. Now they've appointed Vitya the Wolf Hunter, so it will pick up again. They may be capable of going on a family or leasehold contract. Otherwise the only ones who would be strong enough to take the risk are 10, 5, and 8.'

I understood why all these brigades were afraid to take on a contract that would leave them liable when reindeer disappeared or died. It was a minute refraction of the dilemma that tormented everyone at every level of the country: how to live with the requirement of self-management, without experience, insurance, or social security. The fates of Vladimir Nikolayevich, Manchary, Ivan, and Kostya were tied by powerful threads to Sebyan, Yakutsk, and on to Moscow. But here, as we sat in the white gleam of dusk with a candle at our elbow and the glow of the stove burning our eyelids, the tranquillity felt utter and deep.

'I won't manage to fix that sledge-runner,' Vladimir Nikolayevich commented. 'It needs a thorough overhaul. So we'll have to go gently tomorrow, to avoid smashing it. We'll start early – it's a long haul, with traps to check on the way.'

He went out and tugged at the weak sledge to test the weight of its cargo. Our caravans had stopped with the sledges facing the way we planned to set out. No matter how we arrived at a site, the reindeer were always led around to the next day's starting point, even if this was less convenient for unpacking, since they would not start up unless there was a clear run in front of them.

DAY 5, 26 NOVEMBER 1990

We were up by 7.30. Vladimir Nikolayevich's dog came in during the night to be near the stove and pushed the tent-flap open. We woke up cold, but actually the weather was much milder, perhaps −22°F.

'God knows, Bayanay knows!' said Vladimir Nikolayevich. 'It's warm today.'

I was puzzled that the tent was left standing when we seemed ready to leave.

'I'm leaving it up,' Vladimir Nikolayevich explained. 'I've got two more tents and two more stoves. I'll set them up as bases along the river. It's useful for others, too. Old Starostin sometimes passes this way, and the occasional herder from camp 12.'

After yesterday's cold, so sharp that it seemed painful to touch the air, the haze of ice crystals suspended around us today felt balmy. At one point, Vladimir Nikolayevich's sledge fell through an ice bubble on the river, which had frozen to the bottom, leaving a hollow space between the underside of the ice and the stones of the riverbed. He laughed. 'A good thing there's no deep water underneath!'

He stopped to point out the impressions of tiny feet. While I was adjusting my eyes to the way the tracks wove and darted under fallen branches, he had already dated them and calculated their trajectories: did the animal intend to return this way, or along a parallel route, or was it moving onward altogether? Was it travelling with a partner, so that they could both be caught by traps laid in parallel? Traps were in short supply and could be obtained only through the hunting coordinator who no longer existed, so each had to be placed where it was likely to be most effective. We hollowed out a snow cave a few inches deep on a rising slope, angled so that the sun would not throw up a relief of shadows where we had disturbed the snow. Then we laid the little trap with its jaws facing upward and pieces of meat and fish as bait, using a wooden spatula free of human odour to cover everything with fresh, soft snow and brush over all traces of our approach.

The movements of small fur-bearing animals were more capricious than the migration of a reindeer herd. They moved individually, not driven by changes in vegetation but following further movements of other vegetarian creatures on which they preyed. Sometimes an entire species could drift out of a territory altogether. So far there were very few sable tracks, but Vladimir

Nikolayevich had a hunch that they would migrate this way after the new year. However, we did see many elk tracks.

'Why don't you change your quarry?' I asked, thinking of the rich, soft meat on the elks' massive bodies. He had not used his gun at all.

Vladimir Nikolayevich laughed heartily. 'You have to stalk them for days, and the metal gun freezes your bare hands,' he answered. 'It's not like the wooden bow of the old days. Last winter I wasted five days chasing an elk and still didn't get it. And anyway, it's only food – to live, we've got to make money!'

I pointed out some smoke up a tributary: who would light a fire, and how could it move like that? It was Manchary's team of reindeer, puffing out frozen steam like the funnel of a ship. I realized we must leave a similar trail. We turned up the tributary and caught up with him where he had stopped to saw a gap through a huge tree trunk that was blocking off the narrow valley, its shallow roots scraping defencelessly where it had slithered down against the base of a black shale cliff. The fall must have been recent because we were sure old Starostin had just been here. This stream was his hunting ground and it would lead us to his hut.

Starostin lived alone in a sheltered site, surrounded by huge parallel larch trunks which loomed up against the indistinguishably drained white of the surrounding hills and the snow-bleached sky. This hut was not as it would be if Vladimir Nikolayevich were in charge. It was cosy but dirty, the shelving crudely hacked from tree trunks as if by trolls, the floor littered with wood chips and scraps of food. Part of the roof had caved in, filling a quarter of the room with debris from the thick insulating layer of earth and moss above. The hut was very hot. Starostin served us tea, bread and butter, and we contributed a slab of pork fat. Smoked fish lay beside the stove, ready to garnish our meat-heavy supper later in the evening. Toothless and with a permanent cigarette in his mouth, the old man was almost unintelligible in Russian. He had an inscrutable, inner alertness which I saw in many older people who stayed away from the village. As I watched the outwardly oriented alertness in Vladimir Nikolayevich's eyes, I realized how cosmopolitan he was and how much his house in the village was the centre of his life. There was a further layer of solitude in a dedicated hunter like Starostin which lay far beyond the companionable isolation of a camp of herders and which I had not even begun to penetrate. Starostin had a son who moved between these states, working as a herder in summer and visiting his father here in winter. The son had a sharp, intelligent face. He asked me about power-sharing in Northern Ireland and what effect the Channel Tunnel would have on agricultural tariffs. Starostin then started a long, animated conversation in Sakha with Vladimir Nikolayevich about traps and sites. I did not follow much of it, but understood when he confirmed that there were few sables this year. The father played cards and draughts with Manchary, and the rest of us crowded around to advise them on their moves. When the radio chimed in with events from national and international politics, they would break off from their game to answer back with irony or laughter. I had many questions, but did not want to spoil the charmed atmosphere.

DAY 6, 27 NOVEMBER 1990

The radio said −17°F in Yakutsk and 0°F, or 32° below freezing, in Moscow. The stove was like a furnace and the hut was so hot that we had to prop open the door. Though no one had a thermometer, people were very precise about temperatures. Despite the day's mildness, Starostin's son insisted that global warming was not happening here and that in early December the temperature would drop to −72°F.

We left at 10.30 a.m., returned down the tributary to the Tumara and turned right to continue downstream. The low midday sun ahead of us was yellowy pink, surrounded by misty clouds, like a Victorian painting. The splayed willow-fans of islands and beaches were heavily encrusted with aerial frost, as were the row upon row of feather-frond larch branches which jutted horizontally over the edge of the ever widening river, with little drifts of snow lodged in every roughness of the bark on their trunks. We stopped while Vladimir Nikolayevich cut a rowan stick.

'Peter asked me to get it for him,' he explained. 'It grows really well down here.'

Now we were back on our own territory, and our caravan and Manchary's overtook each other as we stopped to set or check traps. For much of the way, we drove through melted water, which flowed over the ice or dived beneath ice and snow. Even when the temperature fell to −80°F, this water from warm springs would continue to flow into the Lena, which also kept moving all winter, deep and dark, 6 feet below the surface.

Once, Manchary doubled back and passed us in the opposite direction on a single sledge hitched to a pair of reindeer. No words were exchanged, but Vladimir Nikolayevich somehow knew he was looking for a trap that he had lost when his sledge overturned.

'The ermines must be happy,' I said. Vladimir Nikolayevich laughed heartily.

Manchary overtook us again in the opposite direction, mission clearly accomplished.

'Not happy any more!' He laughed again.

Vladimir Nikolayevich was looking for an old couple, pensioners who once worked in brigade 12 and told him last winter that they would be here this year. The awareness of who was where seemed almost paranormal, as information and messages moved around the hushed forest like radio waves, unseen and unfelt until they reached a human receiver through whom they would take conscious form. Hunters were not issued with bush radios like brigades, but everywhere one went, there were signs: tent poles had been stacked, supplies had been cached, twigs had been bent to say 'I was here'. Travellers made marks on cliffs or in snow; they leaned three upright sticks together, with the longest pointing where they had gone, or else they made a triangle on the ground with one angle sharper than the others. Other designs could point to routes of animals. Some people wrote a note on a piece of paper and left it wedged in a tree. Each clue took the follower only as far as the next, which had to be read in turn. A hunter could judge how recent each sign was and calculate how fast a person was moving and where they might be by now. I have seen someone gaze at a panorama of forest and mountain stretching for 30 miles, and find a traveller within a day.

We were led to the old couple by the thickening tracks of their daily movements, and pitched our tent near theirs. Courteous but self-contained, they did not sell furs, but simply fed themselves off the land. They had not been to the village for several years, but had a very detailed knowledge of everybody's movements on this side of the mountains.

DAY 7, 28 NOVEMBER 1990

During the night it snowed, and when Vladimir Nikolayevich called the dog in the morning, a hunched little mound under the snow started to heave. The radio said Kobyay −31°F, Verkhoyansk −54°F. We were in the Verkhoyansk area, but Vladimir Nikolayevich reckoned it was no colder than −28°F.

'Bayanay is being kind to you!' he told me.

Today, we crossed a vast ice sheet. This was not a lake but an icecap which never melted, even in summer. Vladimir Nikolayevich stopped the caravan and got me to listen to a river below, like the rushing of a bloodstream under the skin of the earth. He did not talk lyrically about the beauty of nature like Tolya, but I felt that he quietly loved every little shift in the mood and texture of the land. Later, we drove through meltwater which froze instantly and encrusted our runners as soon as the sledge came up again into the air.

'Excess weight!' said Vladimir Nikolayevich, stopping to tip the sledges on their side one by one and bang off the ice with the back of an axe.

In the afternoon our path was barred by another ice sheet which was too awash with water to pass. We could go no further, and hoped for a cold spell tomorrow.

Ice is one of the land's main resources, facilitating movement by its lack of friction. But this invitation to travel also makes it a great killer. Though the subsoil is frozen all year round, it takes a long time for the ice on the surface to become secure. One can walk on a lake or river when the ice is only an inch thick. But it is not until December that it becomes strong enough to bear the trucks, mobile homes with wood-burning stoves, which provide the only land link to remote settlements as they creep bumpily along the uneven rivers while the current still churns below. Even after January, when the ice is 6 feet thick, it may not always be safe. The soundest and most level ice is usually on the still water of lakes. But by May, this, too, thins and cracks, and the last platelets dissolve in June. Spring is a terrible season for accidents.

Aviators read the history of the ice's formation from the air before choosing a spot to land. In winter, they exchange the wheels of their biplanes for skis. Now, they are no longer limited to a few village airstrips but can land on any sizable body of frozen water. One year in March, I sat in the spring sunshine on the bottom rung of a biplane's access ladder, fishing through a 6-foot-thick hole which I had drilled through the ice at my feet; the same year on 3 June I stayed in camp 2 with a biplane parked on the lake, trapped by low cloud. The aviators took off from the last of the ice during a slight chink in the weather; otherwise, their plane would have sunk to the bottom within days.

Every day followed a similar pattern: wake, light the stove, discuss the temperature, eat breakfast; dress moderately, gather the reindeer, and eat an early lunch; dress heavily and go hunting near the same site, or else load the sledges and move on, setting traps as we go; arrive, unload and pitch our tent; supper, sleep. Meals consisted mainly of meat, bread, and butter. At any time the tasks might include: repair equipment, discuss the movement of animals – and drink tea.

And then there was the radio. Being merely hunters, not a herding brigade, we had no hand-cranked transmitter. But we listened to the ordinary radio at least once a day, nursing our batteries and monitoring the temperature: −20°F one day, −70°F the next – though we could feel the changes more intensely on our skins than the meteorologists with their double glazing and steaming radiators. 'Midnight in Moscow' was the call sign of the Russian-language station, while the Sakha station used an electric organ dubbed in with the twang of a jew's harp, the Sakha national instrument. The radio gave us harp music by Charpentier and Schubert's C major quintet. These meant nothing to Vladimir Nikolayevich, who switched off the music without a thought.

Later, long after dark, he would switch it on again to catch some talk: 'Six o'clock, comrades. Shall we listen to the gossip in Yakutsk? Are they telling us the truth or deceiving us?'

The news always progressed from the Sakha Republic to the rest of the Soviet Union, then Eastern Europe, China and Mongolia, and finally Everywhere Else. People were short of food in Russian cities – would there be famine this winter? President Todor Zhivkov had fallen in Bulgaria. A war would start in Iraq on 15 January. Once there was a programme about me, and we heard Tolya's voice saying that I was out on a hunt with Vladimir Nikolayevich at this very moment.

'Quite right, too,' said Vladimir Nikolayevich, 'it's because you're the first foreigner, especially in winter.'

Another time there was an Eveny lesson, which puzzled Vladimir Nikolayevich with its eastern dialect. During a Yukaghir lesson, he and Manchary pottered round the tent mouthing the strange sounds for 'Where does your grandmother live?'

I learned new things about reindeer. On one journey, a white *uchakh* which had been behaving aggressively towards the others finally became intolerably disobedient. Vladimir Nikolayevich tied it to a tree and sawed off every scrap of antler, right down to the pedicle, before releasing it. Each of the other reindeer came up in turn and kicked it, rearing up with hooves flailing. Wherever it turned, they would follow and attack it. In the end, the chastened reindeer moved off alone, the others following menacingly in a pack until it stood in isolation far away. I had a sense of reindeer sociality, of a team that had become exasperated with the bad behaviour of one of their colleagues.

'I've been meaning to do that for ages,' said Vladimir Nikolayevich, who was helpless with laughter, 'but each morning I kept forgetting!'

'Does that reindeer understand that it's being punished?' I asked.

'It's a big punishment. That one's a notorious hooligan [*izvestny khuligan*]! If you cut off the antlers of any reindeer, the whole group immediately starts fighting. It's a struggle for power.' He used the classic Bolshevik phrase *bor'ba za vlast'*. 'It's the same as when you introduce a new animal. Supposing I'm a reindeer and I've come to this camp for the first time, of course they'll start fighting with me. Maybe they're getting to know me, maybe they're saying hello or the devil knows what, but they'll definitely fight, male or female. It's their law. Whichever one wins can't be touched. That one becomes the leader, but as soon as a new reindeer appears it starts all over again.'

A grin spread across his distinguished elderly face, like a naughty boy. 'Once I was working with Misha in camp 9. There was an old lady milking her reindeer and we were driving some other females across the area when they began fighting. So I said, "Come on, Misha, let's cut off their antlers. Then we'll really see them fight." Meanwhile, Grandma was milking away, milk milk milk – we didn't say anything to her. So we sawed off all their antlers and released them again. Then they fought even worse, only this time they were kicking each other with their hooves and Grandma started shouting at us, "What do you think you're doing? They'll kill each other!" The old lady was furious but Misha and I were doubled up in laughter. The reindeer looked so funny with their tongues hanging out and lolling about!'

Laughter, like heat, was hard to generate on this landscape and faded quickly. One very cold night, close to the stove, I learned more about Vladimir Nikolayevich's childhood. He was born in 1932, four years after the headman Baibalchan was killed by the Communists in their own *bor'ba za vlast'*. By 1937, Vladimir Nikolayevich's father had become president of the Collective Farm while Tolya's father was chairman of the Village Council, exactly the same position that Tolya now held. During the war, Vladimir Nikolayevich's father was selected to drive reindeer caravans for the Gulag labour camp at Bear Mountain near Verkhoyansk, and little Vladimir went with him. His father died there in 1943, and the 11-year-old boy continued working.

'At the end of the month, the grown-ups would lead me by the hand to receive my pay,' he said. Then putting on the face and voice of a puzzled child, he chirped, 'What pay? Have I earned money?' Resuming his normal voice, he continued, 'When I grew up, I realized that Sebyan was my native village, and that I had a mother here. So I came back. They could see I knew a lot about reindeer, so they appointed me brigadier of herd 9.'

As I worked out who his mother was, I realized that I had met her during the summer in camp 10. She was Kristina's aunt, who had had several husbands and now lived with old Nikitin. Vladimir Nikolayevich called Nikitin 'Father', though he was almost 59 himself and Nikitin was only 61.

'How did a young child come to be living with his father?' I wondered, still thinking of the little boy on his own in the labour camp.

'My mother left me when I was a baby and went off with another man.' His voice was even and did not reveal the hurt. But he knew his next remark would be shocking. 'I don't have any respect for my mother.' He paused. 'You can take that how you will. She didn't even give me her milk.'

Vladimir Nikolayevich was a good parent himself, and worked hard to look after his children and grandchildren. His mother was only 16 when she abandoned him. She must have had her own sorrows.

'*Tak!*' Vladimir Nikolayevich looked around for the next practical task, the self-sufficiency of the orphan combining with the eager mind of the fixer and innovator. He once visited Bulunsky district and saw that the local herders had a better method of harnessing their reindeer to sledges*. Vladimir Nikolayevich learned from them, improved on their design, and revolutionized the harness technique of everyone in Sebyan. Professionally, he had risen to be general manager of the Farm's herding and hunting divisions (*upravlyayuschiy otdeleniem*) until the director took over in 1976 and sacked him for his independent views.

Over the next five months, Vladimir Nikolayevich and Manchary would go further down the Tumara, working their way several times up and down each tributary, checking for tracks, setting traps, returning to collect the deep-frozen little bodies, noting changes in the drift and texture of the snow, leaving tents and stacks of firewood for common use, and reading signs left by other humans. Vladimir Nikolayevich was convinced that the animal tracks would be thicker in January. Certainly, where he was going, the human signs would grow thinner: beyond our tent, there were no further huts. On the 20th of any winter month, if they had collected enough furs to make the trip worthwhile, the hunters might travel to the Farm office to hand in their catch to a representative of the State fur corporation – if such a person continued to visit the village.

In another week, Kostya's kind, round face would appear to collect me on his sledge. He was driving me back up the Tumara, which would be familiar yet changed by each shift in the weather: one day to the Verkhoyansk Pass, the first travellers to make an overnight stop in the tent Vladimir Nikolayevich had just left behind, another day to join the three or four men crowded into camp 10's winter hut, and a final day to reach the village.

'Every evening I'll come home and expect to find my good Dr Vitebsky!' said Vladimir Nikolayevich. 'I'll be bored when you go. I'll remember you all my life. I'm sure you'll remember me, too. I'm going to name this camp site after you!'

Late that night, I stood outside looking at the moon, which was shining with an extraordinary blue light over a crust of snow which was so vast that it seemed beyond comprehension. We were camped on a river bank near some massive crags of layered shale, each horizontal contour etched in sharp lines of black and white. By the yellow light of a candle, Vladimir Nikolayevich was bending, cutting something, fixing something, whittling Peter's rowan stick, his shadow projected onto the translucent membrane that encased the glowing bubble of warmth that was our home.

By now, the electricity generator in the distant village would have been turned off. The other hunters around us would be asleep, their tents and huts 30 miles apart behind winding rivers, the embers fading in their stoves. An aviator or a shaman could fly around these mountains all night without seeing another light. At any moment, this landscape could snuff out a clumsy, hairless human crawling across its rough surface. I suddenly understood the feeding of the fire in a new way. It was an act of acknowledgement of the flame's concentrated intensity, but also an act of support for its puniness. There was not another tiny island of heat in this ocean of −60°F, except in the metabolisms of the mammals that we tracked for the hairs sprouting densely out of their skins so that they could later contain the warmth of human bodies.

The depth of silence was beyond anything I had ever known, made all the sharper by the occasional snorting and scrunching of our reindeer on a darkened slope nearby, each one named, trained, and bound to us by a loyalty that was ancient, but could be easily lost.

PART III

BEADS FOR THE NATIVES

7

Frightened children and disdainful women

Winter laid a hush over everything, and removed all sense of haste. On the way back from Vladimir Nikolayevich's tent to the village, Kostya and I stopped to visit his men in camp 10, where the herders had settled into their winter hut with only books and a small snooker table to keep them amused during the twenty hours of daily darkness. Once, in the half-light of the midwinter noon, I noticed Peter sitting naked on a log in the snow, smoking a cigarette very, very slowly.

Shortly before the village, as our reindeer skimmed across Lake Sebyan and returned to the bank, Kostya pulled the caravan to a halt. The six months of winter offered the animals little to eat but lichen. Any pasture closer to the village would have been trampled by human feet and crushed by the huge tyres of tractors and trucks. We scraped away the snow in the midday twilight, two silent, padded figures in thick reindeer-fur coats, gloves, and hats, and crammed a sack full with dry handfuls of this brittle, primitive life-form.

As we came nearer the village of Sebyan, other tracks cut across ours, weaving back and forth, merging for a while and then separating again. Fresh from the single scent-trails of the taiga, our dogs ran from side to side, confused by the contradictory smells. I have often sensed that herders, too, feel this each time they are drawn in by the village, becoming agitated by the memory traces of so much conflicting human activity and communion. I have watched them press on in anticipation, tinged with apprehension. Some of them would get blind drunk, and their wives would threaten yet again to leave them. Men would quarrel with women, men would quarrel with other men, knives would be drawn. I suspected that they were wondering, as I was: Who might die this time?

By its mere lack of movement, the existence of the village changed the experience of space. When out on the landscape, one might think of travelling as the old people travelled, indefinitely in every direction, as over a web. Our approach to the village was a reminder that now the State Farm commanded the surrounding taiga, and that space radiated out from the village in concentric circles. Ivan's and Granny's territory in camp 7 was no longer just a place in relation to other places: it was more 'remote' than some other camps because it was further from the village.

The very oldest people I met, some even a generation older than Granny and Tolya's mother and aged over 100, recalled riding thousands of miles* over the hunting trails of diverse clans and peoples, encountering their spirits, adapting to their ecologies, and speaking something of their languages. But recent generations became cosmopolitan in quite a different way. Their attentiveness to the world was drawn away from the land and funnelled through the village airstrip to a life that could not be reached on the back of a reindeer. During the day there might be the distinctive drone of a biplane turning over its propellor just beyond the radio hut and the last houses, though by the early 1990s flights had been reduced from two or three a day to two or three a month. More often there was the distant buzzing of a grinder from the machine workshop or the straining engine of one of the only trucks for 150 miles around. Having once trekked slowly from the plains along the corrugated surface of deep-frozen late winter rivers to join the Farm's stock, these trucks now lumbered and bucked around the village in all seasons over the pitted ruts left by the drying mud of summer.

Kostya's *uchakhs* could stay tethered outside the door of his cabin for only one night before being returned to the taiga. Standing in the dark next to the mound of lichen tipped out of our sack onto the snow, they looked as alien as if they were parked on a city street.

In a world where a camp site could take days to reach and the interpretation of an animal's footprint could demand hours of attention, the approach to this cluster of cabins felt like a rushing together of strands that were too brutal to merge coherently. After the lair-like intimacy of a winter hut or the thin, permeable film of a summer tent, the single-storey wooden houses in the village seemed withdrawn behind their picket fences, insulated double doors, and blank double-glazed windows raised above eyelevel on earth platforms above the permafrost. Each window, never opened, was decorated between its two layers with vegetation that had died long ago: larch cones, tufts of lichen, or sprigs of cranberry in autumn tints, unchanged from one year to the next until something in the window needed replacing. Outdoors lay parts of reindeer, also dead: skins were draped on fences to dry and pairs of antlers, the unifying shard of the skull still attached, adorned porches, outhouses, and privies.

While individual houses were heated with large log-hungry stoves, the public buildings towards the centre of the village were fed by hot-water pipes running from nearby boiler-houses, insulated under earth embankments surmounted by planks laid along the top to provide walkways over the mud. The doors of the bakery and the food store swung open and slammed shut on heavy springs behind women in elegant boots and smart coats, carrying large bags as they shopped in quest of bread, butter, tinned fish, and bottled Bulgarian cucumbers. In the clothing and general store they foraged among the stock, fascinating to me but surely predictable to them, of slacks, socks, exercise books, saucepans, nails and screws, tea services, and sets of vodka glasses with gold bands around the rim.

In dark December, the seasonal change from a mere three months ago was complete. Soon after the start of school, I had followed the children back from Kostya's camp to the village, where I had watched them burst out at playtime into the September sunshine, huge bows of pink and red ribbon from the shop holding together the long hair on the girls' bare heads, black echoes of the blonde Russian plaits in children's storybooks. In the street I had seen Shura's kids and the competent young herder Sergei, who looked like a child again in his dark blue school uniform.

This was when I saw for myself what awaited those children in the happy helicopter photos, in the days when helicopters were plentiful. I was invited to watch a class, one of the three hours a week now taught in Eveny to imbue children with a love of their language and traditional culture. After the pre-war creation of alphabets and textbooks, native languages had been neglected in the 1960s and 1970s to the point where some had become almost exinct. Among the Eveny, too, the 1960s generation was the one with the least knowledge of their language. But by 1990, in a new spirit of cultural revival throughout the country, pressure from native activists had brought language learning back into the schools.

The teacher had the short, permed and hennaed hairstyle that I nicknamed 'Soviet Woman', in contrast to the undyed tresses that were still worn long by Granny and other old ladies in the taiga, tied up under a headscarf.

She began class with a poem: 'Quickly, find page 197. Have you found it? Where have we got in our reading?' Her voice rose sharply. 'Sanya, have you found it? What is this poem?'

'Autumn,' murmured the children, barely audible.

'Autumn,' she confirmed briskly. 'Now look at the picture. What is drawn there? What is drawn? What season is shown?'

'Autumn,' they repeated, *sotto voce*. They were aged about 10, and resembled baby sparrows faced with a hawk.

'Autumn. How did you know it was autumn, Vadim? How are the trees drawn?' There was silence. 'Huh?' The children muttered something and she said, perhaps amplifying them, 'The leaves are yellow.'

Then a child read out some lines. When playing the tape later to Tolya I realized how great the obstacles were to language revival. The textbook was written in an eastern dialect of the Eveny, 600 miles away on the Pacific coast. Though a native speaker of the village's own dialect, Tolya could barely under-stand the words, but he thought they said, 'Behind the cloudy sky where there are rain clouds, the sun cannot be seen.'

'Well done,' said the teacher. Opening up a new topic, she bellowed, like a Russian, 'What happens when September begins? We go to school, in order to learn. What do we do, children?'

'We go to school.' This was said more audibly.

'That's right, Vadim. Why do we learn?'

'In order to know more,' several voices spoke in unison: they knew this line.

'We learn in order to know more,' the teacher confirmed. 'In summer, you lived in the camps. What did you do there in the camps? I'm sure you didn't just sit there doing nothing!' She pointed to one child. 'What did you do with the reindeer?'

'I worked.'

'How did you work? Louder, Igor. What did you do with the reindeer? You looked after the reindeer. And you, Arsen, what did you do in the camp?'

'I helped with the herd and looked after the reindeer.'

'All right.' Then with a renewed shout, she said, 'And you, Pronya, what did you do? Did you look after reindeer?'

'I looked after them during the day.'

'Pronya looked after them during the day,' she repeated.

Looking around the room and raising her voice again, the

teacher asked, 'Who else travelled? Sanya, what did you see in the camp? How many wild sheep did you see? Don't look over there.'

'A few,' he answered uncertainly, turning round to face her.

'A few or a lot? Speak up!'

'A lot,' he decided, catching something in her intonation.

'Who else was in a camp?' she asked, surveying the class. 'Huh? Nobody else apart from these? Eh? You were all here in the village?'

Yes, all. Until now, I had not realized the full demographic impact of the policy of settling the nomads in villages. Three or four children filled a summer camp with life, but with only thirteen camps it still amounted to a small proportion of a generation. The rest had never set eyes on the activity that sustained their existence and defined their identity.

'Now look at the board,' she said, still speaking in Eveny. 'What is the Russian for *mengen boloni*?'

'*Zolotaya osen'* [Golden Autumn].' This phrase was a Russian poetic cliché, which had been translated into Eveny. The children's grandparents would have talked of 'autumnal thought-feelings' (*möltenseri mergen*), their point of reference in an internal state of mind rather than a visual impression, in the psyche rather than in the colour of a precious metal. But the children were not going to learn this phrase.

'Golden Autumn, Golden Autumn. Now we'll draw a picture of *mengen boloni*. We'll draw by ourselves without copying from books. Everybody draw what you think about *mengen boloni*.'

The children concentrated for a few minutes. 'No copying!' she shrieked into the silence.

What was there to copy? All the children I could see were drawing identical yellow triangles on stalks. These were larch trees copied from textbooks of the mind, perhaps based on Russian drawings of Christmas trees. The larches outside the window were not at all triangular, but the children were not supposed to look out of the window. They were not taught how to talk about the shape and character of each tree, its suitability for firewood, fencing, house timbers, or the flooring planks beneath their desks, or how the twigs sometimes bunched together to form a witch's broom, a sign that the tree may be dedicated to the spirits of the upper world. The children were also drawing book-tents, which were triangles not homes, and which most of them had never inhabited, and book-reindeer, which most of them had never stroked.

The school was the main filter that perpetuated the position of the herders as a permanent underclass, under a barrage of teachers' remarks like: 'You'll never come to anything, you're just fit for herding, I don't know why we bother to educate you . . .' 'Your parents are insignificant, illiterate herders, drunkards, and you'll be the same . . .'

At the end of the school day, I watched these timid children pour out of school to run home to their mothers. Until earlier that same year, many would have stayed behind in the village boarding school, which had only just been closed down. Lidia, Kostya, Tolya, Ivan, virtually every native adult of their generation across the Russian North, had been rounded up by helicopter at the age of 5, 6 or 7 and taken away from their parents* like a herd of little airborne reindeer calves. They would hardly see their parents again until the age of 15, except sometimes when they were released from their corral for summer vacation.

Tolya's parents were nomads and he, too, was taken away like this to the village boarding school. But even his schooling was caught up in factional politics. The school principal of the day did not nominate him for special training at the Hertzen Institute in Leningrad, and even warned him with threats not to further his education. To enter college in Communist times one needed a certificate of political reliability from the authorities at one's registered address. Tolya got around this requirement with characteristic inventiveness by stopping off at the university on the way back from two years' naval service and giving his address as Pacific Fleet Headquarters in Vladivostók*, where he had a good relationship with his commanding officer. The director was furious, but it was too late. Tolya completed his degree in history in the 1970s, and this provided a foundation for his PhD in anthropology and for the foreign lecture tours through his association with me in the 1990s. He is now the most educated and cosmopolitan person ever to have come out of the village.

When mothers still lived on the land, boarding schools were inevitable, even desirable, since northern native families were often destitute, and this was a way for the State to rescue their children from poverty by providing free board, lodging and education. But the policy went further. Prosperous parents, poor parents, good or bad: from the 1960s all would be deprived of their children. If it was too late or too difficult to induct the parents fully into Soviet civilization, at least the Government could work on the next generation. The habit of family separation became so ingrained that even in the late 1980s, when only thirteen of Sebyan's women lived out on the land officially as tent workers, many of them elderly, the village boarding school still contained some sixty children.

Even so, the children of Sebyan were lucky to have their own boarding school. Some natives were sent hundreds of miles away, far from anyone of their own ethnic background. A Sakha friend whose father had been the headmaster of a boarding school in a Sakha area told me about his childhood in the late 1970s.

'We were told that we were going to receive a group of native orphans from an Eveny area,' he recalled. 'They seemed tiny, terrified. I felt such pity for them.'

'Were you the only person who felt that way?' I asked him.

'There were two other people who felt sorry for them, too,' he replied, 'the cleaner and the night watchman. The other children laughed and bullied them, and the teachers mostly ignored them – or scolded them when they couldn't understand the homework.'

'Could you do anything for them?'

'I had two of them in my care. I was a monitor. I took them home, so they could watch cartoons on television. They never talked about their past, or their feelings. But they were obviously miserable and frustrated. Their behaviour got worse in the spring.'

Was this his civilized Sakha imagination casting them as wild Tungus? Surely not. There were stories of native children all across the North who had escaped from their boarding schools, sometimes dying of exposure while trying to rejoin their parents. Most of these incidents happened in the spring, when the sun was up but the temperature was still far below freezing. Though the boarding schools had taught native children to read and count, it had left them unfit to survive on the land.

'What happened to those children in the end?' I asked my Sakha friend.

'When they reached 14, they were sent to a special technical school for low achievers. I don't know where they went after that. They'd been told that their parents were dead, though I'm sure it wasn't true. Maybe their parents were just drunk.'

So the children sent to this boarding school might never have seen their families or ever known that they still existed, and would be shy, rootless strangers all their lives.

I was swapping boarding-school atrocity stories* with an Eveny friend in the city apartment of my Sakha student Tanya when Tanya's mother passed through the living room and over-heard us. She stood with her tray of fine Lomonosov teacups quivering with indignation.

'How can you say such things!' she exclaimed in her firm matriarch's voice. Then she launched into a passionate diatribe about how the boarding schools had saved thousands of street children from starvation. 'I was a poor orphan myself,' she testi-fied, 'and if I hadn't been rescued by the boarding school, I'd have perished along with so many others.'

After this outburst, I checked with other Sakha people living in larger settlements on the plains. Nobody came up with the terrible tales that were told routinely in the northern native settlements. Some found them beyond belief and were insulted that I could even think that such things would happen in their republic. The stories of child abuse in boarding schools in northern reindeer-herding communities are so consistent that one must believe them, and I wonder whether the isolation amplified any pathological tendencies.

Around 1990, most boarding schools across the North were closed down and the children reunited with their mothers. This move was a response to public opinion, but it was also an acknowledgement that the de-nomadization of women was complete. There were virtually no children left of school age whose mothers still lived out in the taiga. The scope of womanhood there had been greatly reduced. The intimate space that women had controlled around the fire related not only to warmth and cooking, but also to childbirth and motherhood. By now, giving birth in the taiga, being a mother, even just being a wife, had become almost impossible. With the move into the village, much of the community's womanly activities had been collectivized as women became teachers, nurses, and cooks for each other's children. Their only surviving function in the taiga was that of tent worker who was there not as wife or mother but as paid housekeeper and dinner-lady.

Most young girls in the village had no inkling that the taiga could be an arena for the exercise of a woman's judgement and courage, nor that they could follow in the footsteps of the great female hunters, like the mother of Vitya the Wolf Hunter, who could point with pride to the fur and claws of a bear, her greatest adversary, spread out against the peeling wallpaper behind her plastic sofa where other women would have had a factory-made mock-Turkish carpet.

With my love of the taiga, I have often asked young women why they do not want to live there. Their first answer is always fear of the cold, away from the warmth of the solid log cabins of the village. But they are also reacting negatively to the circumstances in which that cold is experienced. Life in the camps, they say, is not 'civilized' (*tsivilizovanaya*) or 'cultured' (*kul'turnaya*). These two words lay at the heart of the Soviet ideal of how one should live, and persist into post-Soviet consciousness. The continuum from wild to civilized*, from wilderness to village to city, is reflected in all aspects of conduct, dress, and comfort. The frontier of metropolitan elegance lies, not between the city and the village, but between the village and the camps. In their short trips between house, shop, and office, women in the village can pick their way through the mud in fashionable long coats and high-heeled leather boots, rather than the massive reindeer-fur coats and oversized boots that keep a herder alive in the open air for twenty-four hours a day.

The policy of civilizing the well-groomed village women has worked too well. The grandeur of the crags, the fresh air and reindeer milk still beloved by Kristina, the smell as one spreads fresh larch branches on the floor, the meticulous packing of baggage for the next migration, and the competent life-or-death self-sufficiency with only the thinnest thread of radio or helicopter backup – all of these hold no attraction.

Their own women see the herders as coarse and uncouth as they tramp into the village on their home leave in their dirty clothes and hit the bottle. When the herders recover two days later, they sidle into the Farm building to petition other women with fierce office hairstyles as they disdainfully click the beads on their abacus to compute each herder's earnings and his worth on the Board of Honour.

The sight of a drunken herder in his home, surrounded by sober women, is made all the more painful by the knowledge that this is the twisted outcome of a systematic policy to undermine the family. The role of male herders was fatefully changed by their redefinition as industrial workers. The Russian engineers and Ukrainian miners on whom their job description was modelled had deliberately travelled to remote northern frontiers for adventure or money, and would return home in triumph to their families elsewhere. But for the reindeer herders, there was nowhere else to go: this was home.

In Sebyan, these outsiders included Sakha vets, Ukrainian truck drivers and Estonian mechanics, men who might take Eveny girls with them when they left, over the heads of local boys. With cost-accounting, direct air links from one village to another had been completely severed, and what few flights there were led only to Sangar or Yakutsk. The village was now so isolated that young people were becoming too closely related to marry. Tolya's mother, the Nikitins, Lidia's father, the mother of Vitya the Wolf-Hunter, and many more of the older Eveny in Sebyan had migrated or married in from other regions as part of a way of life in which they rode regularly over hundreds of miles to other camps and communities, mingling at the great marriage marts of the spring festivals, showing off their skills in sewing, reindeer racing, wrestling, and dancing to prospective partners and their parents. But most people under 30 were so dependent on aviation that they did not know how to travel such distances alone on the land without dying on the way.

As with collectivization in the 1930s, nothing in the changes imposed on reindeer herders was derived from their own relationship to their land. Their purposeful and responsive nomadism was dismissed as primitive rootlessness, while the factory metaphor transformed the skilful, self-sufficient hunter into a wilderness proletarian. While the going was good, he could also earn bonuses and medals as a socialist hero of productivity, but by the early 1990s this compensation was rapidly disappearing.

As the frontier model of the unattached male industrial worker continues to wreak its destruction, life in the taiga has become ever more masculinized and unattractive to women. Young women's culture has been changing faster than the culture of their men. The weighty brigadiers involved in planning and politics can speak for themselves. But their young herding lads are silent. Village girls have told me that they could not imagine marrying a herder, because of their lack of conversation. Television and magazines portray life as a torrent of words and easily revealed passions, and they came to see themselves in similar terms ('They're just like us!' exclaimed one girl, explaining why she liked soap operas, dubbed into Russian on Moscow TV, about the affairs of the super-rich in Mexico). Whether one thinks these passions are real or artificial, it would be hard to imagine a more opposite ideal from the taiga ethic expounded by Lidia and still lived out by those who feel comfortable on the land.

For these young girls, taiga values are inverted. To them, the communication with animals, rivers, and mountains, which makes much speech unnecessary, seems maladaptive. Instead of a talent, it is now interpreted as a deficiency. Young herders with their shy eyes are perceived as blank people and the taiga, with its finely grained texture which only they can read, as a dead space.

Men fulfilled and men in despair, camp 8

The man who combined family and reindeer herding more satis-
factorily than anyone I knew was Kesha, the brigadier of camp
8. He lived with his wife, his children, his parents and his father's
father, making this four-generational camp one of the most complete herding families anywhere in the region. This unusual
arrangement was made possible by the charismatic personality
of Kesha's wife Lyuda, a Sakha woman who was not merely
a tent worker, but shared the responsibility for the management of the herd as a family enterprise, as Granny did in camp
7 a generation earlier, but unlike any other woman in this
community of her own generation.

Kesha and Lyuda were in their late thirties. With his long,
straggly hair and black wispy beard, Kesha was very much his
own man and looked like a hero from a mythological epic.
Lyuda also wore her hair long. She had come to this remote
northern location from the Irkutsk veterinary school for a practical exercise, never having worked with reindeer, and had stayed.
She handled the reindeer as if born to it: though Kesha was
a good rider, she had won her way through a succession of
regional competitions to become the reindeer racing champion
of the entire Sakha Republic.

Kesha's father, Dmitri Konstantinovich, was a prominent
elder who had been an opponent of Tolya's attempts at reform.
He was allied to the director politically and closely related by

marriage. Tolya had told me in 1988 and 1990 that it would not be opportune for me to meet this family, so I had regarded them as out of bounds. But Dmitri Konstantinovich was revising his political views. In September 1992 Tolya decided to use me as a prop in their rapprochement by taking me to stay with him in camp 8.

'As an exotic living exhibit [*zhivoy eksponat*]?' I teased him.

'And what was I when you took me around England?' he countered. Of course I had shown him to my friends: he was colourful, extrovert, and in 1989 he was the first Siberian Arctic native anyone had ever seen.

We found Kesha and Lyuda after two days' ride, sheltered in a grove of huge larches rising vertically more than 150 feet above shallow roots which splashed over the permafrosted earth like knotted veins. The lower 100 feet of each trunk were almost bare. The breeze, imperceptible at ground level, was picked up and amplified by the feathery branches which swayed and brushed against each other in the top third. Their movement sometimes rippled down to the base, but on the ground it was completely still between their enormous, shuddering trunks and we walked like insects between stems of grass while an agitated climate hissed above us. Larch needles, reddish beige in the dull light, fluttered down like clouds of small insects. Each stone rising up out of the stream to meet them was encrusted around the brim with little splinters of ice.

Kesha's parents, his grandfather and the bachelor herders occupied three small tents, while Kesha and Lyuda lived with their two toddlers in a large, square tent of the same type as Kristina's restaurant. Instead of larch fronds, they had paved the floor with rough-hewn planks, and their bed was on a platform raised behind a wall of logs. These were both refinements I had not seen elsewhere. The tent was well ordered, with three small suitcases serving as chests and sliced logs as stools. There was a magnificent dried willow tree standing in the corner holding saucepans and metal mugs on its many prongs.

Sudden gusts of larch needles would beat down on the canvas outside, sounding like rain. But this was a cold, dry breeze. The last of the mushrooms were already blackened by hard frosts, and the reindeer would soon be moving towards their winter pasture. In the afternoon, I approached the family tent in the Eveny way, drawing the flap aside and hestitating at the entrance. Lyuda invited me straight in with a laugh. Kesha had gone for a walk with their 4-year-old boy Dima, but she was not shy of being alone with me in the tent. She was sitting on the planks by the stove holding little Diana bare-bottomed on her knee and tapping her feet together. When she was not being held by her mother, Diana would run around the tent touching, exploring, picking up a scrap of food, carrying it, dropping it. Lyuda never stopped watching her child and repeatedly drew her away from danger, especially by the hot stove. She talked to her in a special baby voice like a ventriloquist, the voice coming out of a doll that I had brought as a present.

'Diana, Diana, Diana!' she chanted, sighed, whispered.

Diana was not yet 2, but Lyuda was very proud of her child's alertness and self-sufficiency: Diana insisted on getting into her boots by herself, and she did not cry like other children when riding through the rain and the snow.

'She doesn't want to be covered at all against the weather, but sits up and watches everything,' Lyuda ended proudly.

Once when Diana cried, Lyuda explained, 'She's thinking about her papa. Whenever he's out and she thinks of him, she cries!' Of all the places I stayed, this camp was where I felt the most tension between the happiness of being there and the pain of missing the family I had left behind.

'My daughter is also very self-sufficient,' I said. 'She's 7 now. The first words I ever remember her saying were, "*I do it!*"'

'Then she should come and live in the taiga!' approved Lyuda.

The tent was filled with the sounds and smells of simmering reindeer meat and batch after batch of round, home-made bread which Lyuda tended throughout the afternoon. At one point I went out and split some more logs; at another, Kesha's grandfather, who was over 90, brought in another stack which he had split himself.

Lyuda kept several churns of water hot by stacking them against the side of the stove. When Diana fell flat in the dust, she used this water to wash her face and head in a small zinc bowl.

'Diana loves baths and loves having her hair washed. With Dima, it's the opposite. He screams and struggles! Diana usually cries with strangers,' she said, 'and even sometimes with her grandparents from the other tent. But she's not at all worried about you.'

I stayed, enchanted, until Kesha came back with little Dima in the evening. Tolya had returned from another direction, probably from talking politics with Kesha's father. Tolya and I had caught some grayling at our overnight halt on the way and I produced vodka from my bag. While Kesha cradled his Diana, we spent the evening filling in missing links in my knowledge of genealogies, charting the rise and fall of various brigades, and discussing traces of shamanic fits and trances among people whose ancestors had been shamans.

'What a pleasant evening it's been,' Kesha said as Tolya and I rose at last to go to our own tent. 'We so rarely get a chance to talk to new people!'

Another day, Kesha took me to the site of their next camp 20 miles away, where two young herders had replaced some of the beams of the winter hut and were renewing the insulating layer of moss and soil in the space between the ceiling and the pitched roof. We helped the lads to transport the earth and re-fit the windows, which had been removed for the summer so that bears would not rip them out when looting supplies. There was no glass, so we double-glazed the windows with polythene sheeting.

While we were hauling earth up a ladder I told Kesha about

my own little girl, and asked him why his child was given what in Russia is an unusual name.

'She's the goddess of the moon!' he declared, as he heaved a sack of moss into the loft.

'Did you know the Greek name for Diana was Artemis?' I said.

A big grin came over Kesha's face and he leaned over from the top of the ladder and took the knife from his belt. Looking down at where Diana was scampering around in the long autumnal grass, he incised the Russian form ARTEMIDA on the main beam of the hut, like a declaration of love carved on a tree-trunk.

Kesha had changed his surname to match Lyuda's, a reversal of the usual custom. When I asked Lyuda about it, she replied, 'He said there are lots of Krivoshapkins in the village, and he wants to be the first Burtsev.' Then she added, as though there were more to it, 'But you'd better ask him.'

I did not need to ask Kesha directly. The change of name was a way of expressing his devotion to Lyuda. After an acrimonious earlier divorce and an estrangement from his first daughter, he had finally found happiness. Because of Lyuda's commitment, camp 8 was the only family in the region with children in the taiga all year round. They had Ivan's independence in camp 7 without his bachelorhood or his conflict with the Farm, and Kostya's productivity in camp 10 without the enforced separation from his wife and children.

Kesha and Lyuda had created just what the Soviet regime had tried to destroy. To have a family life in the taiga was a political achievement, and this was made possible by the safe political space created by Kesha's father, Dmitri Konstantinovich. Whether out of conviction or as a strategy to protect his family, he had joined the director in opposing Tolya's proposal for regional reform. In this, he had aligned himself with his own brother, who was in charge of marketing in the director's Farm management, and his sister, the stern teacher whose class I had attended in the school.

Dmitri Konstantinovich was a striking figure. He had an uproarious sense of humour and his round face was lined from continual laughter, but his body was so swollen that he could not even shake my hand for the swelling in his arm. As he sat cross-legged at the back of his tent, his legs almost disappeared under his Buddha-like stomach. He blamed his illness on heavy-metal poisoning from a childhood spent herding near the lead and tin mines of Endybal, now abandoned, on the territory of camp 5.

I sat along the side of the tent while his wife, a handsome middle-aged woman who had lived all her life in the camp, sat slightly hunched on an upturned bucket in front of the stove, supplying us with tea. She was Kristina's sister and I could see the resemblance in her face and gestures. Dmitri Konstantinovich read aloud some of his Eveny poems, one about a shaman's flight into the cosmos and another about Russian cosmonauts, which he kept in a small exercise book, and warmed to me when I showed that I could understand a little Eveny. He also read out articles that he had written in *The Leninist*, the four-page local newspaper published in Sangar, in which his nostalgia for the order and discipline of Soviet times was tempered with criticism of the management over the prices that herders received for meat and antlers. He was coming to think that his stance on the region was mistaken, and to agree with Tolya that the director was dangerously incapable of responding to change.

While Dmitri Konstantinovich was reciting these articles, the young lad next to me was silently studying a magazine serialization of Wittgenstein's commentary on *The Golden Bough* by Frazer, a founding father of anthropology.

It would be two and a half years before I returned to camp 8, in March 1995, with Tolya and his wife, Varya. Our sledges headed for the winter hut which I had helped to repair. The layout was typical for a hut: a large stove standing on a sheet of metal in the centre of the planked floor, a table for eating under the window, and sleeping platforms of raised planks around the walls, with an extra-wide one for Lyuda and Kesha. The children were older now, and ran squealing around the bunks and over the heaps of bedding. Dmitri Konstantinovich was there, but his wife had gone to the village. Lyuda was the only woman. All the occupants of the spring, summer, and autumn tents were under one roof in one large room, and the young herders Boris, Leonid, and Ganya gave up their platforms for us and slept on the floor by the stove. Little Dima shared a sleeping bag with one of the herders, while Diana slept with her great-grandfather, Dmitri Konstantinovich's father.

Dmitri Konstantinovich talked about politics and was now open about his disenchantment with the director's regime. The bachelors spoke of Mexican soap operas, *Wild Rose* and *The Rich Also Weep*. Tolya was playing cards with Leonid. He had won seven rounds and if his score went up to ten, Leonid would owe him a reindeer. The very old man talked only about reindeer and their movements, except to say that his lifetime's collection of labour and productivity medals were now valueless, and he had given them to the children to use as toys.

'It'll soon be the fiftieth anniversary of victory over the Fascists,' he told me. 'And I've got nothing to wear. I wouldn't care otherwise.'

Dima was gorging from the communal frying pan. Lyuda told him to stop, and then turned to me, saying, 'He only listens to Papa!' At that moment, the door opened, and she called out 'Papa!' But it was only Tolya, and Dima jeered and pointed a finger at her. When Papa did come in he scolded Dima, though not harshly.

Diana had managed to open a jar of bear's fat. It is a powerful medicine, but Diana would eat it all if she could. She started pulling things out of the medicine chest, and when Lyuda told her to put down the tonsillitis medicine, Diana ran around the hut shouting, 'I've got tonsillitis, I've got tonsillitis!'

Later that night, Tolya and Kesha tried to cap each other's stories about their childhood punishments in the village boarding school while their parents were away herding. Kesha reminded Tolya of how they would be woken up three times a night, just to ruin their sleep.

'Ah,' said Tolya, 'but don't you remember I was so naughty that I would be hung by my collar from a nail on the wall and left there all day? And I'm still a dissident!' Kesha laughed. Turning to me, Tolya added, 'When we go back to the village I'll take you for a meal with Maria Ivanovna. When I was little she would follow me to the toilet to make sure I wasn't running away!'

The children's lives between lessons were like a labour camp, pulling logs and doing the physical work for which their taskmasters were paid. The denial of comfort to the children was integral to their boarding school's institutional culture. They were not allowed to watch the weekly film that was flown to the village hall, and once when they managed to sneak in a teacher became so enraged that she hacked through the electricity cable with an axe, so that no one could see it at all.

'Didn't she get electrocuted?' I asked in amazement.

'No,' replied Tolya, 'those people are immune to everything!'

Kesha roared with laughter again, then said compassionately, 'The principal at that time was a Russian brought up in an orphanage. That's why his regime was so harsh.'

That's too kind, I thought, angry on their behalf. What about the ones who were your own uncles and aunts? What about the native headmaster, your own relative, who would wake the children in winter and force them to crawl around the playground in the snow on their bellies, wearing only their pyjamas, and who was denounced to me so passionately by Tolya's sister, Anna, that she broke down in tears several times during her narrative? What about the children who killed themselves? Why did children who died in the boarding school remain unhappy ghosts and continue to haunt the building? Who was the child with the wings of an angel whose appearance in the school always meant that another child would die soon? The institutionalization of people also institutionalized cruelty on a level that the teachers would not inflict on their own children at home.

It suddenly became very quiet. Diana had fallen asleep on Varya's bed face down with her bottom in the air. Varya stroked her back tenderly until Kesha scooped her up, still asleep, and deposited her on his own bed.

A viable family brigade depends on continuity. Kesha had taken over as brigadier from his father Dmitri Konstantinovich in 1978, and, like all herders approaching 40, he was talking of retiring from the position if not from herding. But his own boy Dima was still a little child. Of the young men Boris, Leonid, and Ganya, Kesha was pinning his hopes for a successor on Boris, a close relative who had been brought up by Kesha's parents as an adopted son, like Gosha in camp 7. Boris had recently married, though his life as a herder would keep him apart from his new wife for most of the year.

'He's my expert deputy,' said Kesha expansively, gesturing towards Boris. 'He knows it all.'

Ganya had been married, too, to Kesha's younger sister. But she had just left him for one of Anna's sons, who worked in the village. I sensed that Ganya was agitated, but it was still startling when he and the bachelor Leonid crashed into the hut raging drunk, fists flailing, one with a bloody nose. Fighting men can seem very large in a crowded hut full of boiling kettles. Dmitri Konstantinovich, immobilized on his bunk, remonstrated in the tone of an elder. The other men, who were sober, grappled with them. Kesha wrestled them to the ground, slapped their faces and turned them out of the hut. Diana had woken and was watching with big eyes.

Dmitri Konstantinovich looked embarrassed and turned to me: 'It happens [byvaet].'

Tolya was laughing: 'Young people having a lark!'

But it was not just a lark. During the night, we were woken

by the sound of a tussle on the floor and the gruff shouting of obscenities. The stove was out and it was completely dark. The door was suddenly flung open, letting in an implosion of cold like the release of an airlock. The constellations appeared sharply for an instant in the open frame and were blotted out by two huge angry shadows chasing each other into the snow. The heavy, reindeer-fur-lined outer door slammed shut.

A moment later we heard a shot, and then another – six in all. Lyuda screamed and ran out into the night. She seized the gun and Kesha dragged the young men back into the hut, still pummelling the air with their fists and yelling abuse. We crammed each of them back into their sleeping bags. I spent much of the night sitting on Ganya as he writhed frantically but ineffectually inside his bag.

In the morning I was exhausted, physically and emotionally.

'A writer works even in his sleep,' Lyuda teased. 'That's why you're so tired!'

The two brawling herders were slumped in disarray with heavily bruised faces, but Kesha was very playful. While the other men sat around after breakfast reading and rereading a stack of old newspapers, he sang little ditties to his children as he cleared the dishes, improvising to match each situation: 'Now I'm washing up the teapot, now I'm washing up the teapot!' I had given him a big knife and he tucked it into the back of his neck and ran around doing Ninja stunts for Dima. He dressed Dima in a reindeer-fur coat with the sleeves sewn together at the end to give built-in mittens and sent him out to play in the snow. Before he left, Dima said he would grow up to be a helicopter pilot, so that he could bring us provisions.

'Will you take the boy with you to England?' Kesha teased me as Dima ran out.

'Or will you give me Diana?' I teased back.

He laughed. 'You'll never separate me from Artemis, the goddess of hunting and the moon!'

In the middle of the day I went with Kesha, Lyuda, and their children to the nearby river to fish through holes in the ice, bathed in spring sunshine and family affection. That night, while Varya massaged Tolya by holding on to a roof beam and walking up and down his back, Lyuda and Kesha made sausage by pouring reindeer blood into intestines. They explained how the taste varied with the age of the reindeer, and even with its character. We ate the blood sausage with raw frozen reindeer liver and the raw fish we had just caught, all sliced into thin sheets and dipped in salt.

'On the hottest day in summer,' Kesha said, 'you dream of this!'

While supper was being prepared the battered Ganya sat hunched on the floor and the other men sprawled on the bunks with their newspapers. I remembered Kristina's remark that quarrels in the taiga flared up and were soon forgotten. But Kesha's family happiness exposed the pain behind the Robin Hood archetype: a band of brave men lived in the wilds, but only their charismatic leader had a consort while the rest were angry young men, unable to find or keep a wife. In a very few years Dima and Diana would reach school age, and then even this family would begin to separate. Lyuda would stay in the village to look after the children at the end of each school day, while Kesha would visit them from the camp when he could.

A few days later, Tolya returned to the village with Varya and I joined Lyuda, Kesha, and their children for the 25-mile migration to their spring hut. We used two snowmobiles, each with a caravan of sledges tied behind. Lyuda carried Diana on her lap while Kesha took Dima and sat me on the first sledge behind him. Kesha's machine had no windshield, so we made a detour to a storage platform to pick up a spare. Lyuda's machine broke down several times and Kesha repeatedly removed and cleaned the spark plug. Snowscooters were not very practical in this region, as they could not traverse rougher terrain and fuel and spare parts were scarce. Several machines lay abandoned, probably forever, beside the main low-lying trails near the village.

Kesha's deputy Boris was supposed to catch up with us on his *uchakh*. But he did not appear, and as we approached the crest of a pass, Kesha decided to turn back with Dima and look for him. I waited with Lyuda and Diana. A wind had arisen and was scouring the flat surface, revealing beige grass-stalks, each of which created a little funnel of disturbance in the snow. Kesha was gone for a very long time. In the near-darkness of the early afternoon, the three of us huddled silently together on a sledge in our huge furry overcoats. Apart from our family and the other herders from our camp, there was no other human within a two-day ride.

The journey and its risks were routine, and Kesha's 'expert deputy' Boris was simply delayed. But the danger behind the routine was revealed the following winter, when Boris froze to death. He had disregarded the first rule of staying alive: always leave your heavy reindeer-fur coat outside the tent. While collecting firewood at −50°F, he stepped out of his tent wearing just his indoor clothes. In itself this is safe for a minute or so, as you have your body heat as well as the tent and the stove to return to – and a fur coat in reserve. But Boris had left his overcoat inside the tent. In the short moment while he was outside, the stove had somehow set the tent on fire and the coat was destroyed. It was just after the birth of his first child, and though he was only a few miles from his family in the safety of the village, he died before he could reach them. Kesha's careful grooming of his successor was reduced to naught, and Boris's wife and baby joined the numerous widows and fatherless children of men who die on the land.

Violent and premature death touched every family I knew. Among the 186,000 native people across the Russian North comprising over thirty ethnic groups, one-third of all deaths are through accident, murder, or suicide*. In 2000, ten years after making my first documentary film, I watched it again with Tolya and we updated the litany: this one died this way, that one died that way; even the playful toddler from Kostya's summer camp seen strapped into a reindeer howdah, who always draws an affectionate laugh from the audience when he peers into the lens, drowned a year later. But it was mostly men who died. In winter they froze; in spring and summer they slipped through the unevenly thinning ice; in autumn they fell overboard from boats on the region's deep, beautiful lakes, never more than a few degrees above freezing, like Kostya's predecessor as briga-dier of camp 10 who died shooting ducks with two other men when their boat capsized. Men died sober; or they died drunk, when they killed each other intentionally or by mistake, not even knowing the difference as they reached in confusion for the knives at their waist.

'My eldest brother was killed in camp 2 by the brigadier of another camp,' Tolya once told me. 'They drank, they fought, he hit my brother with a stick and the brain damage killed him. The other man got three years.'

'My husband went with two men to a lake to fish and hunt mountain sheep,' Kristina recalled in her throaty voice on another occasion. 'They got drunk, and the others came back alone. They said he'd drowned, but his body was found in shallow water with his mouth stuffed full of grass.'

Ivan the Fence-Builder lost his father when he was a baby, like Boris's child.

'It was June,' he told me, 'and my father left us in the camp to fetch supplies from the village. They looked for him, but just found his horses on the bank. We never knew what happened. My brother also died, the spring after he came back from the Army. He went with some men to fetch wood and they beat him to death. He was a boxer, they had an argument. There was a trial in Sangar, but it wasn't a proper one.' Then, perhaps implying supernatural retribution, he added, 'The killers died later.'

The president of the Village Council was obliged to report

Gosha surveys the late August site of Tal Naldin (page 126). The tents are at the foot of this mountain, by the river. The lower slopes across the valley are stippled with larch forest.

The family of camp 7. Photographs taken in September 1990, when they already knew that their herd would soon be exterminated (page 72).

The Old Man.

Granny preparing a reindeer hide for sewing.

Emmie contemplating the fire.

The younger men of camp 7. Gosha (*top*), Ivan the brigadier (*middle left*) and Yura.

Gosha's wife Lidia. The reindeer she rides in dreams reveal the destinies of the men in her life (page 285).

Camp 7 preparing to migrate in late summer. Ivan lifts one pair of saddle bags, while Yura adjusts another (page 118).

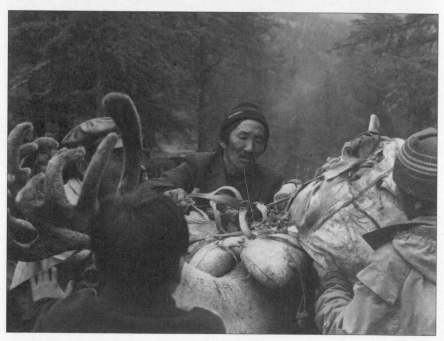

Granny leads the caravan downstream towards the autumn pasture (page 120).

Camp 10 in the perpetual daylight of summer: Kostya the brigadier with riding stick, surrounded by saddles and parked reindeer.

Arkady and his pupil Sergei.

All are fed by Arkady's mother Kristina in her restaurant tent.

Camp 10: Nikitin (*left*) and Tolya in the winter hut. For Tolya, laughter is a way of life; Nikitin is usually more reserved.

Children on their summer holiday in camp 10, while Kostya saddles up for the Communist Party meeting, which was interrupted by his reindeer's mushroom madness (page 138).

December: Vladimir Nikolayevich saws off the antlers of an obnoxious reindeer, which is then also punished by its vengeful sledge partners (page 175).

Our reindeer turn to watch while Manchary catches up with us on the frozen River Tumara. Behind his caravan lie smashed tree trunks marooned after spring floods. The horizontal line of trees further back marks the opposite bank.

Midwinter in the Verkhoyansk Mountains, the coldest place in the Arctic. *Top:* Watching over the herd in the twilight of a December noon. *Bottom:* Vladimir Nikolayevich ten days out from the village. Dependent on reindeer transport and animal furs, he repeats a scene that has hardly changed for 2,000 years.

The village: spring thaw with tracks of human movement (page 183).

Below: Antónov-2 biplane on the village airstrip.

The State Farm office:
a manager. The telephone
leads only to other managers
inside the village. Accounting
was done by abacus until the
end of the 1990s.

Middle and bottom: Men of
power on the steps outside
(page 360). In the lower
picture, Tolya is on the left,
Kesha on the right.

Village women and children crowd around a helicopter.

The woodcutter who worked hard for his medals.

A toy reindeer made from an antler on a hide tape.

The village hall. A reindeer calf prances across the stage to celebrate the abolition of shamanism (page 232). The boys are dressed in mock-traditional costume.

The villagers absorb this complex political and spiritual message.

His wife Lyuda, all-Yakutia reindeer racing champion.

The exceptional family life of camp 8. Kesha in 1996 with his little Diana, perhaps a reincarnated shamaness.

The rotund Dmitri Konstantinovich with a recumbent Tolya.

Patrick bringing a marmot back to camp.

Catherine and Tolya: the herders have got through yet another sack of white death (page 77).

Sally and Catherine reach camp 3 (page 358). Around them are brigadier Valera and the women of his family.

Camp 8 on a carpet of larch needles, September 1992 (page 197). *Middle row:* Kesha, Lyuda and their baby Diana. *Front row, middle:* Kesha's mother, with grandson Dima on her lap. *Left:* Kesha's father Dmitri Konstaninovich, with his own father on the right. The author is in the middle of the back row.

Above: Tolya on his father's knee, next to his mother, around 1950. Note how many women lived in the forest earlier and how few live there today, even in Kesha's close-knit family.

Right: Eveny of northeast Siberia in the everyday clothing of the time, photographed in 1891 by Jochelson, a political exile turned anthropologist. Identities and location unknown, but they bear a striking resemblance to some of my friends today. The centenarians I met around 1990 would have been children when this picture was taken.

each suspicious death and then the diminutive Sakha policeman called 'Little Bird' (in Sakha, Chuuchaakh) was sent across the mountains from Sangar. On aircraft throughout the North, one could encounter policemen and detectives on similar homicide investigations, surrounded by tight-lipped local passengers.

The most likely to have accidents are young men, by taking foolish risks or through inexperience and miscalculation. Late one spring, three youths from Sebyan rode through the high pastures of camp 7 and crossed the highest ridges of the Verkhoyansk Mountains. They descended through the deeply drifted snow of the uninhabited, ungrazed western slopes, normally visited by only the most experienced older hunters, until they reached the maze of sandbanks and deep channels of the River Lena, several miles wide. There, it seems they ventured out to fish on the June ice, a thin skin buckling above the surges of meltwater pumping 1,000 miles downstream from the warmer south along one of the world's biggest rivers. Nobody knows what happened, but a long search later uncovered their bodies washed up against moraines of smashed trees and ice-rubble*.

Young men also suffered from rage at their celibate condition, a rage unleashed by alcohol. A week after his fight in the winter hut, Ganya went to the village. Herders from every camp were gathering in anticipation of the spring reindeer races and were drinking heavily. During an altercation, Ganya was stabbed and wounded so seriously that for the time being we were not certain whether he would live.

But the most terrible deaths of men were not their accidents, nor even their murders, but their suicides. Everyone had suicide stories: Tolya's relative wandered by a lake for three days and then shot himself with a double-barrelled shotgun. Kostya remembered a father of three in camp 12 who was in agony with a swollen stomach. He waited for three days in the vain hope of a helicopter, then shot himself while everyone was asleep. The first time he missed his heart; the second time, as the others were running up to him, he aimed at his forehead.

Two days after returning from a trip to the city, a young man from camp 1, with whom I had once drunk tea, drove a snowscooter to the lakeside, smoked a huge number of cigarettes, ate a whole bag of oranges (a rare and expensive city food which had only just become available), folded his clothes neatly, and hanged himself.

'The authorities usually blame it on alcohol,' Tolya told me, as we were discussing a sequence of four suicides in eighteen months, three of them by bachelors. 'But that's too easy. None of these men was a drinker. These aren't weak people, these are people with strong characters, fighting their circumstances, making a protest! Though drunkenness can be a form of protest, too.'

In this last remark, added almost casually, Tolya painted a terrifying picture of a generation of young men who could see no scope for taking control of their destinies, and for whom drinking and killing themselves were equivalent. It was as if their drunkenness were already a kind of living suicide.

Humans are so few and far between on this land that I have only once been nearby at the moment of a suicide. One night, the electricity generator in the village continued to run after the official cut-off point of ten o'clock. I assumed that the privilege granted to the director and his associates had mistakenly been extended to the rest of us, so I continued writing up my notes while the children of the house used the additional electricity to watch rapists and serial killers from Los Angeles, made newly available through liberalization on the channel from Moscow (where it was only four in the afternoon).

The next morning I learned that the late-night electricity had been to help with the preparation of the body of Motya's brother, who had just hanged himself on his twenty-third birthday while his family were out of the house. Suicides were thought to run in chains, as each victim reached out from the dead to take the next. Motya's brother was the fourth in yet another sequence of suicides. The first boy in the chain had married a girl in another village and shot himself when she gave him up after her previous husband reappeared from jail. The next was also having girl trouble and hanged himself. The third was sent back from the Army for mysterious reasons before he had served his full two years, shot his own *uchakhs* in the taiga and then shot himself. The fourth victim in the chain was today's boy, who had been on bad terms with his wife's family.

I passed the boy's house. Motya came up to the gate, embraced me (a Russian gesture) and led me silently through the other women who were standing in the yard, red-eyed with suppressed weeping. Inside, the boy lay in a coffin lined with cloth striped in the funeral colours of maroon and black, surrounded by strands of blue plastic convolvulus. On a small table by his head lay offerings for his soul: a cup of vodka, a cup of tea with the sugar-spoon still standing in it, a plate with morsels of reindeer meat, cake, and biscuits, and several coins. Behind the cups stood two posed, expressionless photographs of the boy, one taken in childhood and one while in the Army. They revealed little about their subject. The blank face on the body combined the mystery of individual motive with the despair of a generation. Perhaps Motya sensed a danger in the fascination that held my gaze.

'You should go,' she murmured gently.

9

Landscape with Gulag:
brushed by White Man's Madness

Until recently, the Russian public was unaware of the distress of the native peoples living in their own far North. Most people knew only of happy 'children of nature', who danced in exotic costumes and supposedly lived in harmony with their harsh but beautiful environment. As soon as press controls were relaxed in the late 1980s, it emerged that the northern natives suffered from cancer, depression, and a poisoned environment, and that their life expectancy was eighteen years lower* than that of Russians. Journalists and academics started publishing articles with titles like 'Big problems of small peoples' and 'Before it's too late!' Natives, too, began to write about their experiences, and newspapers and magazines were filled with testimonies about the misery and humiliation that they had endured in boarding schools.

An Association of Northern Native Peoples was formed in spring 1990, and I was privileged to sit among the rows of suits and medals* at their inaugural meeting in the Moscow Kremlin. We were addressed by the man who had initiated *perestroika* and ended the Cold War, the Secretary-General of the Communist Party of the Soviet Union, Mikhail Gorbachev himself.

There was no one from Sebyan in the Kremlin, but I saw how close the connections were when I accompanied the Sebyan delegation a week later to the inaugural meeting of the association's regional branch for Yakutia. The meeting was held at the little port of Chersky, where the River Kolymá meets the Arctic Ocean nearly 5,000 miles to the northeast. In the Soviet Union, the Kolymá was the most terrible place name it was possible to utter, as it had been the cruellest part of the entire system of Gulag prison camps. But to local northern natives, and to some Russian settlers, it was home*. For everyone else it was still a closed zone, and even Soviet citizens from outside the region required a special permit to go there. With its modern apartment blocks of cement finished in dingy yellow and pink stucco, Chersky looked like any Soviet frontier town, with streets wider than the buildings were tall. Entire walls were painted with stylized representations of natives riding reindeer and surrounded by the benign but superior Soviet technology of aeroplanes, cranes, and oil rigs. These smiling natives were faced from other walls by huge stencilled Lenin heads and quotations from Lenin's speeches.

The meeting was to be combined with the spring festival of native peoples from the whole of Yakutia. Earlier, each community had run its own festival, but for several years the Government of the Sakha Republic had sponsored a combined festival, held in a different location each year.

'The reindeer have come to town!' one Russian bystander exclaimed to his neighbour, amid the crowd that had gathered to watch the exotica. 'I've lived here eighteen years and I've never seen a live one!'

The town sloped down to the river, an overcast, frozen expanse which stretched a mile or more wide towards the opposite bank hidden by a fog of suspended frost crystals. Far out on the ice, red and white biplanes were parked in rows, their wheels replaced for the winter by skis, some painted with the symbol of the elite Polar Air Service. This was the last time I saw such a richness of planes anywhere in the North. Their passengers had travelled from up to 2,000 miles away, at the opposite end of the Sakha Republic. They had pitched tents on the frozen river, where they boiled kettles of tea and pans of reindeer meat, their collars turned up against the stinging curtains of ice crystals that swept up the open river from the Arctic Ocean just beyond.

To a crackly brass-band recording of 'The Slavic Maiden's Farewell' (Stalin's favourite march), a procession of herders filed past a politicians' podium set up on the ice, holding placards aloft with the names of their districts and wearing their finest coats of brown reindeer fur, inlaid with elaborate designs in white. The patterns were mostly geometric and each one was distinctive to a particular village or woman, but some were experimental and one represented a space rocket, complete with strips of exhaust in white reindeer fur. The Sakha leader of the republic's parliament, who would soon become president as the republic seized further autonomy from Moscow, announced in ringing, epic tones, 'From all the corners of our republic they have come, the Eveny, the Evenki, the Yukaghir, the Chukchi . . .' Reindeer were raced on sledges and in the saddle, while young men wrestled and competed in the long jump over a line of sledges laid side by side. These were the traditional sports by which girls and their vigilant parents would judge a potential bridegroom at the pre-Soviet spring festival.

The meeting of the new local association began the following morning. The front rows were occupied by men in suits, many wearing glasses, probably managers of State Farms and other native officials. At the back, figures were streaming into the hall in the bulky reindeer-fur coats of true herders, the chinstraps of their hats untied and ear-flaps hanging down, stumbling over those already seated. The attendance had overwhelmed the lady with pegs and tickets in the little cloakroom outside.

On the platform stood a middle-aged native man, leaning forward intently over the shiny lectern of synthetic wood which he shared with a gigantic plaster bust of Lenin. He was bareheaded but still wore his outdoor coat of reindeer fur, perhaps as a sign that he was not a bureaucrat.

'First, I'd like to comment on the arrangements for this meeting,' he shouted over the rows of suits so that the people at the back could hear him over the shuffling of their own coats. 'We were told the meeting was scheduled for nine thirty, but we find that it started at nine o'clock. The natives arrived half an hour late, and now we're sitting at the back.'

It seemed the officials did not count as natives, though they could well have been his own relatives. 'So in the end, it's a gathering of bigwigs and there'll be endless talk, talk, talk, with no results,' he continued. 'This is a very special meeting, but it's being run in Russian. If the native representatives had been here from the beginning they would have insisted on running the whole meeting in native languages. This isn't good enough!' Like every subsequent speaker he spoke in Sakha, the lingua franca of the region. Few Eveny, Evenki, Chukchi, or Yukaghir would have understood each other's languages, but at least Sakha was better than Russian.

One speaker after another subjected the audience to an impassioned harangue, reiterating local variants of the themes I had heard in Moscow the previous week from delegates living across the entire Russian North: pollution from mines and from oil, the destruction of reindeer pasture, high mortality, poor health, alcoholism, and the need for education, self-government, and the preservation of native languages.

'I'm an old man,' said one speaker. 'I have five grown-up children and ten grandchildren. Three of my children work in industry, one is a reindeer herder, my youngest daughter is studying at the Hertzen Institute in Leningrad. All my children were educated in Sakha or Russian-medium schools. But it makes me angry that my children don't know the Eveny language. So I collected all the literature in Eveny and I taught my smallest daughter a little bit. Now she's graduating from the Hertzen as a teacher of Eveny.'

'I'm putting together a list of everyone whose ancestors were Yukaghirs,' said another, 'so we can know who has what blood, before it's too late. Over half the population in my region are from minorities, yet in their identity document [*pasport* in Russian] they're written as Sakha. The last person who was officially registered as Yukaghir died last year.' There was an ironic twist to his intonation as he concluded, 'We're living in a Yukaghir area, but the last Yukaghir is dead! Well, that's that, isn't it?'

'Our reindeer herders are walking inside a huge radioactive mine,' declaimed another speaker. 'Uranium mining at the mouth of the Moma River started in 1941, and to this day the mines are open to the air. Now it turns out that 70 per cent of the local Eveny are dying of cancer*. We've been asking this question for a long time, and finally we have our answer: "We'll send you four experts in the summer, and if you don't want to die, you'll have to pay for their tests." I find this hard to understand: if I don't want to die, I've got to sell the trousers off my bottom to give money to these people to pay for their tests? That's how I understand it – but you work it out for yourselves!'

At each point the audience was nodding and murmuring assent in Sakha: '*Söp söp*, Right right!' If the government sponsors had hoped to see a happy folkloristic affirmation that all was well, this year something was going very wrong.

'And I want to talk about land, too,' the speaker continued, echoing Tolya's idea for a redrawing of boundaries. 'We must establish an Eveny administrative territory. Scientists from Magadan and Ust-Nera tramp across our land and won't even allow us to live! Some faceless Kozlov sits in Abyi and issues orders forbidding us to fish in our own lakes, so we can't even catch our own supper. When are things going to get better?'

'*Söp söp!*' nodded the listeners. Even when they agree with an impassioned speaker, a native audience remains restrained. The native faces arrayed in the meeting hall radiated a Soviet aura of medals, badges, and party membership. While some of their number supported the speakers on the platform, others were uneasy. But now they were becoming bolder. When the next speaker enumerated the people responsible for native affairs in the administration, they greeted each name with derisive laughter. Very few of the names were natives.

'It was the great Lenin's successor Stalin who said, "Everything depends on personnel,"' the speaker declared, quoting a well-known saying and glancing ironically for confirmation at the plaster head of Lenin next to him, which was twice the size of his own. The Russian slogan on the wall behind him was from Lenin himself, and read: *Partiya – Um, Chest' i Sovest' Nashey Epokhi*, The party is the mind, honour and conscience of our epoch. Probably nobody present imagined that the seventy-four-year Soviet epoch would end altogether in twenty months. 'Unsuitable people must be dismissed [*snyat'*]!' he thundered. 'The Department of Native Training must have a representative of the northern peoples themselves, because there are tough matters to be decided. We need our own representatives!' His voice rising to a crescendo, he delivered his peroration: 'In the department there should be people who know our language and culture, people whose heart aches about the future. Only people like this can make things work. Let our descendants in two centuries' time be speaking their own native tongue on their own native land!' He stepped down to huge applause.

In the evening there was a concert in a sports hall. Young men in suits and girls in folk costumes danced disco-style to the native singer Eduard Klepechen and his rock band, dressed in stage versions of traditional costumes which combined fur with glitter. To a syncopated throb on the second and fourth beat, so distinctive of Russian pop, he sang in Eveny:

I've got a reindeer, I've got my tundra,
On this land I gallop my reindeer.
Listen, listen to my song, my friend,
My reindeer, my reindeer gallops far away.

Between stanzas, he pranced from side to side of the stage with a scampering movement of his feet, grunting to the beat and driving his *uchakhs* to a run, 'Oi, oi, oi, oi!'

> *My friend, listen, listen to me,*
> *My friend, listen, listen to my song.*
> *Gallop, my reindeer, gallop!*

Shifting to a higher pitch at the top of the tenor range, his voice soared through the hall:

> *O wonderful tundra!*
> *Gallop, gallop, my reindeer!*

He pranced and grunted again through an instrumental ritornello, while a musician in dark glasses plunged his fingers into an electric keyboard. Singing this time in Sakha, he resumed:

> *Who has tundra like this?*
> *Who else gallops his reindeer, but I?* [higher pitch]:
> *Tundra, tundra, wide vast tundra!*
> *Tundra with your swans!*
> *The white cranes and white, white swans*
> *Sing songs of their own.*

During the next instrumental ritornello he twirled on his toes and the fur strips of his coat swirled round him, like the pendants on the costume of a shaman in trance:

> *Who else has such reindeer, who else has such tundra?*
> *None but I!* [higher pitch]:
> *White cranes, white, white swans,*
> *You make my heart leap up!*
> *My reindeer, gallop faster, faster!*

After the festival, the directors of several State Farms, including the director of Sebyan, decided that cost-accounting made such mass meetings of native communities from across the North too expensive to hold every year.

Others suspected them of trying to contain the new mood and to isolate their villages.

The Farm had paid the cost of flying the Sebyan delegation there and back in two Mi-8 helicopters. One helicopter had a corral built inside to hold their four champion racing reindeer. The back cargo door of the helicopter was opened and the reindeer heaved themselves up the step with the help of a push on their bottoms.

I sat minding the animals with a herder called Zinovy, the kind of silent herder who had the taiga soul of a reindeer. Even without the slipstream, the temperature was −30°F, but he latched back the portholes to give his fellow animals fresh air. His calmness seeped into the reindeer, which stood for 1,500 miles without shuffling or complaining.

The journey lasted eleven hours, over a landscape resonant with the evocative names of locations that were still closed to foreigners but would later become part of my personal geography. To avoid a patch of bad weather to the south, we followed a northerly route, overflying Zyryanka, where I would later send my Danish student Rane to study Yukaghir hunters, and Belaya Gora, where I would one day stay with the psychic daughter of a shaman. The weather forced us down at Andryushkino, but just before landing, the pilot suddenly rose again without explanation and went on to Chokurdakh, where he refuelled while I gazed at the blizzard-swept wooden apartment blocks as if at a forbidden planet. Without a permit to be there, it would have been dangerous to set foot outside the airport. We finally cut south to refuel again at Batagay, near the old Gulag camp of Bear Mountain, and on to Sebyan.

Zinovy would eventually die of liver cancer. I saw his slow, weak form in 1999, when he had moved with his wife and granddaughter to join Ivan in camp 7 for the last autumn of his life. He went the way of the people from the radioactive Moma River, of Tolya's mother and stepfather, and of many younger people whose bones and tissues were formed in the early decades of the Cold War and who have since died in their thirties and forties, like the yellow, skeletal man I flew home to die, his daughter fanning him tenderly throughout the flight and hiding her tears behind dark glasses.

Tolya, too, had a brush with death by radiation as a child. One day in the 1950s the sky went red all day long, as if it were on fire, and a strange grey snow fell on the ground. A group of children played with this 'snow', which did not melt. Later, they became burned where it had touched their bodies. The other children died, but Tolya, with his extraordinary life-energy, lived and still has the marks today. This must have been fallout from Soviet atomic bomb testing, but this was not known until the late 1980s.

After Tolya's aged mother died of liver cancer, he told me a dream.

'I was in the city,' he said, 'and in my dream I saw the whole sky on fire above the village of Sebyan. The sky just exploded. Then in the morning, I received a telegram saying she'd died.'

Reflecting later on Tolya's dream, I saw a connection with the red sky in his childhood. When his mother died, he felt that he had lost one of his last links with the old people who lived by the 'true' Eveny culture. He has devoted his new career as an anthropologist to salvaging the memory of this culture, and his book, *The forgotten world of the ancestors,* is dedicated to his mother.

The radiation afflicting Siberian natives was a modern, airborne descendant of the nineteenth-century smallpox introduced by itinerant Russian traders. During epidemics, people would hide in the forest, sometimes seeing no other human for years at a time. The disease was understood as the after-effect of a red-haired Russian woman who sat on a sledge at the back of a visitor's caravan. The only person who could actually see her was a shaman, who would then prepare for a life-or-death battle. If the shaman lost, or if the camp had no shaman, nearly everyone would die. It was said that the smallpox spirit would allow two people to recover in order to bury the dead.

Beyond the overt, conscious violence of the gun, such mass deaths by invisible, unconscious means may be an integral cost of the spread of empire. Siberian reindeer peoples paid for the Cold War twice over, once in areas where uranium was mined and the earth ripped open to spill the tailings, and again in every single community when the bombs were tested from the 1950s to the 1970s above the remote Arctic Ocean islands of Novaya Zemlya, especially on days when the wind would carry the fallout to the east, away from the monitoring stations of Western Europe and over the reindeer pastures of Siberia. Lichens took their nutrients from the air rather than from the soil, and absorbed many times more radioactive isotopes than did green plants. The reindeer ate huge quantities of lichen in the winter, and the radiation was concentrated still further in the bodies of the human population who ate almost nothing but reindeer. The cumulative effect must be many, many times worse than that of Chernobyl in 1986, a single accident that was exposed because the winds were blowing towards monitoring stations in Sweden.

The bombs and the uranium tailings were made possible through the institution of the Gulag. The deportation of millions to these labour camps, which for many also became death camps, began in the late 1920s and continued until some time after Stalin's own death in 1953. The Gulag system was not simply a product of a hysterical mass paranoia in the central cities of Russia, where countless citizens were betrayed to the secret police by their neighbours and workmates. It was also essential to the national economy. Labourers had to be found to extract the minerals that sustained the country's prodigious industrialization and war effort: tin, aluminium, gold, lead, nickel, and uranium. Whereas the sugar plantations of the tropics enslaved other races from other continents, the mines of the isolated Soviet Union could find no one to exploit but their own people. I once went with my family, my Scottish student Seona, a Sakha botanist, and two native psychic healers to the out-skirts of an abandoned uranium mine called Bol'shoy Nimnir, the Great Place of Death. One of the healers had brought a Geiger counter (as well as a revolver against bandits). Throughout the two-day drive there, her equipment registered between 5 and 10 units, but here, still some distance from the mine itself, it oscillated between 2,000 and 3,000. The botanist defiantly ate a cranberry, but we did not linger. Prisoners sent inside this mine could have survived for only a few weeks.

In St Petersburg (then still called Leningrad), I had met rela-tives of Russians who had been taken away in the night, con-demned, and shot or exhausted to death without trace in Siberia. In the mid-1980s it was becoming safe for the first time to pull out surviving mementos of these shameful, long-lost parents and grandparents. I would be shown a photograph fresh from forty or fifty years in storage, or a very recent rehabilitation certificate which said in deadpan official language, without so much as an apology, that these deaths had happened 'incorrectly'.

These other Leningraders did not have the specialist knowl-edge of my anthropologist colleagues who had studied Siberian peoples. For them, Siberia was a black hole. If they had heard anything about the natives, they thought of them as merciless trackers and terrifying accomplices of the Gulag system. Later, as I placed my life each day in the hands of native friends and experienced how they looked out for my safety and comfort, fed me and entertained me, I remembered my old friends in St Petersburg whose relatives had disappeared in Stalinist times. Now I was living with the inhabitants of that black hole, people who might have seen them and talked to them. How had they felt about this? What might they have done, or been made to do?

One day, Tolya stopped me in a level boggy valley on the territory of camp 10.

'This is the Stalin Trail,' he said.

A faint track was just visible in the vegetation. This track had been opened up in 1938 or 1939, and was still used by trucks when the bog froze over in winter. From Yakutsk the prisoners were made to walk for several hundred miles to Bear Mountain near Verkhoyansk, which held 10,000 prisoners by the time the Soviet Union entered the war in 1941.

'Often they died on the way and were left by the guards for our people to bury,' Tolya continued. 'Our parents had strict instructions. If anyone escaped they had to hunt them down, shoot them and present their right arm to the KGB. Our men came across four young men who had run away. One of them begged for mercy. But what could they do? Of course they shot him and cut off his hand. My parents used to scare us with escaped political prisoners when we were little: "Be good, or the *zeki* will get you!"'

Who was scared of whom? This was a strange kind of killing. If you murder someone of your own volition, you might feel guilt, relief, or triumph according to your motive. But what do you feel when you kill a whimpering stranger under orders you do not understand?

This was a new kind of White Man's Madness, a psychotic, industrialized perversion of an older tradition of exiling criminals and political dissidents which was already well established in the nineteenth century, when exile was not designed to kill and the natives could afford to be kinder. In the earlier kind of exile under the tsars, the punishment was that you knew nobody and had to make a new life from nothing. Before aviation, before the Trans-Siberian railway was completed in the far south in the 1890s, it could take two years of travelling to reach one's place of sentence. Some exiles abandoned themselves to despair, some did make a new life and never returned to their families, while still others devoted themselves to studying the local people and wrote classic works of anthropology.

I had not realized how much this system of relatively benign exile had resumed after Stalin, or that one could still fail to return home even now, when any place in the country was only a few flights away from any other place. One winter I was sledging across a lake with Kostya. We were heading for a grubby, tatty tent on the opposite bank, not at all like the usual well-kept Eveny tent. Lakes froze more smoothly than rivers, so we were moving fast and the slipstream was whipping tiny crystals of ice into my face. Yet even through my silk scarf and goggles I thought I saw next to the tent a mound of deep-frozen dead dogs. We pulled up our reindeer next to the tent and a small unshaven Russian came out to meet us. Passing the dogs without comment, we crowded inside for the tea he offered us.

'Go on, ask him,' prompted Kostya. 'You'll find it interesting. I brought you this way specially.' And so the Russian told me his story.

During the 1960s he had been posted at the Berlin Wall, in the office that issued permits to cross to the West. He had acquired an East German girlfriend, who prevailed upon him to issue unauthorized passes to her relatives. And then the KGB, who were watching all along, pounced on him and he was sent into exile in Siberia, to our village.

His exile had long since lapsed.

'Do you want to leave?' I asked.

'I've got a daughter just the other side of Novosibirsk. I haven't seen her for thirty years. I'm sure she'd think I'd be a nuisance if I went to live with her. But anyway' – he laughed – 'every time I save up enough money to get there I drink it away before I reach the airport.'

I was wondering how to ask tactfully about the mound of dogs, but it turned out there was no need. Our host noticed my eyes straying.

'I have a licence from the Village Council to shoot strays,' he explained.

'What do you do with them?'

'Well, I used to eat them,' he answered. 'But then a native spirit with a dog's head started to haunt me. So now I just make them into gloves and sell them in the village.'

Then our tea break was over, and it was time to move on. I met the Russian several times in the village over the next few years. He died without ever seeing his daughter again.

Before helicopters, before even the Antónov-2 biplane, aviation was so limited that the Gulag camps in this region could not have functioned without using natives to bring in supplies on caravans of reindeer. Remembering that Vladimir Nikolayevich had worked at Bear Mountain as a child during the war, I asked him about the trail that came past Sebyan.

'They were driven along this trail on foot like a herd of animals,' he recalled. 'They were marched in groups of fifty with soldiers in front and behind. I don't know what they had to eat, and I don't know about their clothing, but they stayed overnight in superb big tents, much better than ours. They were all sorts: Sakha, very few Eveny, mostly Russians. Some of them looked like cultivated people, probably artists and intellectuals. You can't imagine how many people died on the way, but I saw it for myself. There were skulls lying along the track.'

I was trying to understand the meeting of two worlds. 'Could your people talk with the prisoners?' I asked.

'We couldn't talk to them when they were on the road, only when they were working. Prisoners didn't work only in the mines. They also worked and lived with us driving reindeer sledges, they worked as reindeer herders. The director of the local Collective Farm was a shrewd man, there was a shortage and he would take anybody who was capable, even Sakha and Russians.'

I had not realized there had been this sort of involvement. I also gradually discovered that there had been prisoners from Eveny families I knew. Tolya's mother's brother was one, imprisoned and forced to work as a tin miner at Bear Mountain. He survived and although this was illegal, for the rest of his life he avoided the State Farm system altogether by hunting in the taiga alone with his *uchakhs*. ('I must take after him!' Tolya declared during one of his many clashes with the authorities.)

The master-programme of collectivization was ruthless and the socialist ideal of benevolent sharing was tempered by the 'repression' of anyone who could not be made to fit the abstract ideal of the perfectibility of humanity. During collectivization, natives with a significant number of reindeer might be branded as *kulak*, 'rich peasant', and targeted for *razkulachestvo*, 'de-rich-peasantization'. Though some survived with mere confiscation of any property that they had not handed over, 'repression' could mean anything from imprisonment to execution. Others were attacked because rivals used the climate of victimization to settle old scores. An anonymous letter of denunciation (*donos*) could be enough to cause a disliked neighbour to disappear forever.

Especially dangerous was any contact with the wider world outside the Soviet Union. The grandfather of old Nikitin who knew every reindeer in camp 10 and whose little grandson Sergei was already dedicated to reindeer herding, made an extraordinary journey to Finland in the early 1920s as an inform-ant for a linguist who was compiling a dictionary of the Eveny language.When the political climate hardened in the late 1920s, he was arrested because he had been abroad, and died in prison. The book, a monumental description of the dialect of Sakkyryr*, the village built to command the region where the Nikitins tra-ditionally migrated, took fifty years and a second editor before it was ready for publication as a learned tome, with commentary in German. Copies are shelved in scholarly libraries around the world (though it is very specialized and no one in Cambridge had ever signed it out before me). Nikitin's ancestor was killed for telling someone the Eveny words for the objects and processes that make it possible to stay alive on the landscape: the nouns for the kinds of reindeer, the verbs for handling them, the adverbs that explain how these actions are performed: quickly, slowly, tightly, loosely, like this, like that. Probably neither of those scholars in Finland knew what had happened to the man who was the source of the material that had become their life's work.

As I write, the vocabulary of denunciation has softened, and demands to 'repress' and 'liquidate' people (*repressirovat'* and *likvidirovat'*) have been replaced by demands to 'dismiss' or 'sack' (*snyat'*) them. But such stories made me wonder how safe it might be, even in these liberal reformist times, for reindeer herders to be associated with me. Though their managers have sometimes been wary or hostile, I have never felt any hesitation among the herders themselves. Often they have urged me to pass on to the world what they made it possible for me to know.

'Thanks for coming so far and taking an interest in us in our bear-ridden corner. At least the world will know we exist!' as Ivan once put it. *Medvezhiy ugol*, 'bear-ridden corner', is a Russian idiom for a remote backwater.

Tolya's response was typically robust: 'After *perestroika, perestrelka* [a grim pun, meaning, "After liberalization, an exchange of gunfire"]. But at least we'll have struggled!'

Others pointed out that they did not need foreign contacts to put themselves in danger: 'We can get arrested all by ourselves!' The disappearance of their own headman Baibalchan in 1928 had never been forgotten: 'No foreign linguist or anthropologist that time to serve as an excuse!'

When the local KGB archive was opened in the early 1990s, Tolya planned to find out what happened to Baibalchan, one of the great mysteries in the history of Sebyan. He might have had the same experience as my Sakha PhD student Tanya, who was trying to find out her own family history. Her file in the archives was being declassified at the rate of one page a day, so that she had to break off in mid-sentence and come back each morning to read the next brief instalment. But Tolya's experience was not even that easy: 'Before I could get round to asking for Baibalchan's file,' he said, 'they'd closed the archive again.'

However, one person – Bagdaryn Sulbe, a Sakha historian who had devoted his life to the study of place names – did get into the archives at the right moment to find out what happened. He agreed to see me in Yakutsk. His flat contained the same black plastic sofas, mock-oriental wall rugs, and glass-fronted display cabinets as any other flat. But in addition, it was stacked from floor to ceiling with boxes of handwritten card-indexes exploring the etymologies and histories of every location in Sakha (Yakutia). Huge swathes of the region might see perhaps only one person a year, if that. Yet millennia of nomadic movement, followed by 350 years of Russian cartography, meant that every stream, every slope, every ridge, had been named. Behind each name, the historian had deciphered the description of a natural feature, a chief's marriage, the supernatural feat of a shaman, the migration of a clan, or a change of language as the horse-herding Sakha ousted the reindeer-herding Eveny from the lowlands. The old historian's mission was to ensure that no toponymic trace of significant human activity should ever fade.

'I looked into Keimetinov's case [Baibalchan's official Russian-style name was Pavel Vasilevich Keimetinov] because his daughter gave me power of attorney and asked me to look in the KGB archives,' he began. 'He had taken a prominent part in an early anti-Soviet rebellion at the beginning of the 1920s. After the Reds won the civil war and consolidated Communist rule, they granted him an amnesty. He was free: "Work, live, go about your business – don't mention it, you're welcome!"' He modulated his voice sweetly to match these everyday Russian words of courtesy. 'So he began to work for his people. He turned his house into a school, and when that wasn't big enough he built a special schoolhouse in 1928. You live in those mountains, you know how difficult it is to get supplies up there – well, in the Twenties he organized a cooperative as well.'

The historian's daughter brought us tea and chocolates on a tray. As the door opened in front of her, it released the sound of children's cartoons from the television in the next room. He poured us tea, with milk in the Sakha style, and continued.

'In 1928 the secret police [*cheka*] came to Sebyan, arrested Baibalchan again and took away the goods of the cooperative. "Case number 47R," the old historian intoned in an official voice, "Citizen Keimetinov, Pavel Vasilevich, interrogated 13 April 1929. Charged with taking part in banditry and robbery, and appropriation of state property." Again, he was released and given an amnesty. But then an odd thing happened.'

The historian's tone was precise and even. He must have documented many similar cases. His narrative was moving inexorably towards the arbitrariness of so many arrest stories, into the vortex of personal whims, quirks of procedure, who was on duty that night, whether a message got sent, received, and acted upon, the multiple little turns that determined whether a person walked free or was sent to their death.

'There was a special meeting of the secret police in Novosibirsk that October. The meeting issued a decree to release this Keimetinov from custody. The only condition was that he would be banned for three years from living in Moscow, Leningrad, the Far East, Yakutia, Kiev, Rostov-on-Don . . .' He named several other provincial cities in the Russian heartland where in fact no Eveny would want to live.

'And now here comes the odd twist. When this instruction reached Yakutsk, the local boss sent back a telegram saying, "URGENT MESSAGE STOP CONSIDER NOT POSSIBLE RELEASE KEIMETINOV SINCE BANDIT AND RECIDIVIST STOP REQUEST RECONSIDER CASE STOP.' So on 13 January 1930 they did reconsider the matter in Novosibirsk. They revoked their previous decision and decided to imprison Keimetinov in a Siberian concentration camp for three years. He was sent to somewhere near Lake Baikal.

'The inference is very simple,' concluded the old historian. 'The local bosses wanted to cut off the brains of the minority peoples to keep them ignorant. Even among your tiny tribe living in the wilds, one man had built a school and established a cooperative. That meant he wanted to improve the lives of his people and introduce them to civilization. So they nailed him on the pretext that he had once upon a time been against the new Government. All they wanted to do was capture him and destroy him, so that his people should stay uneducated like cattle.'

'So what happened after that?' I asked.

'That's all. There are no more documents in the file. Obviously he perished there.'

Here was the confirmation of what people in Sebyan had suspected.

'What did Baibalchan's file look like?' I suddenly thought to ask.

'It was a tiny, thin little file – just nineteen pages.'

I did a foolish and pointless calculation in my head: each page in Baibalchan's file was eleven times more deadly than one page in Nikitin's dictionary, which is 212 pages long. One researcher who has studied the KGB archives* told me that the supposed 'verbatim' transcripts of many hours of interrogation would often have taken only a few minutes to read aloud. What might really have taken place on 13 April 1929? In the striving to ensure its inevitable outcome, everything distinctive and revealing about the case was withheld even while the file was being written. So when the memory was recovered it would be rendered empty, like the displaced tribes and forgotten heroes whose souls lingered wanly in the fading Biro of the old historian's card-indexes. Baibalchan's daughter surely understood all this when she commissioned this search. Had it brought her comfort to hear some more details, and to confirm that the ultimate facts – how her father had died, whether he had been thinking about his children – could never be known? Would she bother to press for a certificate like the ones I had seen in Leningrad, stating that her father had been killed 'incorrectly'?

10

Killing the shaman
and internalizing betrayal

Enemies of the people could be anywhere. In the 1920s and
1930s previous kinds of headmen or community leaders were
tracked down in every village of Russia, and either brought
under control or destroyed. The Patriarch and senior bishops
of the Russian Othodox Church were drawn into a process of
conflict, compromise, and neutralization.

But in the remote taiga and tundra of Siberia, there was
another figure, baffling, idiosyncratic, and not answerable to any
synod. With their trances and soul-flights, shamans were an
irritating challenge to Communist missionaries of rational science who denied any other understanding of the world apart
from their own. Whereas native people regarded shamans as
channels to an ultimate reality made manifest through the
actions of spirits, Soviet doctrine downgraded them to cynical,
exploitative conjurers, or psychologically deranged individuals.
Conspicuous by their elaborate robes and the sound of drumming far into the night, shamans were picked off one by one.

Shorn of the intensity of communion with spirits, real shamans were replaced by characters in folkloristic cabaret. Their
ritual dances offered a rich source of material that could be
woven very tellingly into the new propaganda. In theatres across
Siberia, shallow re-enactments of shamanic trances were staged
as a special set piece in dramas. There were professional actors

who specialized in this, and one of them told me how he some-times felt dangerously on the edge of a real trance, which he would not have been competent to handle. The role of the sha-man in any plot was to act mad and wicked, impede Soviet reform, and then be defeated. The stage dramas offered a less vicious reflection of the imprisonment and shooting of real shamans in the world beyond the theatre, in which they might be summoned to heal on a hoax call and then treacherously shot, or dropped out of an aircraft and challenged to fly*.

In remote villages like Sebyan, the sanitized shaman's per-formance survived as an episode in a school play. In the spring of 1990 my documentary crew took in a children's dance perform-ance in the village hall. The culminating scene was set in the middle of a herd of reindeer, played by little girls in white leggings and tutus, their palms upraised or heads crowned with antlers of aluminium foil. In their midst, an evil shaman struggled against a noble young Communist hero, both played by older boys. The outcome was inevitable. The shaman was flung to the ground and the hero planted his fighting stick into his recumbent body. To me, this resembled the pose of a Euro-pean hunter dominating his prey, though I suspected the anal-ogy did not occur to the audience of Eveny, who would never triumph over the body of an animal.

The performance was rounded off with a celebratory *hedje*, the joyous ancient circle dance to the sun. In the panorama of northern peoples at folklore festivals, the *hedje* had become conventional shorthand for 'the Eveny'. This was all that sur-vived of the old midsummer ritual, and it was known and danced by every Eveny schoolchild even if they had never rid-den on a reindeer, far less flown on one. Just like the ancient stones in Mongolia and the tattooed bodies in the graves at Pazyryk, this dance used the reindeer as an idea, but the idea that it promoted denied the animal's spiritual significance. From being a vehicle for the soul, the reindeer on stage had been turned into a cute decorative background for a political tableau.

Underneath this trivialization of the reindeer there was a deeper message about these animals' own collusion in this transformation. In the great Soviet project of transforming the consciousness of the Siberian native peoples, the overthrow of the shaman and the devaluation of the reindeer reinforced each other. As they pranced from foot to foot, these sweet little girl-calves seemed to be celebrating the advent of the State Farm and their own conversion into an economic commodity.

Who was that shaman who lay inert and defeated under their pretty hooves before getting up and slipping off the stage? At the time I took the scene to be generic, standing for the repudiation of an entire way of understanding the world. And surely it was this, too, since similar scenes were performed in native communities all over the North. But it was a long time before I realized that this shaman represented the parents or grandparents of many of the people in the audience who were sitting in silent, inscrutable rows. He represented the executed father of Granny in camp 7, who was bullied at school for being the daughter of an enemy of the people; he represented Kesha's grandmother; he represented the grandfather of a young friend of mine called Yan. Shortly before he died, Yan's grandfather had told me how he used to fly up to the sky in the form of a falcon and was later punished with exile for magically healing reindeer hoof-rot in competition with the official vet. Even the teacher who had choreographed the dance was the granddaughter of a shamaness who had been punished.

This performance was included in the documentary, which was later broadcast many times on Russian television. When video players first reached the village a year or two afterwards I also distributed individual copies. The portrayal of reindeer herders as shrewd and politically engaged rather than quaint and picturesque has been influential throughout the region; while the setting of the dance within the film's narrative of cultural destruction has made it impossible ever to perform this tableau in the village again.

As well as the descendants of shamans, the audience also contained the descendants of collaborators who had betrayed the shamans, or at least averted their eyes as they were being persecuted. The new regime had brought a high level of violence towards anyone who failed to conform to the sudden and alien norms, and offered high rewards to those who were seen to make the transition quickly and openly. How could one have any career, even in a remote settlement, without mimicking these values to the point where one ultimately internalized them psychologically? The proportion of party members in Sebyan was similar to that in the rest of the country, perhaps 10 per cent of the adult population. The director and his close associates like Efimov; his opponents like Tolya; leading brigadiers like Kostya in camp 10, Kesha's father Dmitri Konstantinovich in camp 8, Ivan's father the Old Man in camp 7, Sofia Kirilovna in camp 5, Vladimir Nikolayevich who had been brigadier of various camps: all had been honoured with an invitation to join the party, and all had accepted. The elderly Vasily Pavlovich, son of the Baibalchan who had been arrested and disappeared, wrote to a local newspaper in 1988 to support Tolya's reformist proposal for the redrawing of district boundaries, and bolstered the weight of his signature with the words, 'Member of the Communist Party since 1961'.

Communist Party members formed an elite like the Greek and Trojan kings of Homer's Bronze Age*, fighting among themselves. All were enmeshed in the same project, and all were its implementers as well as its potential victims. Some had seen the party murder their close relatives, and even as loyal members they could never seem to atone fully for these trumped-up crimes. It was sixty years since Vasily Pavlovich's father had disappeared, but one woman in the village dismissed him with the words, 'Who is he to write to the papers? The son of a bandit!' The party elite lived in a state of moral tension in which it was hard to distinguish the State from its indigenous collaborators, to separate coercion from collusion. Betrayal had become internalized within people's lives to the point where it could no longer even be recognized.

The symbols that sustained this collusion had become so much part of people's identity that the new atmosphere of open cynicism sometimes seemed like a painful dismemberment of the self. Here, irony was perhaps the best defence. In 1995 Tolya took me to visit a retired woodcutter in his very neat log cabin on the other side of the village. I had spent years studying reindeer production statistics and the minutiae of how herders' brigades formed and re-formed each season. But I had hardly explored the workings of the ancillary services.

But this visit was also to clarify my understanding of the role of certificates and medals, which continue with an uneasy mock-seriousness to this day (I saw them being handed out in the village hall as late as 2001). The cult of productivity extended from reindeer meat to every support activity, from the truck that brought water from the lake (or a hewn block of ice in winter) to each house when the occupants raised a red flag by their barrel to request a delivery, to the woodcutters who stripped the mature larches from the surrounding hills for 30 miles around to satisfy the village's hunger for timber and firewood.

The woodcutter opened a bottle of the Portuguese champagne that had reached the village shop that year by a freak of the market economy, and showed us his collection of medals and citations. One medal revealed him to have been the all-Yakutia long-distance ski champion back in 1961. But he took greatest pride in his feats of production, and showed us a solemn photo of himself wearing an ill-fitting suit and tie, the final medal of his career on his lapel. The woodcutter had received numerous certificates in Soviet times: on May Day, on the annual Day of the Reindeer Herder, on the anniversary of the founding of the Soviet Union. One said, 'For firm order and discipline' and bore the slogan, 'Strict accounting and auditing of every working minute!'. He had many citations for overfulfilling his work plan: 'For cutting 450 cubic metres of firewood against a plan of 268 cubic metres'. I told him about Henry Ford, who, when a worker overfulfilled his plan, did not congratulate him but threatened to fire the manager who had set his goal so low.

'It wasn't like that here,' he said firmly. 'We respected our workers and they worked harder, also out of respect. When you got one of these citations you didn't sleep, you worked all night!'

Medals lifted the ordinary military tone of the word 'brigade' to new levels, rendering high productivity into military heroism by likening prize-winning 'shock-workers' (*udarniki*) to shock-troopers. Work medals were made of soft, bendy imitation brass, but people still wore them on a military holiday like Victory Day.

'Beads for the natives,' said Tolya, in a virtuoso development of his scathing extended metaphor for white colonialism.

Medals also endowed labour with a sanctity by linking it to the legitimacy of the State. Many depicted Lenin, the founding ancestor of the Soviet Union. The woodcutter's collection of medals and citations had some new versions of Lenin that I had not seen before. As with icons of Christ and the saints, every new version was still traditional and therefore valid. The icons of Lenin merged the sacred with images of political power. Whereas Orthodox saints were portrayed full-face, Lenin was often also shown in profile, as kings and emperors had been on coins for thousands of years.

These medals and certificates were indeed a kind of currency. Like coins and banknotes, they cost almost nothing to create physically, but their face value could extend infinitely upward. They had to be carefully rationed to avoid devaluation or inflation if they were to continue to set the blood racing. This rationing allowed them to be used for political patronage, but after the woodcutter's rebuttal of Henry Ford, I understood how they worked by manipulating those people who would respond to them. The medals and certificates were designed not simply to be displayed (the woodcutter kept his in a folder, deep-frozen in the entrance lobby of his hut where one took off one's coat, hat, and boots), but to be internalized psychologically.

'Respected Dmitri Vasilevich!' began one Address of Salutation to the woodcutter from the Soviet period, which would have been read aloud in the village hall before being presented to him as he mounted the steps to the platform amid applause. 'The Directorate of the State Farm, the Communist Party Representative, and the Village Council warmly congratulate you on your 55th birthday!*' Like the others, this certificate was signed by the heads of all three bodies, emphasizing the single-mindedness of the village's triple authority structure.

'Many worked hard all their lives and got no citations,' Tolya murmured to me, quietly so as not to offend the woodcutter. Yes, I realized afterwards, they were not susceptible: it would waste a scarce resource to award one to someone who would not be fully captivated. The director of the Farm was susceptible himself, of course. Behind his desk, I saw a glass cupboard full of very grand citations with red trimmings from his own superiors, who put him at the receiving end of the same trick that he used on his workers.

'But look! You forgot to sign this one,' I said. The woodcutter was holding out a certificate signed in 1982 by the director of the State Farm and the representative of the Communist Party. Tolya's name was typed in as president of the Village Council, but there was no signature.

'Why don't you sign it now?' I suggested.

Tolya took the pen from my notebook and signed with a flourish, adding the current day's date, '28.03.95'.

'Shouldn't you backdate it, to match the other signatures?' I wondered. They both shrugged their shoulders, and then, at the same time, they both laughed. The scholar in me was more of a stickler for accuracy than these former party believers*. The old woodcutter had spent the past hour lamenting the passing of the regime of discipline and hard work which gave these papers their carefully measured meaning. Yet even he could not help enjoying the irony of the document they had just created.

*

After the Communist Party was dissolved at the end of 1991, the class distinction between party and non-party members suddenly vanished. All the players found themselves with stronger or weaker social capital and carried this forward to the new struggles and moral ambiguities of the deregulated market. As the infrastructure of provisions, transport, and the payment of wages broke down through the 1990s, the misery which this caused to some turned into an economic opportunity for others. Even if no fuel arrived for the boiler-houses, no flour or tea for the village shop, no mercy flights for patients dying of cancer induced by radiation, still the helicopters, biplanes, or winter trucks brought in the vodka.

Gorbachev had introduced nationwide prohibition in the mid-1980s, saving many lives, but this policy had collapsed before 1990. As the 1990s progressed, the village came to resemble a horror movie in which people succumbed one by one to a zombie plague*. Nice people, people who had been my friends for years, people who cared about their families, became unrecognizable. Each time I returned, I found that someone I thought would hold out forever had gone under. I would listen to the new tally, just as I listened to the list of people who had died. Who would be the next, who was still left unclaimed? It was said that smallpox had always left some people alive to bury the dead: would alcohol be so gracious?

When I visited a Native American clinic in New Mexico, I was told that the pun in the English word 'spirit' represented a reality and that alcohol was indeed an evil spirit or demon with its own greedy desires, and must be fought on these terms ('Give our greetings to the Siberian Indians!' was the Native American therapist's parting remark). The pun would not work in any of the languages used in Siberia; and the Eveny understood the destructiveness of vodka in a more ambiguous way, as the dark side of a countervailing life-giving power which they acknowledged each time they fed it to the domestic fire.

The impact was worst among young men, though it was also

increasing among women: one year the Village Council made a list of eighty-two women found drunk on the street. Vodka in the village cost several times the city price during the day and a further two or three times that as the evening wore on. The actual figure varied with the state of the rouble. In 1995 a half-litre bottle cost 25,000 roubles in the city, 100,000 roubles during the day in the village and 200,000–300,000 at night (in 2000, after the currency reform of 1998, these figures were 25, 100, and 200–300 roubles respectively). At the same time, diesel fuel used to operate useful machinery cost 1,900 roubles a litre, so that one could choose between buying 1 litre of vodka and up to 200 litres of fuel.

Reindeer meat, the only tradeable product of all herders' activities, which was also the justification for the helicopter flights and the provisioning, the reason for the village to exist and for a council to run it, fetched 5,000–6,000 roubles a kilo in the outside world. A litre of vodka was worth 20 kilos of meat, or about a quarter of a whole reindeer raised with all its herders' skill, endurance, and thousands of miles of migration. For a few truckloads of this distilled fluid, one could buy out the entire reindeer industry of Russia, half a million years of symbiosis. Like the bomblets of biological warfare, these bottles could destroy indigenous humanity on one-tenth of the earth's surface.

In destroying humans, this demon also thrived on human help. Aviators brought in vodka as a perk, creating a sudden glut in exchange for fresh meat which was worth so much more than money back in the city, but they did not fully understand the consequences of what they did. I brought in small amounts of it and released it gradually over days and weeks, torn between a moral unease and the expectations of my hosts. But the members of the State Farm management and their relatives who ran the bootlegging trade understood the consequences of their actions right down to the dregs.

Some people tried to outwit the demon. Afonya, the gentle man who was Tolya's successor as head of the village's civil administration, wanted to see a shop controlled by the Village Council selling vodka in rationed amounts at normal prices. The money that was now made by speculators and siphoned off to the city would then have been recirculated inside the village. But any such ideals were subverted by the demon's human allies, sometimes using demonic cunning. Since the early 1990s, reindeer herders throughout the country had been paid their wages either late by months or years, or only in part, or not at all. In the mid-1990s, in response to this, the Sakha Government introduced a system of 'arctic cheques' or 'nomad cheques'. These were vouchers which reindeer herders could exchange in shops for provisions. When the first herders, or their wives, tried to redeem them, after flying to a town at the risk of being marooned there indefinitely while they waited for a return flight, they found that shops would honour the cheques for no more than half their face value. Many herding families gave up and sold off their cheques for whatever cash they could get from an entrepreneur.

But the cruellest part was yet to come. The cheques had been an attempt by the Government to fill in for the State Farm managers' inability to pay their herders. In these remote mountains, beyond the oversight of higher authorities, those same managers then bought up the cheques and converted them into vodka. The herders were now paying inflated prices for alcohol which had been bought in the first place with their own hopelessly discounted cheques.

As the depressed and cash-starved villagers drank, and the stabbings, drownings and suicides continued, I passed through various stages of anger at the Farm elite who did this to their own relatives. Yet in the end, I could not help pitying them, too, gripped as they were by a compulsion to exploit which was as great as the compulsion to drink. They themselves seemed pathetic, men with no future like their victims, zombie leaders touched by the same plague from which they profited to no purpose.

I first came to clarify my own feelings about this during a long private conversation with Afonya on a quite different topic. I had been asking him about taiga manners and ethics, trying to clarify what seemed to me a contradiction between people's acute awareness of each other's moods and needs, and their extreme concern, even anxiety, not to trespass on each other's personal boundaries. In his calm, low voice, Afonya was confirming everything that Lidia had been telling me in her more impassioned lectures. At one point, he became abstracted and I felt him hesitate.

'I hope you won't be offended if I say I don't always like the way your people treat each other,' he said at last.

I was so surprised that I could not think of anything to say.

Afonya filled in the silence. 'I noticed it when you brought that film crew, and we were all driving together. When Lindsay fell off his sledge, you and Graham didn't rush to help him up.' I was being reprimanded in the same way that Tolya had scolded me for not helping Vladimir Nikolayevich on the ice sheet. But I had thought that was because Vladimir Nikolayevich was old.

Afonya had been thinking about this for five years since the incident occurred. These hunters who could track an animal for days, unseen and unsmelled, had also been watching us – and making judgements. For them, intervening was not always trespassing, as I had misunderstood, and they saw us as neglecting our companions in need. I thought of justifications: Lindsay hadn't seemed hurt anyway . . . if he had been hurt, of course I would have done something . . . I wouldn't have been of much practical use anyway, compared to your people . . . But I said nothing.

Suddenly Afonya asked a question, one he had perhaps never been able to ask anyone before: 'What don't you like about the way *we* treat each other?'

Almost without thinking, I answered, 'It's so hard to stay alive here, and yet your people are constantly plotting against each other and damaging each other's lives.'

Now it was Afonya who could find nothing to say. But I thought I noticed a tear in his soft, sensitive eye just before he turned his head away.

When I returned to the village in March 1995 after my second stay with Kesha in camp 8, I discovered that the director was due to make a periodic report in the village hall. The herders in the camp had not been aware of this, not even the alert and well-informed Kesha, so I scuttled from house to house to find out when the meeting would happen. Nobody seemed to know.

This paralleled what I had heard about the debates on the regional boundary in the late 1980s, which had taken place in my absence: decisive meetings had been called at short notice when the herders were busy with their herds, repeated if neces- sary until the voting came out to suit the position of the Farm management. Such moments highlighted the split between herd- ers and their managers, even though all were members of the same extended families. Herders needed relatives in the village offices to look after their interests, and office people increasingly needed relatives with access to meat. But the symbiosis was uneasy.

I was having lunch in the house of Tolya's sister Anna, when Tolya rushed in.

'It's happening now – quick!' he commanded.

The politics of excluding the herders from meetings matched the Arctic sense of urgency, the instant when one's own path crossed the migratory route of a wild animal or helicopter, when indefinite inactivity was suddenly transformed into an opportu- nity for decisive action that must be grasped or lost. I abandoned my lunch.

It was 2.20 p.m. and still daylight. A few men were standing forlornly on the steps of the village hall. Some time later, the director arrived. He was sober.

Eighteen of us shuffled into the hall. The boiler was off, the lights had not been switched on, the curtains were drawn, and the chairs were still pushed to the side from the young people's disco two nights before. Women began sweeping up the cigarette ends and vodka bottles, while men pulled the chairs back into rows and started setting up a table on the platform. A woman unrolled the velvet cloth which transformed the plain table into a setting for an official meeting. The cloth was bright Soviet red. This red was a code, just like the contrasting burgundy velvet of the pulled-back stage curtains. Burgundy meant 'culture' in the form of 'theatre'; bright red turned this piece of material into an altar-cloth from an extinct religion, arousing a pale memory of political sanctity.

The director was joined on the platform by other representatives of the Farm management. The entire meeting was conducted in Sakha. About thirty people had turned up by the time the director started talking about how well they had all lived under Soviet rule (*sovetskaya bylast*)*, and how well they were continuing to live even now. I looked around and recognized vets, the school director, the current head of the Village Council, other administrators, and one or two retired herders like Kesha's father Dmitri Konstantinovich. But there was only one currently active reindeer herder: Valera, the brigadier of camp 3.

More people kept trickling in as the Deputy Head of Animal Husbandry (Zamestitel' Nachal'nika po Zhivotnovosdstvu, ZNZ) read out reindeer productivity figures, camp by camp. 'Live weight . . . percentage of calves surviving . . . coefficients . . . wolves . . . percentage of losses . . . hoof-rot . . .'

The hall was cold, and the audience slumped silently in their worker jackets or padded coats.

'The herders' level of discipline has fallen,' intoned the ZNZ.

Suddenly a very old man in the front row muttered, 'It's your fault!' I strained to see who was hidden behind the pulled-up collar of the anonymous jacket. It was the director's old ally and next-door neighbour Efimov. I was surprised, but the director did not seem to be: maybe he had been forewarned.

'Democrat!' he retorted. 'Anarchist!'

In 1990s Russia, these two words were often considered the same. The audience did not flicker a muscle.

I noticed Kesha coming in at the back. The ZNZ continued with his prepared remarks on discipline, saying it was bad that young herders kept switching from camp to camp; bad that herders bought unnecessary goods instead of essential provisions; bad that herders stayed too long in one place. The only brigades that migrated properly were numbers 3 and 8. Was it just a happy coincidence that these were the only two camps whose brigadiers were present?

The theme was picked up at length by the director himself. If a long speech in a listless voice could be said to rise to a climax, this happened when he reached the words, 'If I criticized each herder individually, I could go on forever!*'

'It's your fault!' old Efimov muttered again in the ensuing silence.

'Fascist! Capitalist!' The director adjusted the glasses on his vodka-ravaged nose and continued, looking down at his notes and adapting them pointedly as he went: '. . . workers' collective . . . Communist Party of the Soviet Union . . . Comrade Efimov . . . democracy . . . chaos and disorder—'

'Your cronies have embezzled half the money from the sale of meat in Sangar!' interrupted Efimov, adding very correctly, 'Since you've mentioned me, I consider this gives me the right to speak.'

Like dummies coming to life, the shrouded audience began shuffling in their seats. Efimov kept asking questions: 'Ikkis bopros baar (Another question) . . . and another . . .' He mixed in major issues of life and death for the community with personal ones: 'Why don't I get firewood when I'm a registered invalid?'

I went out for air and found more people hanging around on the steps of the hall who did not know about the meeting. When I told them, the older adults went in. The teenagers stayed where they were. They did not plan to work as reindeer herders – or, in fact, to work at all. The Communist Party and the Soviet

Union had both been dissolved three years earlier, and they were already losing even the faintest memory of the connection between red tablecloths and the immutable one-party State.

Other people started asking their own questions: about leave entitlement, about pay, about veterinary services, about contracts. Experienced vets and brigadiers were using a surprise public meeting to ask questions about the basic procedures of their own jobs.

'To raise the herders' productivity, you have to raise the level of support services. If you simply raise the number of reindeer you'll ruin the pasture,' said Kesha's father Dmitri Konstantinovich.

'Cutting the young antlers in spring to sell weakens the reindeer for the rut in autumn. Also, the herders never received their pay for last spring's antlers,' said a woman's voice. 'So now they've lost twice over: no money, and weaker animals for breeding.'

'Our camp handed in antlers last summer, but where's the money?' shouted another woman from the back. 'I can't be fooled, I'm educated, I know, I understand!'

These women could have been the wives of herders I knew well in the taiga, but I did not recognize them, so separated were their lives. Looking out for their absent husbands' interests in the village had become one of their main roles in the marriage partnership.

Efimov was speaking again in his frail voice, standing up and coughing frequently. The others quietened down.

'It's not only the herders who lack discipline,' he said, 'but also their managers. And the schoolteachers. Everyone is slacking! You could start on the herders by bringing the office under control for colluding with them. The herders from camps 1 and 2 abandon their reindeer to the elements and hang around in the village for a month at a time. They fill in their own worksheets.'

'Where did you get this information?' For a moment the director was so taken aback that he forgot to deny the allegation.

'I know, and I observe. I've seen them in the office fiddling double leave allowance. Even if we're in a new world with new laws, worksheets still have to be filled in correctly! And another thing,' Efimov continued. 'Why do you still retain the services of your Ukrainian sidekick? He hasn't been here all winter, even though he's still drawing a salary. He's got apartments in Moscow and Magadan. How were these paid for?'

'He is our Chief Animal Technician,' replied the director. 'It's thanks to him that the herders have a living. It's because he fought with the buyers in Sangar that we've been able to bring 50 or 60 million roubles back to the Farm.'

'You mean he ate half of it up himself! I know about the percentage he takes on every kilo of meat. And his fake antler receipts from Korean mafiosi. I'm going to tell the herders all about it. And another thing: people are paying ten times the proper price for vodka. You're the one who brings it in – you should sell it in a proper shop! Five or six brigades have already been ruined by alcohol and their herders just hang around the village getting drunk. Young men aren't working as herders, and young girls are working – as prostitutes!' There was nervous laughter. 'Young people, if you're listening, come to your senses!'

Sadly, the only young people around were the ones on the steps who had not bothered to come in. Efimov moved on without a break and began denouncing the new contracts between the Farm and the brigades, saying that they were not properly set up, monitored, or fulfilled. But now the director was shouting, 'Stop this man's speech! Stop this man's speech!' But there were no guards in Sebyan, not even a policeman except when Little Bird flew over the mountains (weather permitting) to arrest the surviving drunk from a homicidal brawl.

The director addressed Efimov directly, using words that ten years earlier would have been powerful: 'You're an Old Communist, but in aligning yourself against us you're committing a serious error.'

'No!' croaked Efimov. 'I see everything with my own eyes.

The Soviet period is over. Don't hang on and drag us all down! Resign!'

There was a hubbub as he resumed his seat.

I had known Efimov in earlier years as one of the director's closest inner circle of comrades. He was an Old Communist of a particular kind, who had deeply internalized the party's ideals of duty and rectitude on top of the old native values of modesty and self-denial. When he held the post of General Manager of Reindeer Production, he had actually toured the herding camps. Such people were very punctilious: his previous hard-line inflexibility, even his hostility to Tolya, had been because he hated fluidity and disarray. His catalogue of accusations seemed too well thought out to have been improvised. How long had he been holding in these grievances? When had he undergone a conversion? What I did not know, but Efimov did, was that he was sick and did not have long to live. Reflecting on the future that would take place without him, it must have taken great courage to reject the system he had done so much to create and sustain.

Efimov died, taking his dissent with him, and the director lived on and continued to direct. Now, in 1995, the number of reindeer in Sebyan had fallen from the 20,000 of 1988 to 16,000, only 11,000 of them still owned by the Farm. Between 1995 and 1997, the summer head-count of reindeer remained around 16,000. Accurate figures were becoming harder to reckon. People were taking great pains to conceal the numbers of their private reindeer, not least because the Farm was trying to charge them for use of pasture above a ceiling of 30 for an active herder, 20 for a pensioner, 10 for a village resident, and 5 for someone living away in the city. By 1999, numbers had risen again. That autumn, I did a careful calculation, brigade by brigade, and worked out that for all herds combined the January figure was 9,925 Farm reindeer and 5,700 private ones, and the June figure 11,800 Farm reindeer and 7,500 private ones, giving a winter total of 15,625 and a summer total of 19,300.

However, this increase was not for a healthy reason. The buyers in surrounding towns hundreds of miles away were hard to reach by truck in winter, while aviation had become so scarce and expensive that few reindeer were slaughtered in the autumns of 1996 and 1997, and in 1998 no meat was sold at all. The herders were labouring to produce a commodity which the Farm could not market and for which it could not pay them. The nearest consumers were in Sangar. The Farm claimed that Sangar could not buy meat because the coal mines were closing and the town had no cash. In 1999 I went there to see for myself and interviewed the head of the Sakha administration and some of the Ukrainian miners. It was true that Sangar was becoming depopulated, like many Arctic industrial settlements, as white workers returned home to the Ukraine and the western end of Russia. But those who remained complained that Sebyan was not sending them meat, and they still had enough purchasing power to switch their custom to the Dutch and Australian beef which was being flown into Yakutsk on large planes and floated downriver to their little town on supply barges.

There had been a moment, around 1990, when almost anything seemed possible. But to act on such possibilities one needed vision, boldness, and integrity. Many opportunities were lost because of an unfocused, debilitating fearfulness, sometimes combined with sabotage by vested interests from within.

The people of the neighbouring village of Sakkyryr saw the point of Tolya's idea of a separate district dedicated to Eveny interests, and in 1991 they formed a district on their own, leaving the village of Sebyan to one side. In 1992 and 1993 Afonya, head of the village's civil administration, proposed to take advantage of new legislation to run his own winter trucks, convinced that he could offer the herders a better price than the Farm for their meat, antlers, and fur, and bring them supplies more cheaply than the Sakha traders who had recently visited the deprived village, reaping a huge profit.

'The director can't pay anyone to do anything, but he won't

let anyone do anything either,' he told me. 'I asked the Farm to sell me 5 tons of meat and 100 leg-skins for making boots – I even offered them a better price. But they said, "It's against our plan." The brigadiers wanted to give me their produce, but the Farm scared them with threats of dismissal. What they were really worried about was that it would show people that it's possible.'

Some of Afonya's ideas trespassed on areas that had previously been controlled by the Farm, even though the Farm was no longer able to deliver, and opposition was inevitable. For instance, he also tried to introduce competition into the supply of firewood, offering woodcutters 10,000 roubles per cubic metre against the Farm's 7,000. Other schemes were more neutral, such as his attempts to establish a workshop making saddles and lassos for younger herders who no longer knew how to make them themselves, or to set up a folk museum of traditional Eveny culture. But all were thwarted, each for its own complicated reasons.

The ambivalent attitudes in the community towards private entrepreneurship raised new tensions along the border between the responsibilities of the Farm and of the administration.

'The director's main fault is that he's so strongly against private property,' Afonya was still saying in 1999, the year when I went to check on the consumers in Sangar. 'He could provide winter trucks and let the herders pay for the petrol. Then they could sell their own produce and feed their own children. Or he could help the school that their children attend by giving the school ten reindeer a year. He says the school is our responsibility in the administration. Well, we get the teachers' salaries from the State budget, but no food to feed the children.'

Help offered from outside was subverted, too. The miners in Sangar might accept the battered and petrol-soaked reindeer carcasses that bumped their way down the frozen mountain rivers in trucks, but the elegant Sakha housewives in Yakutsk were more particular. In 1994 my American student John

Tichotsky persuaded the US Congress to spend a million dollars sending reindeer herders from around the Sakha Republic to learn consumer-friendly meat packaging with Doug Drum, known as the Sausage King of Alaska*. Drum also made a brief visit to Sebyan with Andrew Crow, a Russian-speaking American aid worker. Their helicopter zigzagged them through a microcosm of the community's political tensions: the director sent them to his own relatives in camp 9; Tolya redirected them to Kesha in camp 8 and Kostya in camp 10.

At a demonstration in the village, Drum butchered a reindeer with a finger knife instead of an axe, trussed a joint with string, and roasted the meat instead of boiling it. When he cut the string and opened the joint, everyone rushed forward to taste it.

'What was the most useful lesson for you in Alaska?' I asked one herder who had spent two weeks training there.

'The orientation to the consumer,' he replied immediately. 'We don't have this at all. They think about profit, they calculate ahead: Will people buy this or not, why aren't they buying, could we do it differently? But they have some strange preferences. They like their meat without bones, even without fat. I can't imagine anyone buying meat like that here. How would they make soup?'

The programme was officially cited in Congress as one of the most successful US aid programmes to Russia – and, for many villages, it was. In Sebyan, the success was more qualified. Vitya the Wolf-Hunter went with several others to Alaska, where they were trained to operate a machine for making venison sausages.

'Whatever he learns in Alaska, they won't allow him to do it here!' Afonya commented wryly. Sure enough, a sausage machine was sent to the village, but the director sold it off. The director's people prevented Afonya from creating a company to supply the village, but they created one themselves, buying at a low price from their own Farm, which then did not pay the herders.

'If they were Americans, they'd sue the bastard if he didn't pay them, or burn down his office!' Andrew commented as we were sledging together to a camp during one of his project-fixing visits. The director's company was said to have embezzled US $200,000 owed to the herders for the young velvet antlers which weighed so little but were so highly prized as medicines by Korean buyers.

Perhaps this was related to the fake receipts that old Efimov had denounced with his last breath. But the director's scam was small fry compared to what was happening on the global antler stage. Just as the Farms cut out the herders, so the Russian antler shippers, drawing in produce from across the Russian North, were cutting out the Farms from their 'joint venture' by dealing directly with the Chinese on the US west coast, who had cut out the well-meaning American aid programme. The meagre payment that eventually trickled down to the herders in Sebyan took the form not of cash, but of shoddy Korean sleeping bags, which were so thin that they were nicknamed 'Death to Reindeer Herders' (*smert' olenevodam*).

When reindeer are driven into a corral, they become agitated, circling restlessly for the few hours until they are released to spread out again across the taiga. When nomads were driven into a village, they remained there for ever and social tensions were concentrated. Both reindeer and humans are corralled for the same purpose, and with the same ambiguity: to protect and nurture them, but also to number and control them.

The *buyun*, the wild reindeer, understood the price of domestication, and chose to remain free. The northern native peoples were not offered this choice. Long before the actual word 'contract' became widespread in the late 1980s, the establishment of the village and the Farm had amounted to a contract in which the Soviet State would provide veterinary care, administration, and aviation in exchange for the production of meat.

As part of this bargain, the State arrogated all roles to itself, replacing headmen like Baibalchan with State Farm directors, family production by brigades, and parental child-rearing by the communal boarding school. Throughout the Soviet Union, previous structures were aggressively destroyed or sabotaged, making it impossible to live outside the new ones. The loyalty required was not just one of outward appearance, but a surrender of mind and soul. For those who came dangerously close to the flame of power, this surrender was enforced through confessions, labour camps, or the firing squad. For the remote or unimportant, it was enough to render them incapable of living any other way than as dependants of the State. The crime of Baibalchan, the headman in the 1920s, was not simply one of self-sufficiency, but of modernizing the self-sufficient spirit of the taiga and bringing it into contemporary politics. Like Nikitin, the linguist's informant, he was filling a vacuum that would soon be acknowledged officially, as the foundation of cooperatives and the compilation of native dictionaries became Soviet policy. Nikitin's fate might have been different if he had helped the Soviet linguists who came later, as might Baibalchan's if he had become an official in a Farm like his son Vasily Pavlovich. But this new system was still in the process of forming, and they were working outside it. Their unauthorized projects showed that initiative was still possible, and so they were killed.

The structures set up by the Soviet authorities were not rooted in the community. Oriented away from the land towards an outside world to which they formed the furthest periphery, the new villages required an extraordinary level of outside maintenance and support that after 1990 could no longer be sustained. The literal idiom of formal contracts cruelly exposed this inadequacy, as the Farm became less and less able to fulfil the obligations that these contracts spelled out.

The State's total control had required a commitment to total provision. But by the time the State tried to hand back this responsibility, it had repressed the community's self-sufficiency so deeply that the community no longer knew how to function on its own – or had the heart to do so. The State had tried to break up families and make them factory workmates. Many brigades were formed around a core of male kin, but they had not been allowed to function as families. The factory model had failed, and the family model had been disabled in the process.

Over three generations, as men have been separated from women, children snatched from parents, and communities plunged into a deepening abyss of homicide and suicide, northern native people have become enmeshed in a tangle of betrayal, collusion, and guilt. Many people in Russia talk of despair, inertia, and paralysis, using an apocalyptic vocabulary of *raspad* (decline), *polnyy krakh* (total crash) or *konets sveta*, 'the end of the world', which by an amusing quirk of Russian vocabulary can also mean 'the end of electric light'. Among indigenous peoples of the North, this has taken on a special tone: *my vymyrayuschiy narod*, 'we are a dying people, a people on the verge of extinction'.

'We're like dinosaurs, or the Abominable Snowman [*snezhny chelovek*],' Afonya told me during one of our sombre, reflective conversations. 'Except we've managed not to die out completely – yet. Or like that samurai who went wild in the jungle!*'

Tolya faced sorrowful facts with a different tone. 'There are only 17,000 Eveny in the whole world,' he announced when I introduced him to a dignitary in England. 'We're an endangered species. And you're meeting a real live one!'

In 1990 the director had colluded in abolishing the large regional spring festivals after the Chersky meeting got out of hand. By 1995, his own ally Efimov had brought the Chersky format home to the director's own village hall and reversed the ritual admonition from the platform by attacking the leaders from the floor. The meeting in the village hall, a confused compromise between secrecy and public record, was a discipline session that went wrong, as the director surely feared it would. The Soviet concept of discipline blended accountancy with socialist puritanism. The accountancy was an instrumental concept, expressed through productivity figures, even while all costs and profits were rendered notional by huge, unacknowledged subsidies. The puritanism transmuted discipline into a transcendent moral force, an ultimate touchstone of a person's goodness. 'Discipline' was portrayed as a virtue which struggled endlessly to get inside people, as though they could not perform any useful action under their own motivation. Speeches and medals were catalysts in an alchemy of exhortation which was supposed to turn base human material into socialist gold. The faces on the Board of Honour were celebrated so that they would imbue others with their own higher alchemical state by a chain reaction of example.

Shifts in ideology and personal relationships made it hard to sustain a condition of 'discipline' indefinitely – and, in the end, even to take it seriously. This virtue did not count unless it was directed towards approved goals, so that it eventually stood revealed as little more than an appearance of loyalty to the clique in power. The family of Kesha and Dmitri Konstantinovich survived by supporting the director until the early 1990s, and even afterwards they made their disagreement discreet. But Ivan's father failed to display the loyalties commensurate with his party membership and position on the Board of Honour, and so he was cast down and punished; and it was after several futile clashes with the director while trying to instigate reform that Tolya retreated to the city, where a sympathetic Yukaghir professor of native studies supported him to complete his PhD and become a lecturer in anthropology.

Even if taken at face value, the discourse on discipline by the Farm management was insolent. The herders rode reindeer over ice for fifteen hours a day, survived encounters with angry bears, crawled for help with an inflamed appendix, hunted alone for six months in winter, saved each other's lives in blizzards and avalanches, and took constant care to feed the spirits of the land through fire. What kind of 'discipline' was embodied in the people who mismanaged the Farm's economy and ground their relatives into alcoholic subservience, compared to Vladimir Nikolayevich's astonishing competence, Kostya's dedication to productivity and party work, Kesha's experiments in re-routing his reindeers' winter migration, Granny's watchfulness over her family, Lidia's determination to nurture her frail husband, Ivan's careful planning of each move of his herd, Kristina's everlasting stew awaiting the return of her frozen herders, or Lyuda's fortitude as she bathed her baby in a tin bowl while snow weighed down the outside of her tent?

Whoever strives forcefully towards happiness
Can manage to break strong rope.
When you meet suddenly with unexpected misfortune,
Don't rush: Think! Observe!
Don't turn back!
Stand firm
And you will win in your unequal struggle!
If you want to be a true man,
Rely only on yourself!
It is very difficult to scramble up the steep slope
To your goal,
Even harder to do good to another.
For happiness is not given to anyone lightly,
It is brittle like the first ice of autumn,
It comes only to those
Who are true to mondji!

From a song improvised by old Vasily Pavlovich* (son of Baibalchan) on the theme of *mondji*, the quality of being self-reliant, able to survive in extreme situations, and never giving up

PART IV

SPIRITS OF THE LAND

11

Animal souls and human destiny

Ideas about Hövki, the deity who created the world, are vague. Like gods or spirits of the sky in many religions, he occupies a realm that lies above and beyond our own. But unlike heavenly gods in monotheistic religions, he is no longer closely involved with the daily lives of his creatures. Today he is talked about only in legends about how things came to be as they are; even in the memory of the elders, the annual midsummer flight to the sun on reindeer-back was the only ceremony in his name.

It is quite different with the spirits in the landscape around us. The Eveny live in a modern world, but it is also an animate world in which mammals, birds, fish, rivers, lakes, and forests are alive with their own souls or spirits, giving them some degree of consciousness like our own. Rather than looking to any conception of a transcendent god, this kind of thinking locates the divine inside the phenomena of the world, as part of their composition and nature. Similar ideas recur in shamanic religions around the globe, from the Amazon to Borneo to the Canadian Arctic.

The spirit of a phenomenon represents its essence*, something behind the visible surface that is somehow more real – that which makes a thing what it is. This understanding ranges more widely than most European senses of 'soul'. In animals, the shamanic sense of this life-force may seem similar at first sight – except that humans can sometimes turn into animals, be seduced by them or marry them; for places, it represents their identity, as they are experienced through the cycles of nomadic movement; for tools, it represents efficacy – the spirit of a knife or gun is manifested through its ability to cut or shoot.

Because such creatures, places, and objects have some kind of consciousness, they can also have intention. An animal may cooperate with humans or be recalcitrant; a gun may choose to shoot well or badly for you; mountains and rivers may nourish or kill you. Spirits are the causes of some of the most significant events in your life, and you should strive to be aware of the moods of your surroundings and adjust your behaviour accordingly, in order to achieve your aims and avoid disaster.

Shamans are specialists who are highly attuned to these moods and alert to small signs or omens which emerge into the world of appearances from hidden realms beneath. As they encounter spirits on different levels of the sky during trance and soul-flight, shamans cooperate or negotiate with them on their clients' behalf – and, if necessary, fight them. When a shaman dresses as a reindeer or turns into one, this is not only for its ability to fly. The reindeer is no castrated *uchakh*, but a wild, intact male, and aggressive in its competition with opponents or in the defence of its herd. The shamans' reindeer headdresses often emphasize the spear-like sharpness of the antlers*.

The Eveny word *shamán*, like corresponding terms in various Siberian languages, refers to a figure who draws together many animal powers, represented by the skins, furs, feathers, and beaks that make up the costume, in addition to metal representations of mammals, fish, and birds. In many shamans' initiatory visions, the spirits butcher the novice like a game animal, boil him or her and then reassemble their bones and muscles to form a new person with extraordinary properties. These new powers are also embodied in a range of animal helper spirits which the shaman acquires, giving the shaman the strength of the bear, the power of the eagle to seize its prey in its talons, or the ability of the pike to dive into underwater realms and seize the soul of an enemy in its vicious jaws.

A Tungus (Eveny or Evenki) shaman in the seventeenth century.

The Soviet persecution of Siberian shamans was so thorough that almost no shamans of this old type exist anywhere today*. However, even when there were many shamans, laypersons also engaged in their own personal techniques of divination, working out meanings in the crackle of a fire, the twitching of an eyelid, an event in a dream, or the untypical behaviour of an animal.

The relationship set up at the beginning of time between the Eveny and their animals is different from the relationship in Genesis, where God gives Adam 'dominion' over every kind of creature. For the Eveny, animals are spiritually and psychologically more complex. In the biblical text, wild and domestic animals are not distinguished and the purposes of all animals are subordinated to those of humans. But in the taiga, wild and domestic animals each relate to humans in their own distinctive way. Even domestic animals are not subordinate and their purposes run alongside those of their human caretakers in mutual dependence and cooperation.

The purposes of wild animals are more mysterious. When a hunter kills a wild animal, this is not a right given by God, nor a human domination of a domain called 'nature', but an engagement with the spirit Bayanay, who represents a principle of undomesticated animality. It is under the direction of Bayanay that wild animals migrate, feed, breed, and die. His association with his creatures is intense and multifarious. The Eveny describe him as an old man who is the 'owner', 'keeper', or 'master' (in Russian, *khozyayn*) of the animals and of the forest in which they live. They sometimes say the animals are his pets or his children, but more profoundly that he also *is* those animals: they are his incarnations, manifestations, or refractions. This old man's power is an elemental force that pervades animals and landscape alike and can wax and wane, surge or retreat at different moments, in different locations or for different hunters.

Each species embodies Bayanay's essence in its own way, according to its nature. Like the shaman's helper animals, they engage with humans through their distinctive qualities. Sometimes they pass on something of these qualities, which are themselves just a fragment of the force of Bayanay himself.

'The meat helps your eyesight,' Lidia once taught me, as we were feasting on a freshly caught mountain sheep. 'They're keen-sighted, they live on high crags and see all around. And marrow fat from the hind legs makes you a good hunter and stops your bones aching. Mountain sheep is so strong that you shouldn't eat it if you're ill or pregnant, especially the liver.'

A hunter can kill a wild animal only when it offers itself at the behest of Bayanay, who decides whether to give an animal or withhold it, place it in the hunter's path or send it off in another direction. Like the animals through whom he reveals himself, Bayanay is capricious and hard to fathom. At one moment he may favour you, so that you and your family eat well; at another he may desert you, so that you starve. To encourage animals to give themselves, humans must strive to be worthy of Bayanay's favour by pleasing his creatures. An animal will give itself only to a hunter who treats its body and soul correctly. When it is reincarnated, it will offer itself again to the human who respected it the previous time.

People seemed to understand this in different ways. On my first visit to Kesha's family in camp 8, some years before his deputy Boris froze to death, I saw the carcass of an elk, which Boris had shot, neatly cut up and stacked on a raised storage platform. Its head had been laid carefully upside down, facing east, with a triangular incision cut neatly out of its huge black snout.

Boris said that he had cut out the triangle so that the fat would not leak out. What about the placing of the head upside down? That was to stop the blood from flowing out the wrong way. I did not understand either of his explanations. This young man, brought up in the village school, had performed an action that belonged to the rituals of the taiga but he had given it a scientific sort of rationale in the idiom of the schoolroom – except that the school had never prepared him for this question*, so his response seemed incoherent.

When I asked Kesha's father, old Dimitri Konstantinovich, I received a very different answer. The cut on its nose and its orientation were to ensure that the elk's soul would be reborn in a new body. The elk would reappear again from the east, like the sun.

'But what about the head being upside down?' I wondered.

'That's so it won't see the person who killed it and be angry with him,' he replied. Here was an ambiguity at the heart of Bayanay's gift of life. By performing the correct rites, one hoped to earn the dead animal's appreciation. Yet at the same time the killing was a violation, and the animal's offended gaze had to be diverted away from the very hunter it was supposed to acknowledge.

This ambiguity was at its most extreme with bears, which embodied the most intense concentration of Bayanay's power. Everything about the bear was awesome compared to any other animal. Its paw, fat, gall-bladder and kidneys had the greatest medicinal power. It was in the bear that Bayanay's animal energy was most openly sexual. When skinned, the bear was said to look like a naked human. A little of its dried penis shaved surreptitiously into a woman's tea was a powerful aphrodisiac, and there were stories of women who became too involved with the forest and were seduced there by bears, sharing their winter lair and later giving birth to a mixed litter of human babies and bear cubs.

Many people, even the fearless Vladimir Nikolayevich, were so wary of killing a bear that they would never do so unless they were attacked. If they were forced to kill one, they would not eat the meat, and might negate the act of killing by sewing up its eyes so that it could not see its killer, or by telling the dead bear that it was killed by a Russian.

Lidia and Gosha once gave me a bearskin. 'It was shot last year by Valera from camp 3,' they explained. 'It had just woken up in the spring and had eaten one of his reindeer. They wouldn't have killed it otherwise.'

'Did they eat the meat?' I wondered.

Lidia shuddered. 'Even talking about it scares me,' she said. 'Some old people used to eat it, maybe some people do even now, but most are afraid.'

'I got one last year,' said Gosha. 'It was coming directly towards us. I had to eliminate it [*prishlos' ego ubrat'*]. But I wouldn't have touched it if it had gone the other way. It was fat, but we didn't eat it. I just took the skin, and the gall-bladder and paws for medicine, and buried the meat under some stones. Then immediately my *uchakh* cut its foot.'

'There is a connection.' Lidia picked up the thread.

'It would have fallen ill anyway,' Gosha contradicted her, reassuring himself. 'Herders are always riding their reindeer onto sharp rocks, we make mistakes, the *uchakhs* get cuts, especially going downhill.'

And yet, even while he spoke, Lidia was expressing her doubts. 'Bayanay was angry,' she murmured.

Bayanay is inscrutable. His purposes are obscure and can change on a whim. But clues can seep out and be perceived by an attentive observer. If your *uchakh* snorts as you are setting out to hunt, there is no point in going further. If a crow cries out, you should likewise turn back; but if it is silent, you should follow where it leads you. A male hunter may dream of sexual intercourse* with a young woman, who seems to be Bayanay's daughter. This means that he is favoured and will have a successful hunt the next day.

Such omens and dreams come to us as they will. But you can also take the initiative and seek information deliberately. Kesha explained this to me one winter night in camp 8.

'You can use divination for the most extreme situations,' he said, 'when times are really bad.'

This was not an extreme situation. Kesha and I were sitting with a candle at the table. We had eaten good reindeer meat, and talked and laughed far into the night. The others were asleep: Lyuda and the children slept on benches around the outer walls of the hut, wrapped in blankets and furs, while the young herders lay on the floor at our feet.

'You take an animal's shoulder blade,' Kesha continued, sweeping the palm of his hand across the table to represent the bone's smooth, open surface. 'You take off the fresh meat and dry the bone. I sometimes use the shoulder blade of a marmot or a hare. But the best is the bone of a mountain sheep. It lives high up on peaks all the time, it's keen-sighted, it sees everything.' This was the same idea as Lidia's explanation about wild mutton; but Kesha was going beyond the eyesight of the hunter towards the commanding vision of the flying shaman.

'Now I place it in the embers, not directly on the flame. If it's an open fire that's fine, but if it's the stove in the tent, you mustn't have any kettles or pots on top. The bone will shatter, or the imprint of the bottom of the pot will come through on to the designs on the bone.'

So this clear vision was vulnerable to interference. It was the analogy between the pattern on the bone and the map of the landscape that provided the message. At this moment of heightened openness to analogy, if you failed to set up a pure, well-focused situation, you exposed yourself to the risk of a false connection.

'When you put the bone in the ashes you address the fire in an arrogant way to get its attention,' Kesha continued. 'You say, "Let's have a competition, who knows better – which of us can see better, you or I?"' I must have looked puzzled or shocked, for he went on, 'Yes, of course, it's really a way of saying, "You know better than I do." It's a pretend competition. But in response to this, the fire draws the answer on the bone.'

My notebook was laid out on the table and I was writing his words. He picked up one of my spare pens and drew the outline of a shoulder blade.

'I'm here, and I could go up three rivers,' he said, marking a triple fork inside the outline. 'Which way should I go – the Kumka, the Kelten or the Chokhun? I consult the fire, I look. Lots of tiny cracks appear. I take no notice of the little ones, and look only for the main fissures.' He thickened the three lines, one more than the others. 'These two are weak cracks, there are no big animals there, I shan't go there. But the Chokhun stands out, it's a deep crack – I'll go there and I'm bound to get a wild reindeer, an elk, or a bear.'

Clearly, Kesha felt less unease than Gosha about killing bears. I wondered whether this information was sent or communicated by Bayanay. But from Kesha's next remark it seemed that the cracks were just a sign of Bayanay's intention and that such signs could work by power of analogy alone.

'It also works for finding lost reindeer,' he went on. 'You use the shoulder blade of a reindeer from the same herd.'

The analogy between the design on the bone and the map of the landscape was becoming clearer. It was based on a connection between the animal that provided the bone and the animal being sought. Information about wild animals came via the body of one of Bayanay's creatures, while information about domestic reindeer came from a member of the same herd.

Kesha continued: 'If some of the herd have got separated, I ask the fire, "Where have my reindeer got to?" You don't ask so directly, you never really talk about yourself out loud, but you let the fire know what your concern is.' Kesha was speaking in Russian, but I recognized the Eveny concept of *unuvkame*, the indirectness of 'giving to understand'. 'So here I am wondering, "Where I can find my reindeer?" And gradually these little cracks start to appear, little fissures appear here and there, like this, you see . . .'

He drew another shoulder bone and started to fill it with rivers and streams. The drawing was a process, not a product. Like a photograph in a bath of developing fluid in a darkroom, a map started to appear, faintly at first, becoming stronger as his pen moved back and forth over the bone, until the image was too crowded to read any further.

'Now look, two big cracks have appeared here, you see, you understand? This is where I've been, my autumn route, my spring route, here's the River Tonkochan, here's the pass, here's the Sakhandjá. Look, the source of the river . . . the waterfall . . . the cliff.' Kesha conjured up an entire journey as he started to take a direction through the landscape he had created.'Ah, now it's showing the loop, where our route curves round!' His wrist flicked round to echo the curve in his narrative. 'It's showing me that my reindeer are lost inside this loop.'

Here was a more focused version of the idea of the fire as a source of knowledge. Rather than letting the fire take the initiative and waiting for the vague, chance messages contained in its crackling, you could seek information deliberately. In response, the fire became correspondingly articulate, conveying detailed, explicit information. The bone acted as an amplifier.

We had put no new wood on throughout this long conversation, and the stove had died down to a quiet whisper. The hut was warm, and filled with the scent of larch embers. This was just the moment when the fire would have been ready to heat up a real bone – except that with no genuine problem, we could not abuse this power.

'I use shoulder blades like this a lot,' Kesha concluded, 'it helps a great deal. It hasn't been proved by science, but it works in practice. But you need to take it seriously and know how to understand it.' He looked down at the sleeping herders. 'Some of these lads go out hunting without me, they travel all night and all day from dawn till dusk, and still they come back empty-handed.'

A good hunter like Kesha was said to 'have Bayanay', almost as if this were part of his own character. Having Bayanay was an attunement between hunter and prey that made possible a successful act of killing. Some people had Bayanay much of the time, while others passed through phases of life when they had him to a greater or lesser degree.

'If an animal runs up and gives itself to a hunter, this doesn't depend only on the animal,' Gosha once explained to me, 'or even on Bayanay. It means that this person's consciousness (in Russian, *soznaniye*) is different from the consciousness of ordinary people.'

From conversations like these, I came to understand Bayanay as a vast field of shared consciousness which encompassed the landscape as setting, as well as all the human and animal roles in the drama of stalking, killing, and consuming. This state of super-consciousness was so delicate and precarious that when talking of hunting, especially in the forest, one could not refer to animals by their ordinary names. Instead of the 'crude' but 'true' name *kyaga*, one referred to the bear as *abaga*, meaning 'grandfather' (in Sakha, the identical euphemism is *ehe*).

These were secret names, so that eavesdropping animals would not know that the hunter was speaking about them, and their use became more necessary as one moved away from the village deeper into the forest. I was also told that they were terms of respect which would please the animals. Either way, I think, the meaning was the same: such indirect names were linked to the awe surrounding the moment of sudden, violent action after a long wait, the swift and decisive instant when an arrow or bullet killed, and my friends were transformed in some mysterious way so that they seemed almost afraid of themselves. In this terrible blend of nourishment and murder, in which the animal both colluded and was angry, one had to honour one's prey and at the same time deceive it.

Bayanay did not like any hint of boasting or self-aggrandizement, and the slightest immodesty in speech, act, or thought could lead to failure. A hunter could not talk in advance about his plans, or about any promising dream or omen. The most he might say was that he intended to visit a certain river, or he might invite another man to join him by saying in Eveny, '*Töre hergel*, Let's go and see the land.'

Afterwards, too, the hunter was not supposed to talk directly about what he had caught, and had to avoid any words for 'kill'. If he had caught a bear, he might say, '*Kungan duram*, I have obtained a child.' Even this coded information should be introduced discreetly. As with the conveyance of bad news, the Eveny avoided the shock of a sudden, direct revelation, even about something good. One way of announcing success was to smear the animal's blood on the back of your dog and send it ahead of you into the camp. Everyone would recognize the dog and understand the coded signal. If the hunt was unsuccessful, you might tie up your *uchakh* some way outside the camp and enter on foot, 'so that people won't be upset [*ne rasstroilis'*]'.

The delicate morality of the animal world echoed Eveny ideals of relations between humans – perhaps more scrupulously observed because not clouded by human passions and politics.

The indirectness of relations with Bayanay paralleled the discretion in the face of the mystery of another human being's existence, as expounded by Lidia, Afonya, and other sensitive traditionalists and as lived, unspoken, by older people.

Bayanay also enjoined the mutual support of the Law of the Taiga. Since it was only by the grace of Bayanay that a hunter obtained food, he could not keep it or even call it his own, but had to share the meat in turn with others. As Vasily Pavlovich sang into our microphone*:

> . . . Today I obtain a wild reindeer [buyun],
> I shall share it so that it is enough for everyone.
> The tenderest, fattest portion
> I shall present to my relatives,
> Both close and distant.
> Honour every person,
> Consider them your equal,
> And between you there will be peace and harmony.
> If I deny a guest their share,
> That is the worst offence of all.
> But if your intimate guest is happy,
> Then your domestic reindeer [oron] will be healthy.

But if Bayanay required humans to use meat as a foundation of social relations, there was another kind of hunter who did not share meat with the human community at all. Ever since humans first began hunting wild reindeer, they have been in competition with the wolf.

The way people talked about the wolf was unremittingly negative. In Russian, they used words like *khitry* (cunning or devious) and the rich, buzzing word *zloy* (wicked or aggressive). When *zloy* was used of a bear, it meant 'bad-tempered' and referred to its mood when it was injured or hungry after hibernation. But when applied to wolves, these words implied an inherent quality of the entire species, as if wolves could have no other disposition. Even after Yura watched a Canadian television programme about the family life of wolves, which he found fascinating and sympathetic, this did not affect his attitude to actual wolves on the territory of camp 7. Bears occasionally attacked domestic reindeer, but they were not condemned for this as a species and their killing, even when justified, entailed anxiety. But when wolves did so, they were killed by shooting, trapping, or poisoning, without any concern whatever for their souls.

At first, I understood the herders' wolf-hatred literally, in terms of the impact of the wolf population on reindeer herds. But the meaning of the wolf was not timeless. The domestication of the reindeer around 2,000 years ago must have brought about a reinterpretation in the eyes of their human minders, as the wolf changed from competitor to looter. During my time in Sebyan, the State Farm there generated a new meaning, as the wolf furnished a political and moral metaphor for its decline.

In 1990 there were very few wolves in Sebyan, each of them known. But soon afterwards, the Farm's territory saw a huge migration of hares, which were shadowed in turn by a population of alien wolves. Before this symbiotic double invasion could move on to some other territory, the wolves ate up the last of the hares. Then they turned on the domestic reindeer. In 1992 the Farm's records show that 716 reindeer died of disease (an average figure) while wolves already killed another 282. This was far more than the previous few wolves could have managed. In 1994 disease killed 561 reindeer while wolves killed 2,488, more than an entire herd. By early 1995 wolves were ravaging camps 1, 5, 6, 10, and 11, often killing 20 or 30 reindeer at a time, and it was said that they had wiped out a quarter of all the Farm's reindeer. They were even attacking horses, which no one could ever remember happening before. It was their harassment of the calves in camp 7 that spring which forced Ivan to move the herd on far ahead of schedule.

The most *khitry* wolf in all of Sebyan terrorized the remote camp 2 from 1999 to 2001. The brigadier (Tolya's brother Maxim) sent for Vitya the Wolf-Hunter to outwit this enormous and highly intelligent creature. Vitya caught it once in a trap, but it gnawed off its own leg and escaped. He finally killed it in February 2001 with poisoned bait. A few days afterwards, I arrived in camp 2's winter hut with a television crew to make a new film. The wolf's rigid, shaggy corpse had been stacked on a high shelf and its three legs loomed into the room above our heads as we ate our reindeer stew.

Up to 1990, the Farm management had given rewards for killing wolves, under a routine, low-level programme of control. By 1995 they had increased the bounty, but there were hardly any bullets or poison to give the herders or even to Vitya. Very late that spring they called in a helicopter with the State Fur Corporation's crack marksmen to shoot wolves from the air, guided by herders who knew where their tracks were but whom the State would not license to use such powerful rifles. Because of an argument over who would pay for it, the helicopter arrived two weeks late. There was no room for me on board, especially as I would not have been useful, but the herders told me afterwards that wherever they guided the marksmen, the wolves' tracks had already melted. They did not see a single wolf, though the helicopter gave several people an expensive lift to and from the city.

It could hardly have turned out better for the wolves if they had planned this spectacular failure themselves. Or was that just what they had done? As the State Farm deteriorated economically, so I saw the meaning of the wolf changing in front of my eyes. For all the difficulties of knowing and interpreting him, Bayanay wove the animals and humans on the landscape into a moral ecosystem. Yet he displayed little interest in wolves, or control. As I reflected on this, I realized that the wolf occupied a special place in the modern cosmology: it did not follow any courtesies. Rather, it took without giving, rejecting any moral relationship with its prey's human minders. The increase in the destruction wrought by wolves had run in parallel with the collapsing economy. If previously there had been a coercive welfare state, now the late, decadent State Farm was seen as taking without giving anything in return, failing to honour its own part in any contract with the herders. Like the wolf, at least in remote Sebyan, the Farm represented pure predation*.

The wolf provided a metaphor for other kinds of human predation elsewhere, too. Throughout the North, Russian industrial workers indulge in poaching wherever they come across the domestic reindeer of native peoples, and are known by the herders as 'two-legged wolves'. Even in Soviet times, losses to poachers were sometimes written off in Farm accounts under the extraordinarily frank heading 'Eaten by Russians' (*russkie s'eli*). Nenets herders near the town of Naryan-Mar on the western Arctic Ocean coast told me in 1999 how some members of the Russian population operated like a pack of wolves. Each December, as the Nenets drove their animals to the slaughter house*, town-dwellers would drive snowmobiles alongside the running herd, shoot the reindeer on the hoof and disappear with them into the midday darkness. The herders were furious, but dared not retaliate.

In Sebyan, there was little poaching since the small community was isolated from Russian industrial workers and offered no anonymity to its own people. Instead, the symbolism of the wolf turned inward: the new plague of wolves stood for something rotten within the community itself. The people who gobbled up the herders' wages, who left them without vital supplies, who profited from selling them vodka and then stood by while they stabbed each other in a blind alcoholic rage, were their own relatives. The wolves symbolized an internal predation within the community and a rejection of the taiga ethic of mutual support.

But the wolves and their human counterparts were not the only ones who could be *khitry*. Some of the wolves that dented the Farm's economy were what the herders themselves called 'symbolic wolves' (*simvolicheskyye volki*). The joke went that the local wolves had become *gramotnyye* (literate), because they read the earmarks carefully before deciding which reindeer to kill. The herders were more inclined to eat reindeer owned by the Farm, which had not paid them for years, than their own reindeer; the wolves took the blame.

The dog is an inversion of the wolf. I have not come across an Eveny legend about the origin of dogs, but imagine that such a legend might recount that the wolf, like the reindeer, was offered the option of domestication. Wolves that chose this path became dogs, and today the distinction between them is perceived in terms of human interests: as well as hunting, the dog assists the herder to look after domestic reindeer, while the wolf destroys their joint work.

Legends are recorded from other northern peoples, and often dwell anxiously on the fragile threshold between dog and wolf. A Nenets legend recounts how the first dog was put on earth by their creator god in order to look after his herd of reindeer. But when this god came down from the sky some time later, he was furious to find the dog asleep on a huge pile of reindeer bones – that is, behaving like a wolf. Since then, the Nenets have allowed dogs near their reindeer only under the watchful eye of human minders. In Sebyan, too, there is an anxiety that dogs may slip back to their wolf-like origins. There were repeated accusations over strays, poor breeds, and dogs that turned savage, and these problems were said to be increasing. Kostya once raised a litter of orphaned wolf cubs and trained them to work as dogs, but this was very unusual.

The domestication of dogs, like that of reindeer, removed the uncanny frisson of dealing with Bayanay. Domestication is vulnerable to reversion, but so long as it can be sustained it creates a new kind of relationship between humans and animals: there is nothing secret or dangerous in the language used to talk about domesticated animals. Not only are they aligned to human purposes; by sharing a working life with humans they develop a new kind of individual identity. Among Bayanay's wild animals, one reindeer, elk, or bear is much the same as any other. Even when the hunter carries out rites to ensure their reincarnation, they may know him next time round but he will not recognize them in their new bodies.

Dogs, horses and trained *uchakhs* had their own names. With the help of the specialized Eveny vocabulary, a skilled herder like old Nikitin could also distinguish each of the 2,000 or more reindeer in a general herd. The identity of domestic animals was based on their involvement with particular humans through overlaps and intersections between their biographies. Consistent with the principle of analogy that is so fundamental to Eveny thinking, a domestic animal could stand for a particular person at a crucial moment in that person's life. It might mirror the human, reflecting something that happened to them, by a kind of sympathy; or it might act as a substitute or surrogate, even saving the person's life by dying in their stead.

In the course of my conversations with Vitya the Wolf-Hunter, I came to see that his life was dominated by a particularly powerful recurrent pattern of substitution. He told me of several situations where he believed he would have died if a domestic animal had not taken the full force of the danger and died for him.

'I was hunting for mountain sheep one autumn in camp 9,' he recalled, 'and right on the summit of a mountain I got caught in an avalanche. I lost my footing and slid down the mountainside with the snow. I came to rest on a rock just on the edge of a 300-foot precipice. I broke my ribs and my breastbone here.' He pointed to his body. 'Somehow I managed to use the tip of my knife to dig into the stones, inserting it into one little crevice after another, and hauled myself all the way up to the top again. But my gun, my gloves, everything was down at the bottom – and the dog was dead.'

'Are you saying that you survived because the dog died?' I asked.

'It seems like it [*vidimo tak*],' he answered. 'That was way back in the 1970s. Here's another example. In 1992 I was transporting explosives on horseback to the miners at Endybal, a 42-kilo sack. There was a steep bit with very fine scree, so I dismounted and led the horse up this slope. Then I started coming down the other side. It had rained overnight and suddenly these fine stones gave way in a landslide. The horse was flung all the way down and my hand got burned with the friction trying to hold on to it. The horse slid down 600 feet with the stones and was killed. One of the Russian geologists helped me to skin the horse and we shared the meat. But I was so sorry I almost cried – it was such a fine horse!'

Sometimes an animal would die, not as a substitute for a human but to mirror a person's death. Ivan told me that soon after the director had ordered the liquidation of his reindeer in 1990, some of the director's own reindeer were killed in an avalanche. Though the number of animals killed was smaller, Ivan already saw this as a just retribution.

Some time later, the director suffered a more drastic loss with the death of his wife, a Russian woman who taught in the school. People felt that her death had changed him.

'He really did love her and went downhill after she died,' Lyuda told me afterwards, then added with uncharacteristic sharpness, 'or else he depended on her to advise him!'

Afonya was more charitable. 'Before she died, he was very unfeeling,' he told me once. 'He didn't bother to go to funerals, and people were offended. But since then, he's started coming to funerals. He's become capable of feeling the fear of death, and of feeling compassion for others.'

By all accounts, the director's wife was a coarse person who bullied the herders' children at the school on the strength of a fake teacher's certificate. 'She was very fat, with high blood pressure, and died of a stroke,' Ivan told me years later with great satisfaction. 'During the debate on regional devolution she insulted our Eveny elders and told them to take their dirty

Tungus mattresses off to Sakkyryr. The spirits heard, and she died suddenly soon afterwards!'

I was not in the area when the director's wife died, but when I returned I heard several accounts of animals that had mirrored her death. All involved the people of camp 7. Some people said that the Old Man had shot a female bear in the mountains at the very moment when the director's wife had died in the village. Others told a similar story about a wild reindeer. When I caught up with Granny three years later, I got a richer and more complex story. She was still bitter at the death of her husband and was convinced that he had died of a broken heart. The day the director's wife died, she said, the men of camp 7 were training an *uchakh* for someone who was a mutual kinsman of both the Old Man and the director. As part of this training, they tied it to a tree. Suddenly the *uchakh* took fright, 'as though it were being chased by a wolf', gave a bound and stumbled. In the fall it broke its leg, so that they were obliged to kill it. Just as they were skinning the reindeer, they heard over the radio that the director's wife had died.

Granny reckoned that this must have happened at the very moment when the reindeer was struggling. 'It was her death agony,' she told me drily. 'She took the form of a reindeer.' Then, concerned that I should understand not just the metamorphosis but also the moral, she added, 'For us everything is connected, really, truly. There is a God, who ordains our fate.'

Granny's son Ivan, the professed sceptic, firmly believed this. Vladimir Nikolayevich, a truer sceptic, thought it was just 'a good story!' The multiple versions in circulation, and even the doubts, did not undermine the power of the idea, but on the contrary showed how this kind of thinking was so fundamental to the Eveny that they needed to explore it many times over in several incompatible stories. In all these versions, the animal that died was not a substitute for the director's wife, like the dog and the horse that saved Vitya, but paralleled her death. In their mountains nearly 200 miles away from the village, members of a persecuted family enacted a scene that mirrored a death taking place in the family of their persecutor. They were agents of the animal's death, making it satisfyingly close to an act of revenge, but they did not cause the woman's death and did not even know it was happening: they were simply working with an *uchakh*. It was as if the reindeer's origin provided a channel for their abiding hatred of the director to tune into the distant event, so that a deadly impulse passed into the animal.

Only a domesticated animal, and only a domesticated reindeer, could carry such an intense load of social meaning. The versions involving a wild reindeer or a bear were relatively weak, since wild animals lack individual identity and it is only in passing that they may be made to stand for a human. But when you give away a domesticated reindeer, it does not forget its past association with you and can act as a link between its two owners. Even while the last of the old people who remembered making a ritual voyage to the sun were slipping away, the reindeer still remained far more than could ever be grasped by the minds of the Soviet Communists or of their market-oriented successors today. In the accounts office, a reindeer represented x kilos of meat which could be sold to the coal miners of Sangar. With good contacts through middlemen who knew how to get them to Korea, it could also represent y kilos of medicinal antlers. These commodities cost z roubles in overheads, such as vaccinations and helicopter journeys, to produce. Whatever the permutations of state and private ownership, herders were paid (if they were paid at all) in relation to these notions of costs and returns.

But outside the accounts office, even in the private lives of these same managers and accountants, domestic reindeer were not merely the objects of human plans and actions. They could also be subjects with a will and intention of their own. Among the Eveny, nearly everyone who lived on the land had a *kujjai*, a reindeer that had been specially consecrated* to protect its owner from harm. When you were threatened by serious danger, your *kujjai* placed itself in front of you and died in your place. Each time your *kujjai* died, you knew that it had probably saved you from death, even though you might never find out what the threat was. You then had to consecrate another reindeer to maintain the same level of protection, like renewing an insurance policy. In the course of a lifetime, a person would get through many *kujjais*.

Though dogs and horses could die in their owner's stead, the stories told by Vitya and others did not suggest that these animals died with any awareness or consent. Only a reindeer could sacrifice itself knowingly and intentionally. In the world of the Russians who gave the Eveny Farm managers their bookkeeping culture, animals could behave like this only in fairy tales. Beyond allowing itself to be harnessed, ridden, or eaten, this was surely the deepest expression of the contract that was proposed to us by the first pair of reindeer: to stand for a human, and to die in the human's place, of its own free will. The domestication of the reindeer created a category of creature that had a soul, but which could be killed or allowed to die without anxiety, guilt, or spiritual danger.

The act of consecration was simple. On my first visit to brigade 8 in 1992, Tolya bought a white reindeer from Kesha as a present for my 7-year-old daughter Catherine, whom Tolya had met in England, and consecrated it as her *kujjai*. He muttered a prayer over it and then released it in my direction, telling me to photograph it and take the picture home to her. To live far from one's *kujjai* was now quite normal, and people in the village might keep a *kujjai* in a herd run by relatives. Children might be shown their *kujjai* soon after they were born, and when they were old enough they were sometimes sent out to graze it near the camp. When selecting a reindeer for consecration, one should look for a particularly fine castrated male. This was exactly the sort of reindeer that might otherwise be trained up as a favourite *uchakh*, and I suspect that the *kujjai* was an intensification of the special relationship between a reindeer and its rider. But when choosing a reindeer to turn into a *kujjai*, some people looked for additional signs.

'A *kujjai* is a very special kind of reindeer,' Vitya told me. 'Its eyes aren't like an ordinary reindeer's. I can't really explain it, it's like a shaman, I suppose – it's hypnotic.'

'When it dies, does this always mean you've been in danger?' I asked him.

'They can go in various ways – some get lost, some get eaten by wolves, some just get old and go away to die. You can even kill your own *kujjai* and give the meat to somebody else, but you mustn't eat any of the meat yourself. That's a great taboo.'

Some people said that a *kujjai* should never be harnessed or ridden, but should live a life of ease, 'just walking in the taiga', as Gosha once put it. Others saddled them, but in a way that showed that a *kujjai* was so closely identified with its owner that it was almost like an animal double.

'If another person tries to ride your reindeer or harness it to a sledge, then you'll fall ill,' Vitya explained. 'For example, my wife has a *kujjai* but if *I* harnessed it, *she'd* fall ill. But it's no problem if you ride your own *kujjai*.'

The identification with one's *kujjai* could also make its death alarming.

'Last year I had a disturbing dream,' Kesha once told me. 'I dreamed that a wolf fell into a trap I'd set. So when I woke I went to look, and in the trap I found my own *kujjai*, with its leg broken, so I had to kill it. I was very frightened.'

In Sebyan, only certain kinds of white and dappled reindeer were chosen. Gosha taught me an elaborate vocabulary of Eveny words for a white reindeer with black patches, a silver-grey one, a white one with slight darkenings in certain places, even one 'the colour of white metal'. I did not know what to make of these words, especially as we were sitting in a log cabin in the village with no reindeer in sight. I thought at first that they were simply descriptive colour words, but Gosha insisted that they were also terms of respect or awe, like some of the names used for wild animals when in the forest, except not secret. Another way I understood Gosha's talk of respect was that reindeer of these colours were particularly suited for consecration as a *kujjai*.

But Vitya had a quite different view: 'I'm from another region,' he objected. 'I prefer a brown colour. A white or dappled reindeer doesn't suit me, it gives me an allergy.' I thought this word so bizarre that I asked him what he meant. His answer gave an illustration of a *kujjai*'s identification with its owner that goes beyond allergy into comedy: 'I had a dappled *kujjai*, but I gave it to somebody else because it was making me ill all the time. But that person drove it and beat it so hard that I got bruises on my buttock, here' – he half rose from his seat and turned his back towards me – 'I couldn't sleep at night, it hurt so much!'

The director's dying wife, Vitya's sore behind: these were signs of how a domesticated reindeer could retain memory-traces of its successive owners and link their lives. The connection did not even need to be very close, so long as it could be traced. It might not seem significant until some extraordinary event occurred surrounding the animal's death. The fight in the winter hut of camp 8 led to an event of this sort. Ganya, the angry young herder whom I pinned down all night in his sleeping bag in 1995, was stabbed soon after in a fight in the village. One version was that Ganya had drawn a knife first and then someone stabbed him. Another was that he had fallen on his own knife. However it happened, his spleen was punctured. Ganya had lost a lot of blood and lay in a coma. Sasha the Radio Man tried to summon an emergency helicopter to airlift him to hospital in Sangar. But low cloud and persistent snowfall meant that day after day, as the boy's life slipped away, no help came from the other side of the unyielding wall of mountains.

Meanwhile the reindeer-racing went ahead as scheduled on the frozen lake near the village. This was the local version of the larger festival that I had attended in Chersky in 1990 and which had been discontinued after herders from across the region had turned it into a political protest meeting. Unlike the secretive policy meetings or elections held in the village hall, the date of the festival was made known in advance to the herders, who sent in every man they could spare from the camps. Almost every one of the village's few hundred inhabitants came together on the snowy shore of the lake and spilled onto its frozen surface. Herders and riders wore heavy reindeer-fur coats with inlaid patterns in brown and white, while village women wore the long city-coat which is the winter uniform of women who do not work on the land. Families pitched tents or sat in open-topped trucks, boiling kettles and sharing out meat from the camps and sweets from the village shop. Boys rode around on *uchakhs* and young men showed off on sledges. Others used the knives on their belts to cut blocks of ice out of the lake and carve them into reindeer heads or bas-reliefs of racing and hunting scenes.

The older adults went around shaking hands, Russian-style, pulling the reindeer-fur mitten off their right hand and replacing it each time. I, too, shook many hands, and met the ever moving Vladimir Nikolayevich for the first time since our hunting trip together four winters ago.

'Did you really give up hunting after that time?' I asked him as he released me from his padded furry hug, also a Russian gesture.

'I meant to give up.' He laughed. 'But it's boring in the village – there's nothing to do. This year I went hunting beyond camp 8.' Then he told me where every hunter had been moving throughout the winter and what they had caught in each location. Most of them had still not returned to the village. The taiga telegraph was as finely tuned as ever.

The mood was festive, but underneath there was a subdued anxiety. Without Chersky's Stalinist military marches on crackly loudspeakers, there were just the sounds of conversation, wind, and boots scrunching on snow. The lake had seen many deaths. Far beneath its ceiling of ice, the body of the Russian aviator

Levanevsky was believed to lie in a sunken plane*. Levanevsky had disappeared on his country's remotest frontier in 1937 while pioneering a route from Moscow to Alaska. But Ganya was dying in his own home, among families who were all touched by the rise of violent deaths among young men. As I walked around greeting people, I could know only a few of the past sadnesses which Eveny etiquette keeps unspoken and unwept – and none of the future. Within a year, the spot right in front of where we chatted and laughed would become the grave of Lidia's brother-in-law and his drunken driver as they drove a tractor onto weak ice.

The shorter races were on reindeer-back, the longer races with a pair of reindeer harnessed to a sledge. The most prestigious 5-mile sledge race started invisibly at the far end of the lake behind the freezing fog, out of which emerged tiny dots that would be identified by the people around me almost before I could see them. Muffled megaphone announcements were carried away on the breeze as each race came in, reindeer and drivers both panting, the fur around their faces covered with ice crystals. Spectators ran out onto the ice to greet the winners, who were often my friends from camp 8: Lyuda, the All-Yakutia Champion, and Kesha, his eyebrows, moustache, and long beard caked with ice from his exertions.

Suddenly one of a winning pair of reindeer collapsed and died. The knots of spectators were widely spread out across the scene, but gradually the news spread that the reindeer was connected to the dying Ganya. The connection seemed to vary in different accounts. Someone said it had belonged to Ganya and he had then given it to Manchary. No, I was later corrected when I repeated this story myself, the reindeer that died belonged to one Petr Petrovich, and the reindeer that had been Ganya's was still alive. The connection with Ganya was that his reindeer and the one that died had been harnessed together in the race.

Around me I felt a low murmuring, as people started taking

the reindeer's death as a sign that the boy might live after all. Nothing had changed in the low cloud and the snowy haze on the mountains. But around midday a helicopter clattered over the nearest peak and landed far out on the ice. It stayed for an hour: perhaps it had brought a surgeon, people suggested, perhaps they were doing an emergency operation in the middle of the lake. Finally, the helicopter took off. The first reports that reached the village from the hospital were that Ganya's condition was hopeless. The long, thin, home-made knife had passed through his kidney and bowel. Later it was reported that he would live but that his spleen had been removed as being beyond repair and that he would always be an invalid, hardly able to walk. Three years later I came across him, minus his spleen but working in the brigade as if nothing had happened. The reindeer in the race had not died in vain.

12

Dreams of love and death

'I dreamed I had a very old reindeer,' said Lidia. 'There were lots of mosquitoes sitting on his face.'

She thrust the teapot across the floral plastic tablecloth to refill the cups that Gosha pushed towards her. Instead of the unbreakable tin mugs used in camp, these were made of thick china, and matched the large white dots of the teapot against an orange background, like a toadstool. On the table between us lay a frying pan full of reindeer mince and macaroni, which we would finish as the winter evening wore on. It must have been late, though it was hard to tell, as it had been dark since early afternoon when I first started to ask her about symbolic meanings of reindeer. The trickle of the tea broke a silence that flowed in through the snow encrusted on the double windowpane from the other blanketed huts of the invisible village.

'It was a predictive dream, the kind that happens towards morning when you're half awake,' Lidia continued. Without the headscarf she always wore in the summer camp, her hennaed hair fell in long wisps across her face. 'I wanted to ride this reindeer but I could only sit on him – he was incapable of moving. There was blood flowing from the left side of his neck. I tried to staunch the blood but it kept flowing between my fingers. From one of the reindeer's eyes there flowed a big tear. I felt such enormous pity for the animal.'

She paused, then added significantly, 'Exactly a year later, my father fell ill with liver cancer.'

Here was another dying reindeer standing for a dying person. Lidia believed that the reindeer represented her father, and that its collapse foretold his terminal illness. This dream had given Lidia a rare glimpse of a future event. But it could be understood only later, through the unfolding of her life. Her interpretation of the dream was complete only when it was fulfilled; the meaning of the dream was contained in the subsequent event.

A psychologist might interpret this dream as an expression of Lidia's anxiety about her father, and say that it revealed something about her own mind. Eveny knew about psychology, and used the word 'mind' when speaking in Russian. But they maintained that prophetic dreams offered insights into invisible processes that were at work, not in the mind, but in the world around us. The role of the mind was to be alert to clues, and to reflect on what one noticed. When Granny listened for a distinctive crackle in the fire, or Kesha read resemblances between cracks on a bone and rivers on the landscape, they did so as keen observers of the environment. If they noticed an untypical pattern, or a striking analogy between two forms that were otherwise unconnected, they took this as a pointer to something significant in reality itself.

They were specially alert to anomalies in the behaviour of animals, both in the omens of waking life and in dreams. It was a bad sign when dogs howled too much or when insects came to your tent but to no one else's. It was especially alarming if creatures of the wild came right inside. One hunter woke to find an eagle flapping its wings inside the entrance of his tent, and later learned that his brother had been dying in the village at that very moment. Tolya's mother once found a jay burning in her stove, and later realized that this foretold that her brother-in-law would hang himself.

The eagle and the jay each represented a doomed person, like the reindeer in Lidia's dream and at the death of the director's wife. In other cases, a dying animal was a substitute for a person who would live at the animal's expense, like the horse and dog that saved Vitya on precipices, or the *kujjais* which saved many men, women, and children in ways they might never know.

Despite the Soviet regime's efforts to create a new society of new people, Eveny tended to relate the quirks of fortune, not to State institutions, but to animal psychology. Rather than the bureaucratic arbitrariness of the Farm, their deeper point of reference was the wild inscrutability of Bayanay, tempered by the cooperation of their domesticated reindeer.

Many of the predictive dreams that Eveny remembered and recounted concerned love or death, or their terrible intertwining. These were uncertain areas of personal destiny, which could not be reached by the assurances of scientific rationality and industrial technology. The Soviet State's claim to benign social competence was still belied by their own internal conflicts and the fate of people like Baibalchan seventy years ago, while even the authorities' helicopter magic failed to arrive for Ganya until it was stimulated by reindeer-soul magic.

Lidia was still keen to oblige my curiosity about animal symbolism, and told me another dream.

'I had this dream when I was a young woman,' she started. 'I didn't know it at the time, but now I realize it was showing me my entire destiny.'

The stove had died down and was giving out a low, steady hum. As Lidia started to speak, Gosha rose slowly from the table and pushed on the reindeer-fur lining of the outer door. The door creaked open at an upward angle and slammed shut behind him. It was −30°F outside and he had not put anything over his indoor clothes, so he was going out for only a moment.

'I was riding uphill on another reindeer,' continued Lidia. 'He tried to reach the summit, but just couldn't make it. The reindeer collapsed and died. Then I changed onto a horse. The horse made a huge effort, he really strained, and in the end the horse made it to the top. From the summit of the mountain I looked down on the dead reindeer and felt terribly sad. Then I woke up. Later, as you know, I married twice. My first husband was Eveny, but he died. And now I'm married to Gosha. His health is poor, but I know from that dream that if I look after him well, he'll be all right.'

Lidia's interpretation of this dream was clear. The contrast between a reindeer and a horse was a common code to distinguish the Eveny from the Sakha. The reindeer represented Lidia's first husband Oleg, Ivan's elder brother who was killed in a fight shortly before I first reached camp 7. The horse was Gosha, who was partly Sakha.

At that moment Gosha came back into the house, heralded by a gush of freezing air. He stood still holding an armful of logs as the door swung shut behind him, and gazed at us. He knew the story and I sensed that he had been imagining the telling of it in his absence, like someone who leaves a room during a piece of music but keeps playing it in his head. With the strong folds of skin over his soft, brown eyes, and his unusual drooping moustache, I was suddenly struck by his classic Sakha appearance.

In my quest for animal symbolism I was still trying to work out each animal's role in the story, and the conventional meanings of the horse and reindeer reinforced my habit. But now I began to wonder whether I had been missing the point of such narratives. I had not been paying attention to the way in which people told these stories and listened to them. Animals could be involved in other kinds of indirectness apart from standing for a person. They could allow a speaker to hint at intimate emotions, which for all one knew might never be declared in more explicit terms. This was another facet of the Eveny sense of discretion and awe in the face of other sentient beings, a feature of Eveny sensibility that Lidia herself has done so much to help me understand.

Looking back now, this conversation helps to clarify a moment when I was living with Kesha, Lyuda, and their children in the winter hut of camp 8, on one of our fishing trips to the frozen river. Kesha and I had drilled our holes near to each other and were sitting on the ice in our reindeer-fur coats. We sighted a ptarmigan on the bank and our talk turned to various kinds of birds. He glanced down the river to where Lyuda was guiding little Dima's hand and jigging his line up and down through another hole in the ice, and dropped his voice.

'The first time Lyuda ever came here, it was for her practical when she was a veterinary student,' he said. Lyuda seemed a long way away, but she looked up sharply. I felt sure she knew what he was saying. 'We didn't know who was coming, or when. It was in July. I'd gone off to look for reindeer, it was a misty morning, very early, you could hardly see in front of you.' Kesha closed his eyes and spread his hands out as if feeling his way through a cloud. 'I was riding along the bank by a lake, when suddenly there was an amazing noise and a flurry of wings. And straight in front of me a swan landed on the water and started swimming back and forth. It was the first time I'd ever seen a swan. It just sat there and swam up and down in front of me, and kept on saying something in its own language. I stopped my *uchakh* and sat there entranced. I feasted my eyes on that swan.'

He paused significantly. 'And soon after that, Lyuda turned up from Irkutsk,' he concluded.

'You mean that swan *was* Lyuda?' I deduced.

'That's right,' he confirmed, 'she came to me in advance, in that form.'

How deeply imbued personal narratives are with cultural symbols, and how readily these symbols can be adapted to one's own circumstances! For Lidia, the conventional ethnic meaning of the horse and reindeer in her dream merged seamlessly with the fragility of men as supports in her life, and her concern about their sickliness and mortality. Kesha's swan was a bird of the tundra hundreds of miles to the north, where it symbolized beauty, grace, and freedom, and it came to the taiga as an extraordinary visitor to bring these qualities into the heart of his own life.

Like Gosha during Lidia's dream narrative, I sensed that Lyuda had been following Kesha's narrative from a distance. In each case, there seemed to be a further significance to the half-presence of the other person who was implicated in the story and knew it already. Coming from outside the relationship, I provided a new occasion to repeat the story, but I was not the only audience. The narrative seemed like a veiled declaration of a feeling that could not be told* to the other person's face, or in their presence alone.

When Lidia told me her dream, I realized I had heard it before, from her friend Varya. It seemed that dreams could take on a life of their own in the telling. An oblique confession of intimate feeling was moulded into a public narrative and became widely known as a landmark in someone's life-story. The dream came to resemble a myth of origin in which a clan, species, or custom was established at the beginning of time, like the story of the first reindeer to choose domestication. A situation might be not simply established, but also justified or defended. The public repetition of these dreams and omens served to validate the marriage by repeating the story of how it began. I have heard people do this in other countries, too; but here, the validation was so often based on an animal apparition.

I noticed the combination of ethnicity and male fragility in another of Lidia's dreams about her marriage to Gosha.

'This was after my first husband died,' she recalled. 'I was walking in a meadow with several girl friends. My friend Sveta was carrying a Russian knife in the dream, and she's married to a Russian. Another girl had a Sakha knife and she's married to a Sakha.' I knew these kinds of knife. A Russian knife is one made on a factory assembly line, while a Sakha knife is hand-made by blacksmiths in the Sakha plains, with a handle of birch wood.

'I was the only one without a knife,' Lidia continued. 'I

thought, I need a knife too! I walked and walked along the meadow, and at last I found a big Sakha knife. I thought, How come no one's picked up this knife? It had no sheath [I have heard elsewhere that a knife without a sheath symbolizes an orphan without relatives], and as I looked along the blade I saw that it had little nicks in it. It wasn't a new knife. I thought, I'll polish it up and make it sharp again. Then exactly a year later I met Gosha. I thought, That knife was Sakha, and this man is also of Sakha origin. His health is delicate and he needs looking after. Maybe I'll marry him. And sure enough, I did.'

Varya had told me this dream too, and her account of it was very similar to Lidia's own. But then Varya had also added a story of her own: 'You know, exactly the same thing happened to me, only not in a dream, in waking life! When I was young I was in the forest with my mother and some of our relatives, picking berries. I really did find a knife on the ground – not a Russian one, not a Sakha one, but a small, home-made Eveny one. I picked it up and called to everyone, "Look what I've found!" The women started to joke, "A husband – good find!" Exactly a year later, I married Tolya.'

The resemblance between the two knife narratives seemed astonishing, and the identical imagery revealed deeply rooted patterns of Eveny thinking that flowed freely between people's lives and through sleeping and waking states alike. Again and again, I was to notice how significant dreams and omens were said to be fulfilled exactly a year later, though it was to be a long time before I would be able to connect this to the experience of nomadic time in their annual cycle of movement across the land.

Lidia's sensibilities were darker than Kesha's, whose radiant experience of the swan fitted his bold and optimistic nature. Her dream-animals and other coded allegories of emotional attachment often seemed tinged with a fear of loss. In particular, I felt that the cultural symbolism of her dreams was adapted to express an anxiety about the vulnerability of the men in her life. Her father and both her husbands appeared as riding mounts, but in each manifestation their ability to carry her was outweighed by their collapse or their need for her care and protection.

One light summer evening in camp 7, I sat with Lidia and Gosha in their tent, the Daik stream trickling through the dwarf willow scrub nearby. Several years earlier, she said, in a tent just like this one she had been lying down near her father, from whom she inherited the tendency to prophetic dreams. Suddenly, she had seen him get up and go out of the tent, though she also knew that he had never moved, but was still lying asleep. At this moment, she understood that he would soon die.

Then she started to talk about the recent loss of her beloved brother. As his illness became graver, he had been airlifted to the hospital in Sangar.

'But,' she said, 'just before his death, he came to me in the village.'

'You mean in a dream?' I asked.

'No, in waking life [*nayavu*].'

This seemed odd. It would have taken him days or weeks to get a flight.

'You mean he actually came?' I asked, still failing to understand.

Lidia did not answer, but continued, 'But I didn't see him. He knocked, and I ran to open the door. I called out, but he'd already gone down the steps and left. I ran to the window and couldn't see him, so I ran to the other window – again he wasn't there. And then I thought, But he's flat on his back in hospital in Sangar, how could he come to me? He really wanted me to be there in Sangar. I wanted to get there but I had no money and there was no plane.'

So the sick, dying man had yearned to see his sister so much that he had come to her in spirit! I could not see her face clearly in the pale summer night, but felt a clumsy urge to fill the silence.

'You didn't see him, but I'm sure he saw you,' I said, responding as best I could.

'Yes, I'm sure [*naverno*],' she murmured.

'He wanted to warn her in advance,' explained Gosha.

'He came to say farewell,' she confirmed. It was Lidia who would later explain what I now understand whenever I see someone set out on a journey, the Eveny avoidance of the finality of saying goodbye.

Just as the messages of Lidia's dreams about animals and knives reinforced each other, so her brother's death had been preceded by many signs, all conveying the same message.

'A year before he died,' she said, 'he was lying down on the couch after lunch, alone. And as he lay there, a woman came into the room. She was dressed all in white, with long white hair.'

'And she stroked his hair,' added Gosha. Lidia and Gosha often speak with one voice, but here again was a hint of how private dreams and omens can become public property. Other people, too, probably knew this story and could recount it like a script.

'She stroked him,' said Lidia. 'But when he got up, there was no one there.'

'He woke with a start, in a panic,' Gosha went on. 'Even in his sleep, he had a feeling of panic.' So the script extended to the man's supposed inner state, perhaps reproducing the way her brother had described it himself.

'He realized she'd come to him in a dream,' Lidia explained. 'He wasn't asleep, but he wasn't awake either. He tried to look, but couldn't see her face.'

'Later, when he was lying ill, she came to him again,' continued Gosha. 'The door opened, that's how she came in. He could even hear the sound of her footsteps as she walked.'

'Just like a real, living woman,' Lidia added.

'Who do you think it was?' I asked.

'I think it was a sign,' Lidia answered, 'a ghost.'

'"She's come from the next world,"' Gosha added, quoting the sick man's words, '"she's come from heaven to fetch me!"'

But if this apparition was a portent of Lidia's brother's death, it was not the immediate cause. He was taken by the restless ghost of another close relative who died forty days earlier in a stupid accident. This relative had borrowed Gosha's motorbike and sidecar to fetch a block of ice from the lake.

'He'd brought us our water supply and he was making a final trip to get some for himself. But then he perished [*pogib*].'

'How perished?' I asked. Maybe my curiosity was stronger than my tact, maybe I was supplying the prompt he needed to continue telling the story. Eveny narratives of terrible happenings have a restrained, relentless impetus of their own. The tone remains quiet and even, and is not distorted by any noticeable upwelling of feeling.

'Well, I waited and waited, he didn't come back and didn't come back,' replied Gosha. 'I'm waiting and waiting, till I can't wait any longer.' Or was the strength of feeling betrayed by this unusual switch to the present tense?

Lidia was still speaking in the past tense. 'Then our little Stepa came to the house –'

'– and brings us the news. I can't believe it.'

'"In the lake, on a tractor, he's fallen in" – just like that!'

Once on the lake, Lidia's brother-in-law had parked Gosha's motorbike on the ice and taken a lift with a friend in a tractor. But the driver was drunk and careless, and drove onto an unstable patch of ice. He should have known that the river continued to flow underneath there after it entered the lake, and bubbles of air got caught under the ice, making it very thin. The passenger door had been welded shut because it kept swinging open in the cold, so there was no quick escape. When the tractor plunged through a crack, they plummeted to the bottom in their metal coffin to join the aviator Levanevsky 100 feet below.

'My motorbike was left standing in the middle of the lake,' said Gosha. I visualized the familiar, friendly sight of a motorbike and sidecar parked far out on the ice, seemingly unattended with its rider hidden by a frost-haze as he crouched over a fishing-hole. They were always painted a cheerful orange and offered the prospect of a lift back to the village. The image suddenly became desolate and terrible.

The Eveny have adopted the Russian Orthodox idea that a dead person's soul roams the earth for forty days before moving on to the next world. But this was a bad death. The dead man had become a dangerous ghost and spent his forty days trying to take someone to keep him company. Lidia's brother was just unlucky.

'This man had been appearing in the dreams of all our relatives,' she said, 'they were all having the same dream.'

'What was the dream?'

'They dreamed that he was calling them,' she replied. 'He was determined to take somebody. He even came to my brother's children.'

'But their grandmother knew what to do,' said Gosha. 'Old women sleep beside the children when they get these dreams.'

'They also put scissors under the children's pillow. That's how they kept my brother's children safe. So he took their father instead.'

'He was an open and straightforward man,' Gosha said. 'He'd never been harsh with anyone in his life.'

'He never asked anyone for anything,' added Lidia, 'he was a free man.' She used the Russian word *svobodny*, which for the Eveny also means self-reliant. These words express the most highly valued traits in the Eveny repertoire of character.

The angelic vision came for Lidia's brother a year in advance, and again shortly before his death; the man in the tractor took him after forty days because he was related but unprotected. Other signs, too, had been leading inexorably to his death; but sometimes one does not understand even the clearest of these. Lidia's brother used to go regularly with Gosha on winter hunting expeditions.

'But last winter,' Gosha said, 'Bayanay was *too* generous.'

'They filled four sacks with fish, this big!' Lidia spread out her arms.

'We got elks as well.'

'They got elks, they got everything!'

'Ten mountain sheep,' he enumerated. Then, his voice becoming flat, he added, 'Nature doesn't simply give as freely as that.'

I had often heard about a hunter who was too successful at hunting. He would catch something every day. Animals came up to him, and birds sat on his shoulder or came into his tent. 'Even mice crowd around him!' I was told. All of these were sure signs that the man was about to die.

In trying to interpret this idea, I have come to understand that the soul-force in the taiga is finite, so that each person is entitled to take only a limited share of animal life. When Bayanay gives you more than normal, he may be delivering all your remaining balance in one rush. If you are young, you are being paid in full for a life that is about to be cut off; if old, it is your last fling, a terminal glory, like the huge harvest of Ivan's father just before he died.

I had been so full of these ideas for years, and had internalized them so deeply, that I would have been very alarmed to find myself killing so many animals so quickly and easily.

'Weren't you afraid at the time?' I asked Gosha. 'The two of you, didn't you think of that?'

'We didn't know that it would be like that,' he replied softly.

'He just thought that Bayanay –'

'– was greeting us –'

'– as new visitors on unfamiliar territory,' finished Lidia.

'That happens sometimes,' Gosha explained. 'We thought Bayanay was pleased with us. Look at Tolya, for example, he always has Bayanay. And if he goes to a new place, Bayanay gives him a successful hunt. The spirit of the land helps him, the sun's energy, the moon, the air, everything helps him.'

I thought of hunters I know who seemed to be successful

without putting themselves in danger. All were very careful about reading the mood of the landscape, though in many different ways. Tolya waited for Bayanay to send him erotic dreams, Vladimir Nikolayevich relied on the meticulous observation of animal behaviour, Kesha paid special attention to messages on heated bones. All fed the fire.

As I listened to Lidia and Gosha, I could almost feel her brother being suffocated by omens as they pressed in on him: the woman in white, the sunken tractor, the hunt where he could not stop killing. In other conversations, Lidia also added the knock-on effect of a witchcraft incident in an office, a grouse (*glukhar*) that flew into the village and perched on a post by her brother's house, and a herd of phantom reindeer, which he made the mistake of following in an attempt to greet their phantom herder.

In this dangerous world, how could one know, how could one do the right thing at all times? One's relationship with the environment was strangely like any Soviet citizen's relationship with the State. Just as a citizen might have to break the law in order to get through the day, so even expert herders and hunters could never be sure if they had offended a spirit or broken a taboo. People tried their best: don't sleep with your head pointing downstream, or your soul may be carried away with the current; don't lie in the posture of a corpse, or you may die and take up that posture for real. But however alert you were for signs, you might still fail to understand the meaning of something as obvious as a freak glut of game. Even the warning signs themselves might seem like a trap. How could you insult Bayanay by rejecting an animal that he put in front of you as a gift?

Dreams threw you crumbs of insight and left you scrabbling for them. Sometimes the sign was simple enough that you could understand a situation and take control of it, even within the dream. Motya once dreamed of two of her friends who had died.

'I saw them on the other side of a fast-flowing river,' she told me. 'They were beckoning to me to cross over to their side. One of them was already in the water, coming to help me across. I started wading into the water, but suddenly realized what was happening. So I turned back, all in the dream, and so I'm still alive.'

But often, prophetic dreams and omens gained their force from the very fact that they could not be understood in time, and became clear only retrospectively. Such dreams and omens were made up of several elements, but tantalisingly reflected the uncertainties of waking life by omitting a vital piece of the puzzle. They hinted at an event or situation, but in order to act on this you would also need to know the person to whom it would happen, as well as the place and time.

A dream that was remembered as ominous might even withhold every possible clue – at the time. One of the most obscure dreams I have heard was told to me by Varya.

'I was walking by a river with my brother's wife,' she said. 'It was some place I didn't know. It was dawn and our mood was sad and slow. We were making a garland out of flowers. Then my dream ended. Exactly a year later, the two of us were summoned by the police to a place I'd never been, far from where we lived. They'd pulled my brother's body out of the river. He'd got drunk and fallen in and been carried far downstream. We travelled all night and arrived at dawn. They took us to the river bank, and I recognized that very same meadow I'd seen in the dream a year before. Just as in the dream, I was walking across it with the same woman, it was dawn, our mood was just as it had been in the dream, and we *were* making a garland of flowers.'

This dream put up impenetrable barriers to any understanding before its fulfilment. The place was unknown and the man whose fate was foretold was completely absent. Nothing was specific except the identity of the onlookers to an event that was still hidden – and their mood. Yet the meaning of the dream revealed itself with shocking clarity when it was fulfilled. It was as if the dream were taunting the dreamer with her helplessness.

One did not always have to feel like a victim of destiny. Kesha's technique of divination by fire and bone was an attempt to see through obscurity in advance. While probing into the future, he tried to make it reveal enough about a future event that it could serve him as a guide to action. The bone did not specify the event but if it promised a catch, this was good enough. The question he posed concerned himself, and because the bone turned into a map it revealed the place.

But the bone could also predict disaster. At one point while Kesha was filling my notebook with his delicate maps of the filigree cracks that appear on an animal's shoulder blade, he suddenly drew a heavy cross.

'If there's danger, it always shows like this,' he declared. 'Suppose I'm planning to ride somewhere, perhaps on my own, perhaps with my family. So I heat the bone, and this cross emerges. Here I am, looking at it. Well, I know my own territory, don't I? So I understand where the event will happen. For example, here's the River Sakhandjá with several big tributaries, here are the headwaters, and this is the Kelten. So say it makes a cross somewhere in the middle of this bit here – like this! I look and try to work out all the possible disasters that may be waiting for me in that place: for example, I may meet with an accident, a landslide – it could be anything, even a fatal illness.'

I was startled by his boldness in supposing such dangerous possibilities. I did not think Lidia and Gosha would have done this, even to illustrate a point, though I could imagine Vladimir Nikolayevich doing so, just as he once jokingly lay down on a sledge in the position of a corpse.

'So if you see a cross, does that mean you shouldn't go there?' I asked.

'It means it's warning me,' he answered. 'I should behave very cautiously there. Or go somewhere else – I shouldn't go there at all.'

Soon after their herdboy Ganya was stabbed, I visited Kesha and Lyuda in their house in the village. Diana was playing in the snow and ran inside to announce my arrival. By the time I had taken off my boots and coat in the lobby, my tea was poured. Inside, the house was warm, with Lyuda cooking reindeer tongue against the background of a quiz show on Moscow TV.

Above the epic signature tune of the Moscow news headlines, Lyuda told me, 'If a visitor is going to come from far away, Diana cries for a whole day beforehand. But she never cries before you arrive.'

Diana was playing nearby with a doll. Lyuda looked across at her.

'She gives signals,' she continued quietly. 'We know if someone's coming, but also if someone's going to die. I didn't understand at first, I took her to the doctor: "Is there something wrong with her heart?" But the doctor said there's nothing wrong.'

'She gets it from my ancestors,' added Kesha. 'I've only got it a little bit. I'm a bit of a seer, but only in dreams. She has it far more. She knows about my hunting. When we drink tea in the evening beforehand she purrs endlessly, Prrrr! Then I'll get a lot when I go hunting. If she doesn't do that, it's already certain that I'll get nothing. My daughter is very complicated.'

'Shh,' warned Lyuda, 'not too loud!'

Diana showed no sign of having heard. But she and her father were passionately attached, and she must have imbibed her parents' certainty that she had special gifts. I had no way of telling how she understood the idea that she perceived more than most other people in the world of hidden signs.

Lyuda continued, 'If someone's going to die, she can't stand being near them even for half a second.' Her voice became impassioned. 'She gets aggressive and attacks them. Take Kristina's brother-in-law: she wouldn't go near him, she knew, and straight afterwards he went and shot himself.'

I wondered about Ganya, still lying in hospital on the other side of the Verkhoyansk Mountains. Had Diana given any clues to his fate? Her parents did not answer directly, but Kesha told me about a recent dream.

'It was very inconclusive,' he began. 'There I am feeling very anxious, I'm not sleeping well, and in this dream I don't actually see Ganya but I have a strange feeling. And then I lie there and try to work it out. It means his fate is wavering in the balance, plus or minus.' He stretched out his arms and made a gesture of scales teetering up and down. 'I don't want to say that he'll be all right, but I don't want to say—' He broke off. 'No, it all depends on the person himself. If he can take the strain then it's O.K., but if not then within a day or two he'll die.' He had switched from talking about fate to talking about the boy's own constitution. 'I still feel it could go either way,' he concluded. 'They said over the radio that he's still unconscious. I'll sleep a bit tonight and maybe I'll have another dream. I sleep in a very agitated state.'

'Is it true,' I asked, 'that you shouldn't tell anyone about a good dream until it's fulfilled?'

'That's right,' he agreed. 'If you tell a good dream too soon, then you'll stop it from coming to pass. But if you see something bad in your dream, then you must tell people straightaway to prevent it from happening.'

If talking about a predictive dream prevented it from coming true, then this was one of the few means available to humans to influence the future: to foster a good sign and block a bad one. This confirmed what I had heard many people say. But occasionally I have heard the reverse. Gosha and Lidia once said that to tell a bad dream makes it come true. When I asked Kesha about this, his response suggested that he saw the matter more in terms of remaining attentive and open.

'You have to pay attention to your dreams,' he explained, 'you have to take them seriously, and the next morning you don't say anything to anybody but you just try to analyse them. Everyone has their own style of dreaming, and their own style of interpretation. So for me, in my dreams, money always means some kind of illness, like flu or inflammation. But when I see strong alcohol in my dreams, then I immediately go out to hunt and the hunt is always successful. If I dream of fire then it always means something unpleasant will happen to me.'

Some of Kesha's symbols were conventional in Eveny dream interpretation. A man's dream of strong alcohol or of a sexual encounter with a woman foretold a successful hunt; a girl's dream of a knife said something about her future husband; and any Eveny person's dream of a Russian foretold violence or serious illness. Other symbols were used more idiosyncratically, and I have noticed that people interpret money, fire, water, and bears in diverse ways, perhaps making connections and developing meanings through the experience of their own lives.

Perhaps this is not very different from dream interpretation among people I know in England, but it also matches the Eveny emphasis on self-sufficiency and flexibility which runs so counter to the Soviet emphasis on doctrinal certainty and single explanations propagated in school. Old people, who had had least exposure to schooling, were the most emphatic about this.

When I asked Granny about consistency, she explained, 'It all depends on your life – the interpretation has to fit you. It's better if you interpret your own dream. Somebody else might interpret it differently. You may even come to doubt your own interpretation later. I got it wrong, you may think.'

When Granny talked about dreams, she took a very long perspective. Instead of evasion and control, which are perhaps the concern of a younger person, she seemed more concerned with recognizing connections, as if seeking a retrospective coherence in the story of her life. Even her satisfying sense of revenge when the director's wife died was set in a framework of inevitability that seemed almost karmic.

'I always believe in dreams,' she once told me. 'I had many dreams, and it was only later that I realized that they meant that my eldest son was sure [*dolzhen*] to die.'

This son was Lidia's first husband Oleg, the reindeer that collapsed and died halfway up a mountain. So Oleg's wife and mother had both had dreams foretelling his death, each from her own perspective.

'When I was young,' continued Granny, 'I used to discuss my dreams with my husband. But he didn't like it. "We aren't shamans," he used to say. "It's all nonsense!"' We were sitting in Granny's tent in camp 7, surrounded by Yura and Ivan, who both laughed. 'So later I stopped telling him my dreams. But before he died I had a dream. We were climbing down a steep mountain together and he fell right down, out of sight. He got separated from us and I couldn't find him. He was with our elder son, they both disappeared. At the time I took no notice, but after he died I remembered it and reflected on it.'

The most spectacular attempt to cheat fate that I have heard was told to me by Vitya the Wolf-Hunter. He came to the village as a child, when his mother brought him from another Eveny region several hundred miles away. One winter, when I was living alone in Tanya's apartment in Yakutsk city a few weeks after Ganya was stabbed, Vitya came to visit me on his way home from one of the meat-packaging courses in Alaska (which he thought was like Siberia, only with more food and more bars). We perched on stools in the kitchen while I served him tea, sausage, and pickled cucumber. Across the street, the nineteenth-century wooden houses, buckled and nearly capsized by the permafrost, were due to be replaced by a smart concrete apartment block, to be built on low-conductivity stilts by construction workers from Georgia and Chechnya. A few doors down, the new regional parliament had recently declared economic autonomy from Moscow in order to lay claim to a larger share in the region's diamond revenues.

It was in this little kitchen, suspended between the unprecedented freedom of rootless travel abroad and the waiting entanglements of the taiga, that Vitya told me about his recurrent attempts to escape from a deadly dream. He had worked as a herder all over the State Farm's territory. Many herders changed brigade from time to time during their lifetime, and the

Farm might redeploy them wherever there was a shortage of manpower. But Vitya had a further reason to keep moving on.

'I worked for sixteen years as a reindeer herder in a brigade at Terekhtyakh,' he said. 'The rivers there all knew me, I'd been riding all over for sixteen years. Wherever I went on every river, I would stay the night, feed the fire, have supper, relax. But all the women started getting angry with me in my dreams. A woman in a dream means a river, it's the spirit of the place. And they said, "Get out of here, go away and leave our family." So I moved far away, to the lowlands. I became Sakha and worked as a horse herder.'

He took a sip of the tea I had placed in front of him. 'Then one of my horse herders got into a fight with some reindeer herders,' he continued. 'It was Red Army Day and they'd been drinking in a village. He ended up shooting himself. So I gave up horse herding and became Eveny again.'

I tried to work out what he was telling me. Wherever he went, the spirits of the place rejected him. In the same breath, he said that someone was killed, apparently because of their association with him. It was the beginning of my understanding of a logic that I came to see as pervading Eveny thinking, that humans could die as substitutes for each other, through the same logic as the *kujjai*, or his own horse and dog that had fallen down a mountainside. I tried to probe at the connection.

'Do you mean that he died and so you stayed alive?' I asked.

'It can't be,' he answered. It seemed I was either wrong, or too direct. But why was he telling me this?

'It can't be?' I echoed, playing for time.

'Yes, it can be,' he mused, 'it can be.' From what followed, it seemed clear that he had already thought of this, and believed it.

'After that, the Farm appointed me brigadier in camp 7. The Old Man himself was the brigadier and he handed over to me. He was retiring, he said he was tired of working: "Here you are, you do it!" So I did it.'

I had not realized that Vitya had been the brigadier of camp 7. I did not recall any mention of this by Ivan or Granny. From their family narrative, I had formed the impression that the Old Man had passed on the position of brigadier directly to his eldest son Oleg, Lidia's first husband, before the position came down to Ivan on Oleg's death.

Vitya sliced off a piece of sausage and continued, 'Everything went well for a while. Then I had a dream. There was a woman talking, an old woman. She must have been the spirit of the local river. She said to me, "Go away, you don't belong here, you're not wanted here!" So I left.'

Wherever he went, Vitya made regular offerings to the rivers. But nonetheless, the women who kept reappearing in his dreams reminded him that he was a social and spiritual outsider and showed him that the landscape did not accept his attention. It communicated with him personally, but only to exclude him.

'What would have happened if you hadn't left?' I wondered.

'Well, after me – you know Oleg?' he replied darkly. 'He took over straight from me, and then he died in Sangar under mysterious circumstances.'

Was Vitya suggesting that this, too, might have been instead of his own death? The logic of the unintentional human *kujjui* became clearer as Vitya brought his story up to date.

'This autumn, I had lots of dreams like that again. This time I was working in camp 11. That's why I left, why I decided to get right out of the area and travel.'

Here was someone who thought he knew what a prophetic dream portended, and was determined to forestall its fulfilment. Little did the US Congress know, when they underwrote the training programme in sausage making, that it would also serve as a defence against a recurring bad dream.

'That's why I'm still delaying my return,' he went on. 'I've been wandering from town to town, you understand? Look, I'll draw it on this bit of paper.' He reached across the table for the notepad by the phone. 'This is the village, and this is me.' He sketched. 'It's like a hunt. I fly like this, and like this, from one town to the next, so as not to leave any tracks. Something bad has got to happen, I know for certain.'

Here was a vivid and explicit image of what it feels like to be on the run from fate.

'Who exactly is on your trail?' I asked. 'Spirits?'

'Perhaps.'

'Can you outwit them?'

I had put it more unguardedly than he would have done. 'Ha!' he reflected. 'That's interesting, isn't it, to outwit them!'

'Will you go back to the village now, or wait?'

'I'll hang around here for another week, I'm not sure, I'll wait until something bad happens, then I'll know it's safe to go home. Ganya was only wounded, he didn't actually die.'

A huge revelation had opened up in front of my eyes, one of those moments of connection which make an anthropologist dizzy with discovery. This was how the logic of substitution worked! Only when Ganya or another victim died would Vitya feel that the pressure had been taken off himself. He could not block the impetus of the event, only deflect it onto someone else. Each time he was threatened in a dream he would run away. When someone else was killed he was left wondering whether they had died in his stead.

But also relieved. It was a shockingly frank admission of how one survived under threat, at a level that seemed older and deeper than the backstabbing of village politics. Perhaps every community has an ultimate threshold between social responsibility and self-preservation, and I had just found the Eveny point. I felt an enormous compassion for Vitya.

As with other significant dreams, it turned out that Vitya's story, too, was public knowledge. But I found that other people had very diverse understandings of his case.

'Nye-e-et!' responded Tolya scornfully when I tried out my interpretation on him, with the special intonation he always uses to mean I have completely failed to understand Eveny culture.

According to Tolya, Bayanay was seeking Vitya's death in exchange for taking a quantity of animal souls beyond his entitlement. 'The women in those dreams are sent by the local Bayanay of each place. Bayanay is angry with him because he's been catching too many marmots!'

Back in the high mountains of camp 7, I asked Ivan about Vitya's time there as brigadier and received a different explanation again.

'He's a cheerful fellow, very active, runs around,' Ivan responded. 'I like him a lot. He has good memories of us in camp 7. But he can't live here, it's too mountainous. It's because he comes from a place with low hills. His horse doesn't like mountains, his gun doesn't like mountains.'

Ivan seemed to agree with Vitya in tracing his problems back to his position as an outsider. But his explanation was psycho-ecological. The most basic tools that a person needed in order to live on the land colluded in Vitya's alienation. His horse and gun appeared as extensions of his person, and shared his feeling of a lack of harmony with the environment which was irredeemable.

I think Ivan was being sincere in his appreciation of Vitya, and I have never met anyone who does not speak warmly of him. But given the strong family feeling in this brigade, I wonder whether there had also been a sense that Vitya was interrupting a family succession. Even if the Old Man had handed over graciously, as Vitya said, he had nonetheless succeeded the Old Man as brigadier over the heads of the Old Man's sons. Though nothing was expressed openly, the dream women who chased him off their territory may have been expressing a resentment at his presence. What the Eveny called the actions of 'spirits' ran parallel to what one might call the impact of 'unconscious' thoughts and feelings.

This kind of interpretation appeals to me, but can never be 'verified'. Even if I could explain it to the people involved, they might deny it. But this might not make it any less valid. An anthropologist tries to perceive patterns and repeated connections in people's thoughts, feelings, and actions. Cumulatively, these amount to a distinctive style of relating to the world and to each other, which we call a culture.

The Eveny, too, seemed to explain things that were difficult to understand in terms of patterns and connections, though often of a different kind. Once a puzzle appeared, interpretations of an Eveny sort would start to cluster round it. But one interpretation need not exclude others. Vitya had upset his balance sheet with Bayanay by killing too many marmots (Tolya); he was living in mountains when he should have been living in lowlands (Ivan); he was living on the territory of a family where he did not belong and had taken over leadership of the family enterprise (my reading of Ivan's family history). The explanations were over-determined, like the dreams supporting Lidia's marriage or the omens preceding her brother's death. Even if you dismissed one there were others, and they all pointed in the same direction.

As Vitya and I moved on from sausage and pickled cucumber to salted fish and dark chocolate, he revealed yet another possible explanation for his troubles. As I was rummaging in the fridge, he suddenly said, 'I sing songs. It's as if there's something missing inside me.'

He then told me that he composed songs in Eveny and sang them at the top of his voice as he rode across the mountains, saying that the words and the melodies just rolled off his tongue and that he was unable to hold them in. This behaviour was extraordinary, since everyone knew that the spirits of the land hated the loud, intrusive noises of humans imposing themselves on their territory. Songs were powerful, like prayers and spells, and had to be aired with great discretion.

'That's true,' he acknowledged. 'The old-timers, the old men and old women, if you speak loudly, if you shout loudly, if you sing outside the tent, they think all of that is taboo, they don't like it. Perhaps that's why I keep having that dream. But I'm generally a cheerful person [*ya voobsche vesely chelovek*]. Even when I'm on a bus in the city I sing all the time. I sing in bars and pubs. I *like* singing songs.'

Surely it was safe to hear one of his songs here? In a melody that combined elements of Russian 'romance' and archaic Tungus clan song, his clear tenor floated around the kitchen:

Ajit gu ajit gu,
Ayavalcha bihem gu?
Nongon elle emereken,
Dyulavo ireken.
Nyulten togan duleni,
Dolba nengu ngeritni.

Ajit gu, ajit gu,
Ayavalcha bihem gu?

Seriously, seriously,
Have I fallen in love?
When she comes into the tent, even at night,
It is immediately filled with light.
When she goes out, the birds do not sing,
Even the wind does not play.

Seriously, seriously,
Have I fallen in love?

I thought of Vitya's long and rich love life, and felt as if the little north-facing winter kitchen had indeed been filled with light.

I heard him sing this song again six years later. In 2001 I escorted a film crew to the spring hut of camp 2, which he had just liberated from the three-legged wolf's reign of terror. Our little red and white biplane skied to a halt on a frozen river a mile away from the hut. Driving a team of brown reindeer with not a piebald animal among them, and wearing brown furs and dark glasses against the reflected glare of the spring sun, Vitya sledged up to us and dismounted. While the camera turned, he rested his hand on the plane's flank and serenaded it in a declamatory style, like a Victorian salon artiste leaning on a huge, winged grand piano.

Looking back, I cannot help seeing a deliberate perversity in this man who was so accomplished that he could stalk a cunning wolf that had evaded every other hunter in the region for years. His loud singing across the landscape seemed to contradict his professions of respect for the spirits and the fire. It suggested, not simply that he recognized that he did not fit, but that he was defying and provoking the spirits who would not accept him. A person without a location, he lived a complex life, taking spiritual risks and playing with hidden powers.

13

Sacrificing at a nomad's grave

I came to the Arctic after studying another shamanic society, a pre-Hindu tribe in India called the Sora. Sora spirits were elaborately named and distinguished*, and I counted over a hundred categories. Each category of Sora spirit was said to reside in a particular rock, tree, or stream. To emphasize their distinct identities, each kind of spirit had to be summoned by singing its own signature tune and fed with the sacrifice of a specific kind of animal, the blood of which was poured into a prescribed number of cups made from the leaves of a particular species of jungle tree.

Having learned religion from the Sora, I expected that the Eveny, too, should identify many categories of spirits, each with its own distinct name – and that I might perhaps use these spirits as a guide to their ideas about the essence of human life. But all my enquiries seemed in vain. My friends were uncertain about the age and characteristics of spirits, and sometimes even their gender. The most I was able to coax out of them was a general agreement that the spirits of water were women (some said old, some said young), and that Bayanay and the spirit of the fire were both old men. Though I joined them in making offerings every time we crossed a mountain pass, no one was ever able to *describe* the spirits of these places at all.

'Is the spirit of the fire really distinct from the spirit of the Earth?' I once asked the precise and articulate Kesha.

'*N-n-n-yet*,' he replied hesitantly. 'I don't really know how to tell you. They're all sort of the same [*vse odinakovo, vse budto*].'

I tried another approach. 'How do you know the spirit of a lake is female?' I persisted.

'It must be,' he replied logically, 'because when you talk to her, you say "Grandmother" or "Mother, give me some fish". The same as when you ask Bayanay for game: "Grandfather," you say, "give us an animal."'

For a long time, I assumed that the Eveny must have suffered a cultural decline and lost what they knew before. But after many such inconclusive conversations, I was gradually forced to realize that I was missing the point. As I saw how the Eveny related to spirits, and learned to do so myself, I began to understand that they perceived them as spirits of a place, and that the identity of each spirit derived from the character of the place itself. Most of these spirits had no other name, except the name of their place. The character of spirits was brought into focus, indeed the spirits were made manifest, one after the other as you migrated through a succession of sites. The spirit of a place was said to be identical equally with the spirit of the river or lake beside which one camped, as with the spirit of the fire that one lit there. The act of pitching a tent drew the spiritual forces of fire and water into a partnership to create a habitable place.

As in their attempt to replace the family with the brigade, Soviet attempts to change Eveny life gave rise to a simplified parody or perversion of everything that the regime tried to supplant. The land, too, was rationalized into a huge open-air meat factory, but this did not completely destroy an earlier sense of the land as a huge open-air temple.

Other religions, too, offer a succession of paths for movement and places to rest during an act of ritual, contemplation, or worship. In a typical ancient Egyptian or modern Hindu temple, worshippers progress through ever more sacred enclosures, finally reaching a cave-like inner sanctum; in a typical Christian church, they enter at the west and progress down the nave towards the rising sun, in an orientation representing the transition from death to life in Christ. In such buildings, the most sacred area is experienced as the destination of a journey, the sense of sacredness intensifying as the worshipper moves ever closer to the focal point. If the building is also a site of pilgrimage, this journey can extend halfway round the world.

But in the sacred space of the Eveny, there was no final destination, no one camp site that was more spiritually charged than the others. The herders progressed around a succession of places that never came to an end. The sacredness of each place was equally intense – but the herders would engage with it for only a few days, until they were forced by their animals' onward migration to move on to the next place.

Then the old place would lie unvisited, not enlivened by human presence, as if asleep. The seasons came and went, and in a climate with little wind they left astonishingly little trace. I came back to a camp site after a year and found a lost piece of paper from my notes, my child's unfinished wood carving, or the champagne cork from my wife's birthday; as I migrated with Yura, he suddenly dismounted to pick up a small black lens-cap I had dropped a year ago almost to the day.

These moments showed how intensely the annual nomadic cycle tied an event to a place, and suggested an answer to my puzzle about why momentous predictive dreams were so often said to be fulfilled 'exactly a year later'. This was the moment when a nomad returned to the place where the dream occurred. It was as if the place were a gateway, which was at its most open at the same time each year. The dream was a kind of pre-echo foreshadowing the event, which became fulfilled when one returned to the same site.

If you understood a warning dream sufficiently, could you forestall its fulfilment by not returning to the place where you dreamed it? People did try to thwart bad dreams by telling them to others, staying clear of certain places, or even leaving a territory altogether for a while like Vitya. But it was an unavoidable condition of the reindeer herders' nomadic life that one generally had to come back to the same place at the same moment each year.

But a place was not always exactly the same. One August in camp 7, I went with Ivan's uncle Terrapin to fetch firewood. He led me up an exquisite mossy stream where I had never been before. The narrow valley was overhung with crags draped with a floppy kind of juniper. Down on the ground, every strand of moss held silver air bubbles under a thin film of living water, so that I felt I was destroying worlds with every step of my massive rubber boots. The stream led us through an old camp site. Here were the old storage platforms on stilts and the rectangular lines of stones that had once pinned down the edges of tents; over there, the remains of a trap for the wolverines that raided the stores, which would have buried them under an avalanche of logs when they tugged at a piece of meat. There was also a small corral for the *uchakhs* and milking females, and even some old cardboard which had been wedged under stones without rotting away.

This place was only a mile from our current camp site at Djus Erekit, and was another version of it. Terrapin remembered living here for several years running in the early 1980s, always at the same season around 10 to 15 August. Herders' routes contain many such variations, as they sense that a site has become exhausted and needs time to recover. The effect of these variations is like sketching in pencil, where an outline is drawn repeatedly, but always a little differently.

Reindeer, too, vary their routes as they alternate between travelling rapidly along paths and lingering in places where they graze, mate, or give birth. The reindeer themselves 'know' where and when to go, as each place in their repertoire 'draws' or 'pulls' [*tyanit*] them in turn. But their repetition is likewise tempered with variation. Each migration route contains many alternatives within itself; on any given day, the animals may follow each other in single file or may fan out and create a new path.

Reindeer temperament contains a blend of conservatism and curiosity, especially among mature females. Just as in the migrations of wild herds, there is a tension in domestic reindeer between the animals' instinct towards cyclical repetition* and their response to changes in the environment. Even when their migration routes remain consistent for centuries, reindeer are constantly making fine judgements. In the low-lying bogs of the northwest Russian Arctic, the thrust of the great migrations from the summer coast to the winter forest far inland has not changed for millennia. But here in the mountains, the seasonal cycle is enacted up and down numerous valleys: a herd may even reverse its trajectory altogether provided they can find lichen for the winter, and a spring grazing ground may suddenly become an autumn pasture.

But there are limits to the reindeers' flexibility. Like their human companions, the animals in camp 7 had become accustomed to their own territory over an unbroken period of eighteen years until the slaughter of the herd in 1990. Within a few days, there was not a single reindeer left alive that knew this territory, apart from a few *uchakhs*. The new herd that repopulated the land was drawn from surplus animals taken from other camps. The animals did not feed well in their new location, the calves did not learn navigation from their mothers, and confused or adventurous leaders kept branching out in unsuitable directions. Ivan, Yura, and Gosha had to take turns sitting up day and night watching the herd, even outside the usual period of midsummer vigilance.

Granny was still complaining about this several years later. 'Reindeer always run away in a new place, they scatter helter-skelter in all directions,' she told me. 'We had to train all of them from the beginning. We even had to train them to come up to us to be milked. For them, this was a bare place.' By 'bare', Granny meant that it was a place devoid, not of vegetation, but of habit and memory.

Humans nurture their sense of space by naming places and

recalling family histories. The most coherent single exposition of this came from Kesha. One winter night, while Lyuda washed up the remains of supper, the bachelors read newspapers, and the children scampered around the bunks along the outer wall of the hut, Kesha started to draw me a map of his entire territory. As he marked important sites, he moved seamlessly from the reindeers' need for pasture to the deepest traces of human presence. The effect was like peeling back the layers of transparent paper from an old-fashioned anatomical diagram.

As with the migrations of Ivan's herd, it was easy to see the animals' steady upstream movement through the spring to the high slopes of summer, followed by a descent down another parallel river to the winter pasture. As Kesha explained the complex operation that he ran under the simple name of brigade 8, the most telling way he could find to express his pride was to say that they migrated more than almost any other brigade.

'In fact,' he said as he warmed to his theme, '*nobody* migrates as much as we do. Nobody's keeping an eye on them, so some brigades stay in the same spot for months. They stay in one place for the whole of July and another for August. They make just three camps through the entire summer and then it's back to their winter hut.'

I was used to brigadiers decrying each other for staying too long at each site. The need to move on frequently is one of the few points of agreement between the value-systems of the traditional nomad and the State Farm manager. But this was surely an exaggeration. I have seen Ivan and Granny in camp 7 move three times in August alone. But I said nothing.

'And so the pasture gets ruined,' Kesha continued. 'Let's calculate: camp 10 is completely grazed out, there's nothing in camp 1 or camp 11, half of camp 7 is used up, there's nothing in 5 either.'

This list included some of the best-run and most productive camps. Now I had to protest: 'But surely there are good brigadiers in those camps?'

'Yes, there are,' he conceded. 'But it doesn't depend on just the one brigadier. You have to think long-term. People want an easier life, and that leaves bigger problems for later. For example, camp 1 made a really small, curved fence to hold back their reindeer and now the lichen inside it has been trampled to death.'

Fences were a key symbol of control and management, but at the same time brigadiers also criticized each other for relying on them too much. Kesha turned back to the map. 'Look, we've got a completely ruined area like that too, here, from the mouth of the River Nebichan to the mouth of the Khaldykhchan. Before us, Ivan's father the Old Man worked there, and before him Starostin. They worked the Nebichan area very hard. They used it in autumn and winter. Now there's not a single gram of lichen, just green plants. It's impossible to keep reindeer there in winter any more.'

The Old Man had originally worked in camp 8 and when he handed it over to Kesha's father, Dmitri Konstantinovich, in 1972, the two families exchanged a part of their existing territories. To reinforce his point, Kesha added meaningfully; 'That winter hut dates from 1957.' I understood what he was gettting at: the comfort and permanence of the hut had lulled the Old Man into staying there too long each winter, keeping the reindeer nearby in one area and destroying the precious lichen, which would take decades to grow back.

The fathers of Ivan and Kesha shared the same grandfather, and when times became hard in the mid-1990s Kesha even suggested to Ivan that they might amalgamate their herds. But despite the closeness between the two families, I sensed a tinge of rivalry now as Kesha compared his own father's management of the territory with that of Ivan's father before him.

'This is our current route,' continued Kesha. 'My father Dmitri Konstantinovich worked it out and now I implement it. Before us, the Old Man used only this part, but we extended it to this new summer and autumn bit which nobody used at all – it was completely wild. We vary the route every three years, between the full circuit and the short cut, to preserve the pasture.'

I knew that the routes allocated by the State Farm were fixed year after year, until the State Farm decreed a change for its own purposes. How did Kesha square his method with the instructions of the State Farm?

'The Farm doesn't know about any of this. They have maps and plans. But I've been brigadier since 1978 and they've never given me a map – they just keep it in a safe. It's a formality, no one ever checks up on anything, so we do it with a clear conscience [*na chistuyu sovest'*]!'

Reindeer moved around the Arctic for millennia before the recent arrival of humans, and long before there were herders to impose any direction. As we talked, Kesha gradually punctuated his map of where the reindeer went with focal points of human action and control: huts, storage platforms, and corrals. These are new places, constructed where nothing may have existed before, so that reindeer behaviour could be crosscut by human intention.

Kesha dealt systematically with each kind of place, turning the physical landscape layer by layer into a complex human memoryscape. He started his account of human structures by marking a winter hut that he had built in a traditional Sakha style called *balagan*. In the grandiose tones of a dignitary inaugurating a monument, and using his own exotic adopted surname, he declaimed: 'Burtsev's Balagan, constructed September 1994.'

Then in a normal voice he added, 'This migration route was handed down by my ancestors. The only innovation I've made is to build that *balagan*: Burtsev's Balagan, the place is called now – my place in history!'

I wrote the name on his sketch.

Kesha pointed to another place. 'This is the hut my father built when he was brigadier, in the mid-1970s.' He withdrew his hand, to allow me to write on the map.

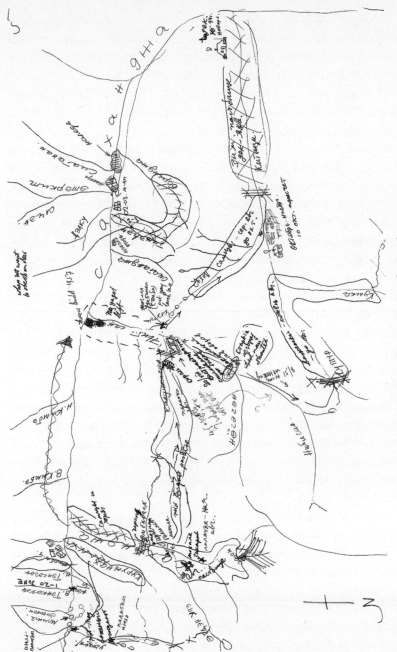

Kesha's map of the territory of camp 8.

Moving back upstream from winter to autumn, he said, 'Here's the corral where we do the autumn head-count, around 10 October. First we count the males, then we release them into their winter pasture over here. The next day we do the females and the calves. We choose the ones to send for slaughter and drive them off to the village – in that direction.' He swept his pen off the page to the northeast.

'These three huts and their storage platforms are our base during April and May.' He had moved his hand to the other side of the page. 'The females come here to give birth, and over here we keep the males separate to give the females some peace.'

Moving back to the autumn side of the page, he said, 'Here's where you visited us the first time, in September 1992, on the River Nimichan, remember? We were still living in tents but making trips ahead to repair the winter hut.'

'I remember how you carved the name Artemida on the doorway of the hut,' I replied. It was a tender memory.

'You told me it was Greek for Diana,' he answered, looking up at me. 'I knew you were thinking about your own little girl you'd left at home.' He turned his attention back to the notebook. 'Next we come to the Lower Tankachan River, meaning "Mighty Grandmother". It's named after an old woman who was very little but very powerful – perhaps a shamaness. No one knows anything about her, except that she was little. And just here are the poles of an old tepee. That's probably where she lived.' Kesha drew a trianglar tepee shape, and I wrote in the details.

'Here's the ruined *balagan* of Chief Mikele, from our Zakhárov clan,' he went on. 'He was one of the richest Eveny. He had reindeer along the Sakhandjá, at Endybal, along the Essiyu. He had so many he didn't even know how many he had. After the revolution they confiscated his reindeer. Around 70 per cent of the State Farm's reindeer came from his herds – his and two others'. How else would the State come to have any reindeer?'

By now, Kesha had moved on from his own recent presence

to the threads that linked him to the landscape through people who had lived there before. Mighty Grandmother was a lost person who had become no more than an obscure name and a heap of sticks. But Mikele still had an identity, because he had a living story. His traces reassured Kesha that the arbitrary relocations of the State Farm's brigade system had not cut him off from the pastures of his ancestors. One derelict hut, a few fallen beams, were enough to pinpoint the episode of collectivization and to recall the ruin of a prosperous family.

Not all identities could be known, but even unknown people long dead could remain intensely attached to a place and might reveal themselves as ghosts. At one point in our conversation, Kesha made a mark on the map.

'One of our old men was riding at dusk along the River Nebichan, just here, looking for his reindeer,' he said. 'He reached an old, old fence. There's no grave there or anything like that, just the site of an ancient encampment. Suddenly he heard a dreadful human shriek: "*E-e-e-e-e-e!!!*" His *uchakh* dropped to the ground and refused to get up. The old man was so terrified that he just abandoned the reindeer, retrieved his saddle and walked for 20 miles waist-deep through the snow.'

The place contained a residue of a forgotten human consciousness which was crying out for a recognition it would never again receive. For a living passer-by, unable to engage with its desperate inarticulacy, the only response was to run away.

Hut beams, tent poles, fences: these little arrangements of tree-trunks were left behind as an incidental by-product of a human presence that lasted for a few days each year, for a few years, and then ceased. The only kind of construction that the nomads built intentionally to leave a lasting trace was the grave, which concentrated the accumulated experiences and memories of a lifetime into one unmoving point. Before the coming of the village, each person would think for years about where they wished to be buried, in the forest, on a lake-shore, or on a commanding rocky outcrop, and they would pass the point at least once every year. Perhaps the early reindeer hunters, too, did this as the glaciers retreated and they became the first humans to occupy new territory, perceiving spirits and rapidly giving it the feel of an ancient, used land.

Most graves were set near a path, because the dead enjoyed visits from the living. But while a grave could protect you as you travelled across the surrounding territory, it could also harm. You must never look back when walking away from a grave. I still find this a difficult compulsion to resist, like the taboo on looking back as you migrate from a camp site, because of my liking for lingering farewells and my hunger to keep taking in each scene to the last. For the Eveny, the long, backward-turning gaze is not comforting, but dangerous. An extreme attachment to one place is suitable only for the dead: a living nomad must keep moving.

When passing a grave, you should dismount and add something to the pile of coins, vodka bottles, cigarettes, bullets, reindeer saddles, and snowscooter drive-belts that lie for years under the windless sky. I have only once failed to leave an offering, when our party decided that it was getting towards nightfall and that we had paid our respects there a few days earlier when passing the other way. Within minutes, our horses had bolted and scattered our baggage across miles of bog. It took all night to retrieve the horses and find the sodden saddlebags. My companions, ironically using KGB terminology, called this a 'low-level warning'.

Some of the graves that Kesha drew on the map were of his own ancestors, others of unknown people.

'Here's the grave of an old woman, an ancestral grandmother of our Zakhárov clan. All this was Zakhárov territory. And here's an old man, old Afanasy, he's one of ours too. He died not long ago, during the war, in the 1940s. This grave still has power [deystvuyet]. And here's another grave. This one does only good, never harm. We've no idea who it is, but we give them offerings and they always help us to catch moose, wild reindeer, sables, bears. And just a bit lower, half a mile down-stream, is the River Ambardakh. Right at the edge of the forest there's a skull lying on the ground.'

'Just lying there?' I asked, trying to picture it. 'No skeleton?'

'There must be a skeleton,' he answered, 'but the skull is all you can see. It's the same type of grave as the one in the Yaros-lavsky Museum in Yakutsk, not up on pillars or in a tree, but buried direct in the ground. It's been preserved all this time. No one knows whose it is, but it's very old.'

Herders were drawn to museums, but their feelings about them were complex. Native laymen held graves in awe on the land, and felt yet another kind of fascination when they saw them safely contained in the scientific space of a museum. Communist education encouraged a respect for archaeology as science, while disregarding native sensibilities about graves. Native scholars were uneasy about excavating them, and there were many stories of the terrible consequences.

The most extraordinary concentration of power, intensifying the grave's potential for both protection and danger, lay in the grave of a shaman.

'This place is called Chorkhon,' Kesha said, as the writing on our map grew ever thicker. 'There's a platform with a shaman-ess's coffin high up on it.'

'Is she still powerful?' I asked, already expecting a story.

'There was someone called Starostin who used to work in that area,' he replied. 'He cut down one of the larchwood supports to heat his stove. He simply didn't realize. The coffin was sus-pended high up on posts, he didn't realize and just cut one support. It was an old, weather-worn post. Then he went mad, completely lost his mind.'

'How could he not know? Wasn't it obvious?'

'No, you couldn't tell, everything had fallen down ages ago. First he went mad, then he killed himself. So my grandmother used to tell me that you shouldn't just leave cigarettes or matches or vodka on that grave, but you have to leave her a post – cut down a young larch sapling and leave it there. So whenever we pass that way, that's what we do.'

Kesha had shamans among his own ancestors, though they lay on what was now the territory of camp 7, a testimony to another part of his family history.

'My maternal grandmother was a shamaness, and her father was a shaman too,' he once told me. 'His grave is on the River Munne, on the right bank of the Belyanka, by the mouth of the River Djotondjá, where Ivan migrates now. I remember seeing his drum still hanging on the grave when I was little. There was no skin left on the drum, just the frame. There were also spirit effigies carved out of wood lying there. Nobody touched them, they were too frightened. If you touched them, an ulcer would appear on your skin.'

The way Kesha and others talked about shamans' graves suggested that they were a point where a shamanic impulse was stored while waiting to be actualized in a new shaman. Understandings varied about the residue of power in a shaman's grave. Another time Kesha told me about the shamaness, 'Before she died, my grandmother prophesied that she'd be reborn within three generations, and sometimes, I think – I just wonder – maybe my Diana . . . I just feel it, I can feel it already.'

I once went with Ivan, Yura, and a caravan of *uchakhs* down the bare Daik Valley, to collect firewood where the stream descended into forest near the grave of the apprentice shaman who gives his name to the area. Daik had not yet been fully initiated before he was killed by a landslide that pushed him into the river. Ivan told me, with characteristic but unverifiable precision, that he had died in the seventeenth century.

'Daik River, Daik Pass [Daik Okat, Daik Högnök]: the whole area is named after him,' he said, and then added, 'before Daik, these places didn't have a name.'

'Why was Daik buried just where he died?' I asked Ivan, thinking that it might be because of his freak death.

'Perhaps it's because he didn't live long enough to choose a place,' he answered. 'Normally you choose somewhere you love, so that you can see the land where you used to migrate.'

In this kind of grave, the body was laid on the ground and the grave built over it, some of the logs eventually working loose. Tolya recalled a time when you could see how the body had been wrapped in birch bark as a shroud, while Daik's bow still lay on top of the grave. Ivan remembered when the corpse still had hair. But now there was nothing more than one leg bone, and they suspected that the grave's power might have been weakened by this decrepitude. There seemed a curious analogy between Daik's interrupted initiation centuries ago and the execution or persecution of Granny's father, the old man who was punished for healing hoof-rot, and others, cutting off the transmission to their successors. I wondered whether this was a characteristic of the shamanic impulse that it was repeatedly blocked by circumstances, yet continued to push through, like Kesha's idea that his grandmother's power might be reincarnated in his little Diana. Was this perhaps a test of its power? I could not work out how to ask this, and in any case, the dour sceptic Ivan might not be the right person. I saved it up for another time.

'Did Daik have a successor?' I wondered, deciding to ask him something more mundane.

'I don't know exactly,' Ivan replied. 'But anyway, all our dead shamans are very powerful. They take the form of eagles and perch on the crags all around Sebyan. They guard our passes against intruders. Not long ago, a Sakha psychic came from the city to the village hall. She said she was a shamaness and wanted to give a demonstration of healing. But she was only pretending to do healing – really she was trying to suck the soul-force out of strong young Eveny people for her own evil purposes. But our ancient shamans protected the youngsters and made them strong. She accused the audience of blocking her psychic waves, then she ran out of the hall and jumped on the next plane out of the village!'

Coming from Ivan, this seemed astonishing. I had often witnessed the tension in him between rationalist dismissal and fervent belief, but usually his position was less fantastical. Spiritual battles between rival shamans formed a classic scenario in Siberian legend, and this motif was later developed into stories of the shaman's triumph over intruders and unbelievers like Russians and Communists. The most powerful grave in the entire region was that of a shaman on the territory of camp 5. This grave protected Kesha's family from a downpour and saved Kristina's sister from certain death in a blizzard, but most spectacularly it rendered a helicopter-load of Russian geologists incapable of heaving themselves off the ground until their Eveny guide had persuaded them to leave an offering.

A few elders were still buried in the old way, especially if they died in the taiga ('Why drag a corpse all the way to the village in summer through the bog?' as Vladimir Nikolayevich put it, with his mixture of practicality and dry humour). But for most people, burial had been removed from the land and centralized in the village – an arrangement that some felt led to an unhealthy clustering of ghosts in one place. The village cemetery lay beyond the last of the houses, between the space of the living and the start of the taiga. The dead were grouped alongside relatives, and their graves were identified with inscriptions giving their dates of birth and death, along with photographs that deteriorated gradually through many snows and thaws.

For the old, birth dates were often guesswork. For the young, they were accurate. The first time I studied the inscriptions in a cemetery, I came to the shocking realization that many young men died either on festivals or on their own birthdays. Here was a new and sinister continuity with the cycle of the seasons and with annual repetition. Some drank vodka and fought while celebrating, or staggered outside lightly dressed to relieve themselves and then collapsed into a snowdrift and froze to death. But for others, like the boy whose body I visited in his house, their birthday was a moment when they reached a nadir of despair and killed themselves.

Regardless of the age or circumstances of death, the few funerals I have witnessed have been grim. In tribal India, I was able to make a detailed study of death customs, tape-recording the ritualized lamentations and jokes and serving formally as a shaman's funeral assistant. But at an Eveny funeral I have never felt I could so much as take a photograph.

The dead person would lie at home, while the women of the house placed offerings of vodka, oilcakes, a cup of tea, and a knife at their bedside. They also laid out a plate with seven pieces of raw reindeer liver, one for each day of the week so that the dead would never go hungry. Inside the house there would be only hushed talk. Some said that one should not lament audibly, while others said that women could weep but men could not. I was also told that a person who wept too loudly would soon follow the dead.

The burial was attended by only a small group of close relatives. In the old days when everyone lived in the taiga, the body was laid on the ground like Daik or placed in a wooden coffin and lodged on a platform or in a living tree, suspended between the earth where the person used to live and the sky to which their soul would soar. Following Orthodox Christian influence in the nineteenth century and Communist sanitary regulations in the twentieth, a hole was now dug in the ground to a depth of 6 feet, using burning logs to melt the permafrost. The coffin was covered with logs, then a blanket, and then earth. Everything that had been contaminated by the funeral had to be thrown away. This included the crockery used for offerings and the dead person's bedding and clothing, which were usually burned in a bonfire beside the grave. The mourners would walk away without looking back, for fear that the dead person would interpret the most fleeting glance, the slightest betrayal of longing, as an invitation to take the mourner with them.

The burial of suicides was a specially secret and shameful

business and I did not see how this was actually done while I was there. I was told that they were buried face down in the earth with no coffin, especially if they killed themselves for no obvious reason or died with their eyes open. This was to prevent one victim from seeking further victims and triggering a suicide epidemic.

The grave was a palimpsest of all the doctrines that had touched the Eveny during the last 300 years, clustering around the mystery of death to offer competing certainties. The bland mottoes on the plastic wreaths served as neutral captions for a montage of shamanist offerings, wooden Orthodox crosses, and Soviet stars snipped out of red-painted sheet metal.

People seemed unsure about what happened to a person after death. Most agreed that for the first forty days the dead person would fly around and revisit all the places they had ever been on earth, 'to complete their work here before flying to heaven'.

'Will you go back to my house in Cambridge?' I once asked Tolya.

'Of course!' He laughed. 'And that damp seaside cottage in Yorkshire* where we stayed with Plato, and the conference halls you and I went to in Finland and Budapest.'

The older belief was that you would be reborn*, usually into the same family. This was the reason why you should be buried with your head pointing east, just like the head of the elk I saw on a platform in camp 8. But whereas the elk's reincarnation would have been taken care of by Bayanay, ideas about the reincarnation of humans had faded, and not many people followed them through as specifically as Kesha when he contemplated the origin of his little girl.

While traditional Eveny thinking easily substituted animals for humans in dreams and omens and in fending off the threat of death, the effect of these new doctrines was to make humans and animals less like each other than before. Christianity taught that humans had souls but animals did not, while Communism insisted that nothing had a soul, but nonetheless distinguished humans from animals because of their political significance as citizens of the State. While animals, particularly large edible species, were still thought to be reincarnated, humans now made an irreversible onward journey to another world.

Some people told me that the landscape and seasons there were similar to those of this world, with huge herds of reindeer surging back and forth between summer and winter pastures. By contrast, I was told at the funeral of an old woman that her orientation to the east was 'so that our grandmother can fly to heaven'.

I suspect that the topography of the next world was not important because attention was focused so intensely on the grave. Here, while the earth was still being shovelled in above the coffin, a reindeer was sacrificed. The meat was cooked on the spot, and the living and the dead ate it together by the graveside. The fear of contact between them, at least at this moment, was softened by this act of communion. Every mourner had to eat some of this meat, as did the men who thawed the permafrost and dug the grave. As well as the seven pieces of liver placed earlier at the bedside, a further seven pieces of meat were left on the completed grave, along with another cup of tea and a glass of vodka.

The sacrificed reindeer was not any animal from the herd, but the dead person's favourite *uchakh*, which would be reconstituted so that they could ride it over the swamps and precipices of the next world. It was taken as a bad sign if the reindeer struggled and a good sign if it died easily, as this meant that it was going willingly. Even village administrators, accountants, kindergarten teachers, and boiler attendants would be nomads in the next world: after they died, a suitable animal was brought to their graveside from a camp where they had relatives.

The skull and antlers were set on top of a post beside the grave and all the other bones were scrupulously gathered together in a wooden box, which was placed on a ledge under the antlers, out of reach of scavenging animals. Every bone had to be extracted from the flesh and meticulously counted. If even the smallest bone was lost, then the reindeer would be lame and the dead person immobilized. Sometimes the reindeer's skin was burned, along with other personal effects, sometimes it was suspended from a tree, where it preserved the animal's original shape with legs flopping down like an empty sack from which everything – meat, bones and soul – had been sucked out.

To visit a recent grave was to stand in an eerie silence of shattered objects, each symbolically 'killed' so that it could pass into the next world. Vodka bottles were smashed, cigarettes snapped in two, wooden sledge-runners cracked, tin bathtubs punctured. On a child's grave, toys were ripped and dolls mutilated. The grave was a portal from this world to the next which sucked in not only the dead but also everything that the living brought as offerings, and even – if they looked backward – the living themselves.

It was here that death most clearly revealed the essence of the Eveny person as an eternal nomad. One might think that the grave was the place where a nomad came to rest and was no longer preparing to move on. But the violence done to these offerings showed that they were being sent on an onward migration, slipping through the portal after their owner. Transformed irrevocably by their destruction, the reindeer, cigarettes, and bottles would continue to follow the dead person around a succession of camp sites which the State Farm could not reach.

14

Bringing my family

In August 1996 I took up Ivan's invitation to bring my family to live for a summer in camp 7. They were excited but apprehensive, and hurried to learn horse-riding and basic Russian. I, too, had mixed feelings. My quest to understand the Eveny had taken me away from home for months at a time, year after year, and I felt that my family would be less likely to resent these people if they had lived with them. I wanted my wife and children to experience the wonderful journey down the River Munne from the cool, bare slopes of Daik, past the deep mossy floor of the larch forest at Djus Erekit, to the high whispering poplars of Tal Naldin. On the other hand, this landscape was more dangerous than the Asian jungle where I had taken them on earlier fieldwork; and I was older and had become more fearful.

Bringing together people from two separate parts of one's life also involved another kind of risk. Would the herders and my family like each other? Would they judge each other by inappropriate standards? How would I keep my family from becoming anxious or unhappy, and showing me up in a bad light? How would their more direct behaviour find common ground with Eveny reserve, and how might this affect my future research? Having studied the lives of these people who opened up their homes to me, I would have my own intimate life scrutinized, under very demanding conditions. Not only might my family reveal aspects of me that I may have wished to conceal, they would also see through any persona I might have developed in front of the herders.

Perhaps I was more concerned with the effect on our hosts than on my family. On this landscape, even the smallest group of helpless outsiders would put a strain on the resources of a camp; I could always reason with the family later.

Sally, Patrick, and Catherine each responded to the reindeer camp in their own way, with that contradictory alternation of enthusiastic participation and the need to withdraw which I recognized from thirty years of my own fieldwork, but had learned to disguise. My wife Sally was a psychotherapist, and brought new psychological insights to the understanding of my friends' family relationships; for her, the landscape was beautiful but menacing. For Patrick, aged 19 and full of restless energy, it offered a fulfilment that eluded him in the confined spaces of England. Catherine, who was 10 years old, related to the animals, the people, and the forces of the land by constantly drawing them.

We were accompanied to the camp by Tolya, his wife Varya, and their boy Kolya, who was Catherine's age. Ivan was pleased to preside that year over such a large community, which occupied several tents like a small army. Granny, Yura, and Emmie were there as always, along with Uncle Terrapin. Masha the teacher had come out again for the summer. Gosha and Lidia were camped nearby, with the children Nadia, Lialka, and Stepa. There were several young bachelors, including a handsome, brooding relative of around 20 whom Catherine named the Bison because of his massive, muscular shoulders and his way of walking with his head thrust forward on his powerful neck.

Sometimes we would crowd into Granny's tent along with her family and the bachelors for tea and food, exchanging our own supplies for their meat and fish; sometimes Tolya's family and mine would each cook separately in our own tents or on outdoor fires. Sally and Catherine were both vegetarian, but

Sally ate meat out of courtesy. Though Catherine joined in the catching of fish and the slaughtering of reindeer, she would not eat them and her mother cooked for her separately from a supply of pasta, nuts, porridge, and packet soup. She also browsed on berries, and Emmie and Lidia saved her small mugs of reindeer milk.

Sally was not sure when she was supposed to join in the tent work and when she should just sit as a guest. She did not belong with the daughters who tended the bubbling pans by the stove under Granny's all-surveying gaze, but found her place with Varya along the side of the tent – not all the way to the back with Ivan, Tolya, and me as senior males, but further in than Uncle Terrapin and the low-status young bachelors who crowded near the entrance. Her cloud of red hair flamed among the short black hair of the men and the long black hair, tied back, of the women. Emmie was playful and at home, Masha sullen and out of place. Once when Sally made me ask Masha whether the women's life contained its own kind of fulfilment, she replied with an emphatic 'Uh-uh!' Her answer needed no translation.

In the evenings we came together after the dispersed and varied activities of the day. On one occasion, Granny found some batteries for the radio and the tent was suddenly filled with drooly saxophone music. Emmie, sitting on the ground and washing up with lukewarm water and an old rag, started dancing vigorously with her torso, giggling and rolling her eyes. Another time, Sally pulled out a box of chocolates and passed it round. Granny smoothed out the silver paper and folded it carefully. Noticing Sally watching her, she explained, 'It's useful for keeping matches dry in your pocket.'

Another evening, when I was discussing ancestral graves with Ivan (a topic which for me never palled), I overheard Sally talking to Granny at the other end of the tent. Through the screen of Varya's halting translation, they were swapping stories about their children's upbringing. Granny was the key to Sally's acceptance in the camp, and motherhood was their common currency. She was telling Sally about her anxieties for her children, things I had not heard before: how Yura had been invited to take part in an athletics event in Moscow when he was a boy, and how she had refused to sign the papers because she was frightened to let him go.

Meanwhile, Ivan and I continued to talk. For him, the location of graves was not just a research topic. He was beginning to review where he might be buried himself, and his voice had become choked. In almost the same words he had used several years earlier when the family herd was about to be destroyed, he concluded, 'Every site here is linked to my childhood, to my life.'

He fell silent. On the other side of the tent, I heard Granny saying, 'What did you do about warts when your children were little? I used to take my children outside the tent and expose them to the waning moon. I would count the warts down, to shrink them till they disappeared.'

Sally noticed the sudden hush at the other end of the tent, and glanced across at Ivan. His remark had not been translated and his eyes were full of tears. The look on her face seemed to say, 'Why is he crying because he had warts as a child?'

'I like our evening conversations,' Ivan told me late one night. 'They remove tiredness and lift one to the plane of the mind.' I later found out that he had not been making any entries in his diary since we arrived. It was not only a technical logbook, but also a companion in a life of solitude.

Catherine would often sit in a corner, drawing. Sometimes she drew alone, sometimes she would squat in the semi-darkness of the autumn night with the other children, all drawing together by the light of her headtorch. She drew animals, mountains, portraits of herders as they sat drinking tea, and everything she heard us talking about: Bayanay as the spirit of the landscape, or the soul of a dying shaman leaving his body and passing into his successor. The sight of a child from another world fascinated the herders. More intriguingly than an adult visitor, it allowed them to reflect on alternative ways in which a person can be formed.

'I used to draw very well,' Ivan once commented, as he watched Catherine. 'I was left-handed, too. I would mount a reindeer and use the lasso left-handedly. But in school they made me right-handed. I had to stand in the corner all day until I got it right. They tied up my left hand so I couldn't use it. Then in the Army they made me shoot right-handedly. Now I can't even handle a knife properly.' Then he added with sudden vehemence, 'Don't let them change you, Katya!*'

In the 1990s there were barely any left-handers in Russia.

'Here it's been allowed just for the last three years,' Ivan continued. 'It's democracy. The same as they threw away Communist insignia and trimmings. Children can even decide not to go to school!'

'Everything is permitted today,' added Gosha. 'Left-handedness, long hair – anything you can think of!' It was a reminder that it was not only old Communists who equated democracy with anarchy. The paradoxical sense of loss that accompanied increased freedom allowed herders to look back to the State Farm of the Soviet period as their 'traditional' native way of life.

'Catherine should write a book in the year 2020,' Ivan said, 'when all this is gone, saying, "This is how they used to live – I saw it with my own eyes!"'

During the day we would sit outside, stretching dressed hide over the forks of antlers to repair reindeer saddles, or using thread made of reindeer sinew to stitch patches onto punctured rubber waders. Sally would go with Varya to gather *Boletus* mushrooms to dry for winter, or climb the nearer mountains with Emmie to find Siberian ginseng* (*Radiola rosea*) on the crags. Yura took the children to cut down trees in the forest and taught them to saw and split logs. Gosha taught Catherine to carve animal figures out of sheep's horn, and she spent much time

Pictures for a memory game, drawn by Patrick and Catherine.

with Gosha's and Lidia's girls, playing cards or going over their English textbook, all giggling together at their mistakes. Sometimes Lidia would join in with no knowledge of English at all, and they would goad her to ever more bizarre mispronunciations.

But Catherine did not want an Eveny girl's life. 'Those girls used to spend hours helping their mother make bread,' she told me afterwards. 'They were so well behaved! I'd get fed up and just run off into the forest and eat berries, like a feral child!'

I imagined that Lyuda from camp 8 might have been like this when she was little. Little Stepa understood, and made Kolya include her, reluctantly, in their boys' games.

Catherine joined in the men's work, sawing, chopping, and repairing the corral with the Bison. She had left her own dog Homer in England, and would spend hours playing with the women's pet fox, which they were planning to make into a hat that autumn when its coat had thickened, and cuddling the husky-like dogs tied up around the camp, such as the old one-eyed prophet, a puppy called Doghor (Friend), and Sharik (Little Ball), Ivan's daily herding companion. My experience of Sharik was different: I thought him bad-tempered, with mean, opportunist eyes, and once when Ivan was not looking he removed some flesh from my leg even through two layers of thick trousers.

'I suppose it's all right to stroke the herding dogs, Katya,' said Ivan, to whom such cuddling seemed very strange, 'but you mustn't play with the ones we use for hunting. It would make them soft.'

At the end of each day, the reindeer were driven towards the trampling ground to grunt and mill anticlockwise like water round a plughole. As leaders darted out of reach of the spinning lassos, drawing small columns of followers behind them, the children would run around the outer edge of the herd to drive them back towards the centre. The men wrestled each male reindeer to the ground, where it would glare with a bulging eyeball at the children sitting on its flank to weigh it down while the men castrated it with tongs or sawed off its antlers. The young herders were less skilled than Ivan, Yura, and Gosha, but in these scenes of sudden action the Bison's powerful hulk seemed very handsome. The girls were still children, and it seemed sad that there were no young women to appreciate him. But the Bison was playful: as he rewound his lasso, Catherine stamped on it and it jerked to a halt in his hand. He looked round and gave her a big grin. He would not remain a herder for long. 'Girls want more education,' he told Sally, 'they want more noise.' Within two years, he had left to join them in the village.

Apart from Gosha, the person who paid most attention to the children was Yura. Head tipped to one side, eyes twinkling, he was always ready to bring them into a conversation. When little Kolya blundered with his clumsy feet across the women's preparation of meat and flour, they scolded him and Masha slapped him. But Yura drew him aside and explained patiently what he was doing wrong.

'Yura is so good with children,' Sally once remarked to Ivan. 'He's a child himself,' he replied.

Yura adopted Patrick. 'He's naughty like me,' he told Sally on the first day. 'I can see it in his eyes.' He used to skip school, just like Patrick, and would go with a friend to watch films. When Yura was sent out of school, this friend would pretend to be naughty, too, in order to join him. But later, the friend had continued his education and moved out of the village, leaving Yura behind; in Patrick, it seemed Yura had found another kindred spirit.

Day after day, Yura would take Patrick walking across the mountains, teaching him to trap and shoot marmots. Patrick perhaps came closer than any of my family to the Eveny sense of quiet solitude on the land. He and Yura shared no language, but communicated through gestures. They would sit in silence for hours gazing around them.

'Even when we were on different spurs of a mountain, it felt very intimate,' Patrick told me later. 'We were not talking, but we weren't separate.'

Patrick even started to dream of native women. 'That's a good sign,' I teased him. 'It means Bayanay's daughter is taking an interest in you!' As he came home with his rifle and his marmots, Sally perceived her son for the first time not as a boy, but as a man.

For our children, it was sufficient to join in physical activities and to be polite in order to be accepted. For Sally, as an older adult, it was more difficult to be sure what was expected of her. The women did repetitive tasks that were nonetheless highly skilled, and talked quietly (except for Emmie's shrieks) in languages she did not know. Sometimes Varya could translate, with her limited English; sometimes I would interpret, but this took me away from other activities. As a man, too, I would inhibit the conversation, which it seemed was often about medicines for women's problems. They respected Sally for making the journey to see them, for riding, for looking after her children, for laughing. But she sometimes found it hard to live up to what she perceived as their standard of unprotesting endurance. She retained her sense of control by keeping our own tent in very good order, lighting fires, cooking regularly for Catherine, and washing clothes in the icy river. She had found a spot where the river made a deep pool against a cliff, so that her clothes would not be whisked away by the current that raced just beyond the edge of her calm inlet. She would also jump in and swim for a few seconds, a habit that our friends considered extraordinary. The women warned her to be on the lookout for dreams about fish, since these always foretell a chill.

It was through Sally's and Catherine's observations that I fully realized how confined the world of women has become* since they ceased taking part in hunting and family herding. The men's daily narratives are full of far-flung place names, and I had been so preoccupied with tracing and mapping these places that it did not occur to me that the women may not have seen them for years. Most women I have observed never moved further than a mile or two from camp in quest of wood, berries, and mushrooms.

Ivan was also hoping for a visit from Vladimir Nikolayevich, with his well-trained horses. He asked me whether we should wait, or migrate to the next site without him. It was obvious that I was not competent to judge, but I felt there was more to this question than mere courtesy. I told Sally about this during her next washing session.

'Don't you see?' she said, as she crouched on the tiny stony beach and swirled a shirt around the pool. 'He misses the support of his father and he needs the assurance of old Vladimir Nikolayevich's judgement. When he's not around he turns to you even though you're not an expert, because at least you're a slightly older man.'

I had been too aware of my own dependence on Ivan for logistics to think that he might depend on me in other ways*.

'He's anxious to make us happy,' she continued, 'they all are. They have to do it, but it's out of genuine kindness, too. Yura looks after our son, Emmie takes me for walks. Ivan wants us to have a smooth migration, he wants to look after us as well as run his herd. But we cause him a lot of anxiety. He's suffering from PMT: Pre-Migrational Tension!'

That was one joke I would not even try to translate. So Ivan and I were both having the same problem: keeping up appearances in each other's eyes and downplaying the strains.

On the last evening before we left Daik, I sat with Catherine on the mountainside where I had sat alone at the same moment the previous year. Though it would be her first migration, I think she, too, sensed the moment of stillness just before a move. In the camp below, tiny figures composed and re-composed themselves in a living Brueghel painting. Granny in her floral headscarf stomped back and forth with her purposeful forward lean, her movements steady and thorough as she lifted tarpaulins and shifted things around between mounds of supplies. She stooped through the flap of her tent with an armful of logs, and the wisp of smoke from her stovepipe intensified. Masha washed pans in the stream, while Emmie boiled reindeer stew on an outdoor fire. The column of smoke rose straight upward, unbrushed by any breeze. Tolya prowled between tents with a notebook under his arm. The undisturbed line of Emmie's smoke, the stability of the tents weighed down with rocks, the delayed transmission of smells and sounds on the still mountain air, all masked an intense expectation of tomorrow's dismantling of the camp and onward movement.

Sally had hung our sleeping bags on a tree to air and was taking them back into the tent before the night chill could settle on them. She came out again to put more wood on the open fire under Catherine's supper. Patrick and Yura were descending a distant mountain after their last day's marmoting.

'When I'm old, like Granny,' Catherine said, 'I'll never forget this.'

When the caravans started moving off the next day, Catherine vanished ahead of us before we had a chance to give her any last-minute instructions or reassurance. We caught sight of her in the distance, far beyond our reach or help, riding on the back of Cleopatra and lurching up and down every contour a few seconds after Granny, whose reindeer Abdullah was tied in front of her. On the reindeer behind Catherine was a wooden crate containing the fox. The reindeer behind the fox carried a plastic rolling water-barrel we had given our hosts to make it easier to fetch water up the steep bank at the next site. Later, Catherine described how the women had called instructions to her in their three languages, none of which she understood except for the obvious 'Lean forward!' and 'Lean back!' as the caravan hauled itself up an embankment or teetered down an eroded slope.

'I wasn't scared,' she said later. 'My reindeer was friendly and I was excited. I trusted Granny!'

Sally, Patrick, and I followed on horseback. We had brought bulky supplies, to ease Ivan's burden of hospitality; but we had no system of advance depots, so this increased the number of animals he had to provide. The other herders teased us for the bulk of our cargo, but Ivan was dutiful and practical. We perched on top of a huge pile of baggage, with only one stirrup each. It was impossible to reach the horse's flank with one's heel and there was no contact with the horse except through the bridle.

Sally was given the Bear, who had been the Old Man's best horse. She had a back problem and led the Bear to the edge of the camp, where she asked me to help her mount discreetly to avoid the humiliation of being seen falling off. The reindeer-fur mats laid over the saddle made it so hard to gain any purchase that she made me remove the top one. At that moment Tolya came up in indignation. 'If you leave that fur behind, she'll have to sleep on the ground in the next camp!' he fulminated. I took it off anyway, and it somehow found its way onto the back of someone else's reindeer.

If the rider is not firmly in command, the horse will make the decisions. I was often reprimanded for letting this happen, and Sally was in no position to impose her tentative will on the Bear. Often the path was not clear and once I saw him plodding relentlessly to the edge of a precipice overhanging the river.

'I tried to redirect him,' she told me afterwards. 'I thought, This is it, I've had it!'

But the Bear understood that the ground away from the clifftop was boggy, while along the edge there was a dry rim. Whenever Sally fell behind, Tolya would shout back at her to hurry up. When Ivan was anxious, he became gloomy, and Sally sympathized. But she experienced Tolya's anxiety as bossiness. At last, her fury drove her to let go of the reins long enough to raise one arm up in the air and hold the forefinger of her other hand across her upper lip, exclaiming in English, 'You – Hitler!'

I feared an explosion, but Tolya gave her an appreciative grin. They were well matched. For several days, he would enter Granny's tent, not in the discreet Eveny style (sniffing the air, as Sally called it, like a reindeer), but striding in with a Hitler salute. Even at a distance, I knew when this was happening from the sound of Emmie's shriek rising above the laughter of the others.

There had been no bad omens and the mood at Djus Erekit was relaxed. Yura was unable to sleep after a move, as usual, and pottered around all through the first night in the late summer twilight. But everyone else slept well. After breakfast, people dispersed and I found myself alone in Granny's tent with Sally and Varya, drinking yet more tea in the mid-morning silence when one is left behind in an empty camp, the stove still projecting what is now unnecessary extra heat on one's face. Today, the red glow of the stove was almost invisible in the powerful sunlight that streamed in through the weave of the canvas.

Varya wanted to know what we had dreamed. One's first dream at a new site was an important indicator of how one might fare there, and we wondered what to tell and what to withhold. On the first few nights after each arrival, Patrick and I both had dreams of blood and violence involving our local friends. We could not tell the people involved, but we felt we had to forestall their fulfilment and so told them to each other.

Sally had dreamed of a galleon with full sails, 'like something from the Spanish Armada. We were all together, your family and ours, up in these high mountains. I was struggling to get the ship to fly, but it wouldn't, however hard I tried. I suddenly managed to pull in one of the sheets. The wind caught it, and the whole ship took off and went straight up in the air. I could see everything blowing around underneath. It was like a hurricane, but it was also as though there were a huge war down there, and we were escaping. I could see a long perspective right down the valley; there was a tremendous feeling of height.'

This was too complicated for Varya's English, so I repeated the dream in Russian. Varya looked thoughtful.

'That means we'll have great difficulty leaving,' she said at last. 'There'll be a problem with the helicopter, you'll see, we'll have to fight to get out of here.' Sally looked alarmed, and Varya added, 'But it means we'll manage it in the end.'

Where Varya's interpretation was predictive, in local style, Catherine's was psychological.

'It means if you were more in control, you'd do things better,' she told her mother.

Sally came to feel that the land had sent her dreams even while she was still in England. Before she ever reached Siberia, she had dreamed of an old native woman who warned her that it was dangerous to play with wolves. On her first night among the reindeer herders at Daik, the camp was attacked by wolves trying to eat the *uchakhs* tethered near the tents. The men had been drinking the vodka we had opened to feed the fire and it was the women who first heard the dogs' frantic barking. In the twilight of midnight, no one could find any bullets and we had to chase off the wolves unarmed. Later, I found out that Granny had felt a twitch the previous day in her left cheek, on the side that indicates something bad will happen. As soon as we started drinking, she had hidden the bullets under her pillow, judging that to deprive the camp of protection from wild animals was a lesser danger.

'So you see,' Sally told me afterwards, 'we really were playing dangerously with wolves!'

Hiding bullets, weapons, or tools made of metal under one's pillow* was an act with complex meanings. The object might protect you, like the scissors that saved Lidia's brother's children from their uncle's ghost, but it might open up your soul during sleep, making you vulnerable to the seeping in of a negative influence. Whenever Ivan put bullets under his pillow, he would be chased all night by violent Russians. Again and again I was told that it was always bad to dream of a Russian, as this foretold death or serious illness. It was here, in the deeper recesses of the psyche, that the trauma of colonialism continued to be felt. I saw an analogy between the Russians' introduction of guns and metal, and their spreading of smallpox.

'A spirit with red hair just like Sally's!' Granny observed mischievously.

Granny had learned something of the Russian language but had not lived in Russian places like her sons. She sometimes referred to us as Nyukha, the Sakha word for Russians, perhaps the only way she knew to talk about Europeans. It turned out that before Sally arrived, Granny had dreamed of meeting a woman with red hair, though she never made it clear whether this was good or bad.

'Of course you have to tell a bad dream,' Sally commented during one of our sessions down by the river, taking a break from the crowded camp. 'Otherwise it remains in the unconscious and gains power. If you don't tell someone, you make yourself susceptible to its consequences. The shadow has a lot of negative energy if it stays repressed.'

Sally's psychological terminology of 'repression' and 'the unconscious' led to the same conclusion as Eveny talk of spirits and destiny: to tell a bad dream is to break its hold over the dreamer. But when it came to her own dreams, which did not fit Eveny patterns or imagery, Sally received improvised interpretations which sometimes seemed quite western. Before leaving England, she had also dreamed that there was a bear on the roof of our house and that I had told her not to be frightened, as the bear would not harm her.

'That bear represented us,' the Bison told her. 'You were afraid of coming to live with the wild Tungus. But Piers already knew us and assured you we weren't dangerous.'

Tolya's interpretation of another of Sally's dreams seemed Freudian. In the forest near the camp, a naked gnome with dark skin and an Asiatic face was wearing a stack of hats on his head. He stole another hat from Sally and perched it on top of the others.

'That means Catherine will marry an Eveny reindeer herder,' Tolya declared with certainty. Perhaps the gnome represented Bayanay, and the hats were girls he had captivated.

'The boys here are gorgeous,' Catherine confided to me, 'but I'm not going to marry one, because I don't want to spend the rest of my life washing up in a tent!'

Tolya was no stranger to forest spirits himself. Following an erotic dream involving Bayanay's daughter, he knew it was time to go hunting. He decided to teach the boys and called at the pile of branches that the children had made into a den. When he invited the boys to come with him to 'see the land', Catherine understood the code and was furious that she was not included.

'I can ride an *uchakh* as well as they can!' she sulked. 'I wouldn't get tired or fall off!'

That day, Bayanay gave Tolya and the boys six mountain sheep. In the evening, the whole camp gathered round an open fire and feasted on the rich fat and soft flesh of wild mutton kebab, while strips of meat were hung to smoke on a rack over the fire. The firewood was larch and the meat would keep for months. With a certain kind of willow that grew further downstream, it would have kept for years.

'Did you know these sheep were all caught on your mountain?' asked Yura, turning to me. This was a mountain they had renamed Piers' Peak, after we had climbed it together several years earlier and built a cairn at the top.

'I broke a bottle of vodka in my baggage this morning,' I said, suddenly seeing a connection. 'It spilled on the ground. Maybe that's why we got the sheep?'

'Yes,' he agreed, 'that's why they were on your mountain.'

'Bayanay received the vodka* through the spirit of the place,' added Gosha.

While we were talking, Emmie was toiling back and forth from the stream carrying cans of water. Our rolling plastic water-barrel lay idle on a storage platform. Presents of knives, torches, ropes, and down sleeping bags were received almost without comment but conspicuously appreciated by being put to immediate use. I wondered whether the barrel had been a mistake and asked Ivan if they would ever use it.

'It's too good to use now,' he replied. 'We're saving it for later.' Then he added, 'I'll put it on Granny's grave, along with the bottle from that whisky you gave her last time!'

'Make sure you puncture it, so I really get it in the next world!' Granny muttered.

Late that night, Granny's stove gave a strong 'tssk!'. A few minutes later, all Catherine's dogs began to bark and we heard the whinnying of horses approaching our camp, answered by our own horses. Vladimir Nikolayevich was arriving. He pulled the flap aside, not hesitantly but firmly, and entered the tent. The mood had been lethargic, but Vladimir Niklolayevich's air of mastery immediately sharpened the atmosphere, as someone hastened to light a candle. He greeted me affectionately. While Varya interpreted, he was introduced to Sally, who thanked him for looking after me during our winter hunt five years earlier.

Vladimir Niklolayevich's round, red face beamed like a midnight sun. 'He was an easy companion,' he answered. 'He followed me everywhere and always did exactly what I told him!' There was laughter.

We talked of hunting, and the following day Vladimir Nikolayevich gave Sally some practice at shooting. Though arthritic and nearly twenty years older than us in his mid-sixties, he strode in front in his rubber waders, and as we made for the top of a mountain he was always one ridge, one crest ahead of us.

He propped up a flat slate vertically. Sally took aim from 200 feet away with his old rifle and shattered the stone with her first shot. Vladimir Nikolayevich looked at her with a new respect, but then added, 'It's because I've got a very good gun. And I keep it clean – not like the other herders!' It may be that he found it hard to praise a woman. But I am sure he also thought that the gun appreciated his attentiveness and enhanced Sally's performance by helping her to fulfil her intention. Before we came down the mountain, he made Sally leave an offering for Bayanay as an apology, since she had used up a bullet for an 'empty' (*pustoy*) purpose.

Vladimir Nikolayevich took my family on several journeys around the nearby mountains, his horses disciplined, his harnesses complete. When the horse carrying Catherine tangled her in a horizontal larch branch and refused to reverse, she called out several times for help: 'Volodya!' She had worked out that Volodya was the familiar form of Vladimir, but had missed the etiquette of addressing an elder. He ignored her until she switched to the respectful 'Vladimir Nikolayevich!' whereupon he turned back and released her effortlessly.

Another time, he took Patrick and me on a longer trip. We set out on horseback, soon parting company from Ivan and four other men on *uchakhs* as they turned off to gather the herd, which had dispersed across a wide green mountain. The three of us continued up to a mountain pass, where we left an offering and descended right out of the catchment area of the River Munne. The landscape became lunar, the jagged vertical crags devoid of any vegetation. The sharp sun alternated with deep shade through layer after layer of mountain.

One cliff was called Ilbai (Help Yourself!), named sarcastically after the occasion when a hunter shot a sheep at the top of the high precipice but found nothing at the bottom except a splattered mess. Another was called Bába-Yagá, after the witch in Russian folklore from whose hut there is no escape. The cliff began high over our heads and plunged past us, shaley and powdery like the inside of a volcanic crater, to a barren, sunless floor.

'A man was hunting mountain sheep up here,' Vladimir Nikolayevich explained. 'He leaped onto a ledge to get a better aim. But he hadn't thought it through. There was no way to climb up or to jump down, he couldn't get off the ledge. He shouted for a whole week: "Somebody please shoot me!" But there wasn't anybody, so he starved to death. Sometimes you can still hear his cries.'

We rode slowly downhill and entered the zone of dwarf plants, then the sparse beginnings of the forest. On a small rise overlooking a river, we came upon two graves side by side, the trunks of their superstructure interlocking at the corners like miniature log cabins, each surmounted by an Orthodox cross.

One grave had no writing, but was probably the grave of Tolya's stepfather's mother, though Tolya himself had never visited the site. Carved into the bleached, desiccated wood of the other, in Russian, was:

born 190[ ]

[ ]

August 1945

The weatherworn missing line was puzzling, since the length of the inscription was too short for a name, or for any of the usual words for 'deceased' or 'departed'. The word order would be odd, but could it be the month of birth?

'No,' said Vladimir Nikolayevich emphatically, 'they weren't educated then, they didn't understand the names of months.*'

Nobody herded in this area now. Perhaps someone knew who lay under those logs, but would either of us ever trace that person and intercept them through their constant movements around summer tents and winter huts?

Vladimir Nikolayevich did not smoke, but carried cigarettes to give to mountain passes, fires, and graves. He and Patrick left cigarettes, while I offered aspirins. We sprawled on the ground near the graves in the warm sunshine. Vladimir Nikolayevich had no qualms about snoozing here. I had warned Patrick earlier about the taboo on eating berries growing on graves. I glanced towards the luscious cranberries and murmured, 'Forbidden fruit!' Our guide understood the gesture and nodded his approval.

We had already started to move on, when Vladimir Nikolayevich held up his hand. He had spotted seven mountain sheep on a nearby mountain. He gave Patrick his horse to hold, tied up his two dogs, and set out to stalk them on foot. Not having been brought up with guns, I was still amazed at the powerful magic of a person who could point his fire-stick and will it to kill a distant animal that I could hardly see. But this time I was sure his quest was futile, since the sheep had already seen us.

Vladimir Nikolayevich climbed above the last trees of the forest and started to make his way around the edge of an open bog. In his baggy khaki suit he was already hard to keep in sight. Then there was nothing. We must have waited half an hour. Suddenly there was a volley of shots, the echo rolling over and over around the crags. The dogs on their leashes barked furiously. From a far ridge, Vladimir Nikolayevich was shouting and beckoning. Patrick mounted his horse and climbed slowly towards him. They returned on foot, the horse carrying a sheep with a crimson wound staining the side of its fleece.

Our guide unpacked a picnic of surprising richness and variety: bread, butter, vodka, salt, tea, sugar, and a tin of Pacific sea-fish. Then he lit a bonfire and fed it with cigarettes, mutton fat, and tea leaves.

'Do you always feed a fire on the trail, even if you're not camping there?'

The rules could never be known for certain, but Vladimir Nikolayevich was always confident in his judgement as he picked his way between them. 'We're here for the first time,' he answered, 'we're near those graves, and we've just come down from a pass – all good reasons!'

He left camp 7 the next day, planning to spend the rest of August and all of September hunting wild reindeer down the valley. He had no stove, but took a little nylon tent from England that we had been using to store our luggage. He left all the mutton behind for us. Wherever he went, he would pick animals from the land the way we picked cranberries.

15

How to summon a helicopter

We thought about Vladimir Nikolayevich in his unheated tent as the rain set in over the next few days. It soon became torrential and we woke one night to find our reindeer-fur mats afloat and our sleeping bags sodden. A river was flowing through our tent. How could we have pitched it in such a stupid place?* We staggered around in the dark rescuing our most vulnerable belongings and squeezed ourselves into Tolya's tent for a cramped, uncomfortable dawn alongside his family.

Catherine's irreplaceable supply of food had been threatened once before, when Sally had spotted the back end of a reindeer sticking out of the tent. But this time it had been engulfed by a slurry of reindeer dung from 1,600 bottoms on the herd's trampling ground up the slope. A vegetarian child on a land that yielded little but meat! I waited till Catherine was sawing logs out of sight the other side of the camp, then rinsed her food out in the river during a break in the rain and laid it on the ground to dry.

The Munne had risen by 3 feet overnight. We made radio contact and heard camp after camp reporting swollen rivers. That evening at supper, Ivan was morose. Some of the herd were on the opposite bank and could not be reached. He could not start taking supplies down to the next site. He was expecting three men from the Farm to build them a new winter hut where the Daik curves round and joins the River Munne.

'The old hut is rotting,' he muttered, seeing his hopes of a comfortable winter fading. 'And the lichen there is trampled. And the new site would have lots of sable! The Farm is obliged to build us a hut, under our contract. A small hut will do; there aren't many of us in winter. If the Farm won't pay, I'll even pay out of our operating budget. I just want those men to get here!'

It continued to rain throughout the following night and the river rose even higher. In the whole region there was only one civilian helicopter left in working order. This was maintained by the region's electricity corporation and we had negotiated a price for it to pick us up in a week's time, on 24 August. This would be the first in a chain of flights that would bring Catherine back to England in time to start her new school, and Sally in time for her therapy clients. But the helicopter could not fly when the mountain-tops were in cloud. There began an anxious flurry of radio contact morning, noon and night with Sasha the Radio Man in the village.

'Snop 7 . . . snop snop snop . . . !' Sally called this the 'radio torture' and still gives this call sign when she wants to attract my attention at home.

Sasha the Radio Man made contact with the aviators in Yakutsk and Sangar, who were basking in autumn sunshine under a clear blue sky. Flying around the plains, they could see a block of cloud encasing our high range. Each day, the helicopter avoided our mountains and went about its other business in the lowlands.

'*Pogoda ibga!* [The weather's fine!]' snorted Ivan. 'Sangar is deceiving us!'

'They say the pass is closed to traffic, but a bar of chocolate would persuade them* to open it!' added Tolya derisively.

It was harder to divine from the crackling radio than from a crackling fire, as it drew in so many more characters and their plans were so much more inscrutable. An emergency medical flight had penetrated the mountains to take Kesha's father,

Dmitri Konstantinovich, to hospital, but this was before we knew we might need a lift. More hopefully, a political dignitary was said to be on his way to the village. His importance might stimulate an aircraft to take advantage of any opening in the weather. If he came by helicopter, would it make a detour to our camp? If he came by plane, could we put together enough animals for the difficult ride to the village airstrip in time to be picked up? The 24th August came and went. We tried to remain permanently packed ready for a sudden airlift, while still joining in the life of the camp.

'Snop, snop, snop 7!'

Groups of villagers would shuttle back and forth between Yakutsk and Sangar, covering the 200 miles on river-boats in summer or in buses along the corrugated river-ice in winter, hoping for a chance of a flight. Now the radio said that our dignitary was still waiting in Yakutsk and had not moved on to Sangar. Tolya and Ivan tried to decode this move. Sangar was the nearest jumping-off point for Sebyan, and had one functioning biplane but no helicopter. Yakutsk had one helicopter and several biplanes, but the planes were no longer willing to fly to Sebyan. So the dignitary must be expecting to take the helicopter.

Catherine and Sally were not the only ones with a deadline. Masha the teacher and the other children had to reach the village school before 1 September, and Tolya and his family had to reach Yakutsk. We had not brought supplies for an indefinite stay. Tolya still had some bullets left and could contribute meat to the camp, but our children needed flour and they were already eating into the sack we had given Granny to last her into the winter.

Sally had started to go on her own to pick mushrooms on the surrounding hillsides. But Djus Erekit is also an early autumn halt on a migration route for bears. They usually avoid people, but we occasionally caught sight of them.

'Last year there were lots,' Yura said. 'They were attacking the reindeer and we had to kill five of them. But there are plenty of berries this year, so they're leaving us alone.'

Lidia was worried, but Granny was not. At one time Ivan was worried, at another not. As always, there was no one safe opinion. Unsettled by their uncertainty, Sally now perceived a sinister undertone to the autumn beauty of the land, combining her uncertainty about bears with her anxiety about how we would get home. The bright red leaves of the ginseng that she sought against the black shale of overhanging cliffs seemed like far-flung spots of the blood that she felt suffused the landscape through to the crimson growing tips of *Sphagnum* moss and the turning leaves of the dwarf birches.

'Destruction, killing, eating, rebirth,' she said. 'It all feels very savage and cruel. Even the larches are killed and skinned to make fences.'

I had not seen it like this. I knew I was defenceless on this landscape, but had felt safe from the time all those years ago when I had first put myself at the mercy of these unknown people who were to become my friends.

'You make leaps of faith,' she said. 'You entrust your life to strangers when you've only just made eye contact with them. That's why you're an anthropologist. Some people would call it foolhardy!'

Now the radio said that camp 3 was summoning a truck to crawl up the stony bed of the River Nuora and carry their equipment to their next site. Our herders derided them for driving on a long detour along the river to avoid a short migration over a mountain on animals. But the truck could be useful. Maybe we could ride there to meet it; maybe Ivan's hut-builders could be persuaded to come, if they knew that our horses and *uchakhs* would be waiting to take them on to camp 7.

'Snop 7, snop snop snop!'

More messages, and more decoding: if the truck driver was delaying his departure, this might mean he was expecting a plane in the village, and was waiting so that he could do business distributing its cargo of vodka around the camps. But by the time he reached us, the plane would have turned round and left anyway.

Many trucks in the village had been privatized and some of the owners and drivers were our friends. But the director still controlled their fuel supply. At one point, a hostile voice said over the radio*, 'They can all stay there until New Year!' and I heard Tolya and Ivan talking seriously of sledging to the village after the ice on the rivers became safe in late November. I did not translate this conversation to my family.

During the next two weeks, only one aircraft managed to cross the mountain range. Dmitri Konstantinovich had died in hospital, and the Sangar biplane was bringing his body back to the village during a short break in the weather. I stood with Granny watching the tiny aircraft floating like a dragonfly, seeming to stand still while the clouds rushed between us, like a stone at the bottom of a mountain stream.

'It's raining to close up his last footprints on earth,' Granny remarked. People would often say this about rain or snow after the death of an elder.

'He was too fat – his kidneys gave out,' Tolya told me later, more mundanely. 'They offered him an operation last year, but he refused.'

We talked to Kesha on the radio. He had put his herd in the care of Boris' replacement and was setting out for the village to bury his father. I tried to imagine the large, happy family of camp 8 without Dmitri Konstantinovich. Talking of Kesha reminded Ivan of the family life he had never enjoyed.

'I'll be 40 this year,' he said reflectively. 'I've worked for twenty years, so I'm entitled to retire. Yura can look after the herd. Then I'll be able to take you all over the place! We need time to get married, start a family . . .'

'There's no one to marry in the village*.' Yura was thinking of his own longing. 'Most of us are related. But I'd be happy to marry a Buryat, or even a Khokhol.' Buryats are Mongols from around Lake Baikal in southern Siberia; Khokhol is slang for Ukrainian. But then Yura added, 'So long as she was willing to live in the camp.' It was hard to visualize the blonde daughter of a Ukrainian mining engineer cracking reindeer bones to extract the marrow for soup on a remote mountain under Granny's watchful eye. I felt sad for this family with no heirs.

Sally was having similar thoughts. 'I don't like to think of a time after Granny,' she said, 'but I bet you these men will marry after she's gone.'

With our departure unresolved, the brigade continued their migration downstream. The members of the camp moved first, and Yura brought the horses and *uchakhs* back the next day for Tolya and me to follow on with our families. When we had all assembled at Tal Naldin, Ivan decided to leave the herd in the care of the bachelors and use every horse and spare *uchakh* to escort us as far as camp 3.

That evening at supper, Ivan asked, 'Will you all come back another time?' I felt a moment of alarm, as I wondered whether he was checking that he would not have to look after us again. But then Granny added invitingly, 'It's very fresh up here!' and I realized that the assurance they both sought was the opposite: that my family had not been put off by the physical arduousness and the emotional tensions. I felt reassured myself that my relationship with the brigade had survived the experiment of bringing them together with my family.

The caravan that set out on 28 August carried fifteen people and their baggage: my family; Tolya, Varya and their boy Kolya; Lidia, Gosha, their two daughters, and their nephew Stepa; and Masha the teacher, with Ivan and Yura to escort us and bring the animals back. Granny and Emmie hardly looked up to say farewell, as if we were merely going to collect firewood nearby. We would indeed all come back another year, when Granny, who had not expected to live to see Sally again, would spend the entire first night singing in an archaic kind of trance through which women in the old days released intense, inexpressible emotion.

The fox did not last long, and did not even make it into a hat: that autumn, it was eaten by Sharik.

Our short, easy migrations down the River Munne had not prepared us for the shock of crossing with our baggage and inexperienced riders to another river system. We rode beyond the tree line to the bare rock that led to the notorious Hotorchon Pass. There was little talking on the trail, but before we set out, Ivan had told me, 'Several men have died there. I remember two of them. One froze in an avalanche when I was young: his *uchakhs* couldn't get out. The other was recent. His sledge turned over on a steep slope and the baggage fell on top of him. He should have walked instead of sitting on the sledge.'

'I remember when I was little,' added Tolya, 'I went on a winter hunt with my grandfather. One of the men on that trip froze: he got drunk. The spirits on the pass swallowed him up.'

'I'm not a believer,' said Ivan, 'but when two people die in the same place, I think it must be a spirit.'

As we hauled ourselves up towards Hotorchon, it began to snow. Smooth, almost vertical slopes swooped hundreds of feet beneath our hooves, every crinkle picked out by the thin, even cover of the new snow, which at the same time disguised where the surface was firm or loose. The caravan became strung out and we dismounted to lead our animals on foot.

As we approached the summit, the ground levelled off and we were able to ride again, but a bitter wind from the other side was lashing the snow into our faces. Lidia's girls were whimpering. Catherine was in a euphoria of hardship, but her delight in this astonishing experience was broken when Masha the teacher started turning round and shouting angry instructions from the *uchakh* tied in front so that Catherine, too, ended up in tears.

We made our offerings at the pass and started to follow a

stream downward. Where we re-entered the forest, the snow was deep. It was midnight before we reached a suitable site to put up the one large tent that would accommodate us all. The children were sent to cut down trees for cooking, and we ate reindeer meat sitting in the snow round an open fire. Inside the tent we lay on reindeer hides. For some reason, no stove was set up to drive out the moisture. Our combined body heat melted the snow beneath as well as the ice hanging from the roof above us, and we awoke wet and chilled.

The next day, 29 August, the mountains became smoother and gentler, and in the afternoon we reached the late summer site of camp 3. The Nuora was much bigger than the Munne, and even after the recent rain its many channels ran through a wide bed of bare stones. The river led much of the way to the village, and several women and children had come out to live here for the summer. The camp was businesslike and cosmopolitan, and the cheerful brigadier Valera had a factory-made table and wooden stools in his tent. Ivan and Yura looked like wild mountain men next to a confident manager.

A truck was expected. Even out of season, the prospect of transport always leads to a small reindeer slaughter, and the carcasses were stacked up in readiness. The camp felt like a railhead waiting for the arrival of a train.

I was used to the way the faint rattle of a helicopter separated itself from the silence only minutes before landing, but it took an hour from the first strained, low-gear drone of the truck before the headlights of this extraordinary vehicle burst into the communal darkness between the tents. This *mashina*, as it was called, with its huge, ridged tractor tyres as high as a man, was a metal box on wheels, said to have been designed for transporting prisoners. The driver had defied the Farm management to come for us because he was a relative of the sick man I had once flown home to die.

'I heard you needed help,' he said, as he declined to discuss any payment except for the fuel, 'so here I am!'

He would not give camp 3 a lift on their migration after all: the truck would simply turn round and return to the village. Perhaps Valera, too, was returning a favour because I had saved him during his appendicitis. The reindeer carcasses would be traded or distributed to relatives in the village and the driver had brought supplies for the camp, as well as friends for the ride: one act can encompass many motives. By morning, they had unloaded and reloaded the *mashina*, and the driver was doing some last-minute repairs.

Contrary to Eveny custom, Ivan and Yura stood and watched us leave. Ivan's eyes were misty. It was the last time any of us would see the gentle, humorous, teetotal Yura. That winter, Catherine would receive a letter from Lidia's girls to say that he had been stabbed in the village by a drunk, joining the long catalogue of wasted young male lives. He had already taken in his last view of Daik and Djus Erekit with us, and after he had seen us depart he would go back to finish living for his last time at Tal Naldin.

The *mashina* had no windows, just an open door at the back through which we could see a small, framed moving image of the riverbed. The passengers never knew when the next stop would be. Ladies, and people feeling sick, would take turns to ride in the driver's cabin. Mostly we ground slowly over the stones, but sometimes we crept down an embankment into a deep channel, where our tyres would be completely covered by rushing water. Young men on the roof of the cabin would read the parting and merging threads of the river and shout directions down to the driver. We were isolated from any voices outside and our orders were banged on the roof of our metal box: 'Move to the left!' and we would scurry to the left-hand side just before the *mashina* lurched to the right, shifting the mounds of baggage and dead reindeer beneath us. If the *mashina* capsized on this landscape, nothing could right it and it would join the larch poles and occasional flecks of abandoned machinery as mute remnants of a human event.

Sometimes we would approach places where the bank had been scoured by spring floods, or river islands that floated on a bed of stones like pancakes of earth topped with poplar groves, and hear the barking of dogs. The *mashina* would stop near a tent to collect and deliver sacks, passengers, or letters. The passengers would get out, the children would play, and everyone would be served with tea and dried fish from the shoals hung out between the trees like socks on a washing-line. Though we sometimes offloaded meat, it was never served to us. In this world of fish, meat was a scarce commodity. These people were engineers and mechanics from the village with their families, many of them Russians and other non-natives who had no relatives in the herding camps but spent the summer catching what they could for the winter.

The journey lasted for eighteen hours, through the night and into the following afternoon. The driver never slept. The passengers, too, spent the night singing to keep themselves awake because the exhaust was faulty and on the *mashina*'s last trip a child had died from carbon monoxide poisoning. Once during the night we were told to get out and walk, to lighten the *mashina* as it drove across a patch where an immense ice bubble in the permafrost had melted, leaving nothing but a collapsing, hollow cave of soft earth.

The village wrapped me in a familiar paradox of relief and depression, which I think affected my family in a strikingly similar way. Down by the offices, men in rubber boots and heavy jackets of artificial leather would stand around each morning in the slush, shaking hands with new arrivals. These were not unemployed youths, but older men of power who greeted smartly dressed women as they mounted the steps into the office to type a memo through several layers of battered carbon paper or push beads on an abacus. Everyone would disperse to their homes for lunch and drift back for a while in the afternoon. In the taiga, depression was kept at bay by the constant imperative of physical survival and the tidal ebb and flow of the reindeer;

in the village offices there was nothing but paper, and it could always wait.

It was still difficult to contact our helicopter. For two days, the encounter between our cold mountain air and the warm air of the plains created a screen impervious to radio waves.

'I'll try again after lunch,' Sasha the Radio Man comforted us on the second day as he coiled little lengths of spare wire with slow, exquisite neatness. 'It's thawing here, maybe the screen will get weaker.'

I went with Sally to visit the local energy employees, called the Energetics (*energetiki*), whose boss controlled our helicopter. They were relaxing in front of the TV. The Chief Energetic seated us on a worn plastic sofa and offered us tea. For a long time we had seen no European faces but each other's. Our host was tall and lean, with piercing blue eyes and a deep voice. Such men had been pioneers of the Russian empire, yet I found it hard to dislike him on behalf of Asian history, or to feel that anyone would fall seriously ill as a result of seeing this warm-hearted man in a dream.

The Energetics had a better transmitter than Sasha, and were able to make contact with the aviation authorities even through the weather screen. The meteorologists spoke of a cyclone that had been bringing moist air from the Pacific Ocean. Our mountains were holding in the western rim of a blanket of cloud which extended all the way to the coast 600 miles to the east. For a week, this blanket taunted us by peeling itself back from the mountains each night.

'Tomorrow you'll definitely fly,' Kesha assured me as we stood together reminiscing about his father under yet another starry night sky. But by dawn, as on every other day, I stood in the midst of a windless whiteout while a curtain of tiny ice crystals floated slowly down all around.

It was well into September and Sally and Catherine were becoming frantic. Catherine was missing the first week at her new school and the other children were already making new friends and finding seats together without her. Sally's therapy clients, who knew nothing of her journey to Siberia, would find a closed door with no note of explanation.

I decided to phone England and pass on messages. Sasha the Radio Man could not imagine making a connection beyond Yakutsk, but agreed to try. For an hour he coaxed his way through a series of operators, each one grudgingly connecting him to the next, from Yakutsk to Vladivostok, to Novosibirsk, to Moscow . . . Suddenly we heard a British ringing tone.

No reply. Could we try another number?

'You've had one today – that's enough!' scolded the operator in Yakutsk, the gatekeeper to all the others. I translated for Sally.

'I know her type,' said Sally. 'She wears frumpy socks, her husband's a drunk, and she's a queen among her petitioners!'

Sasha pleaded for a further hour, until we heard a ringing tone again. Sally's sister Moya answered.

'We're safe!' Sally announced triumphantly. 'We managed to get out!'

'Safe from what? Out from where? Was there a problem?'

The radio hut was crowded with village women. 'Lon-Don!' they murmured in admiration. Sasha's feat became a legend.

Why was the weather punishing us? It was Tolya's sister Anna who suggested an answer. Dmitri Konstantinovich had been buried only a few days earlier, and he was detaining us because I had not visited his grave. The next morning, there was still no flight rostered. The peaks were blotted out as on every other morning, and it seemed certain that no aircraft would enter the mountains that day. I walked with Sally and Tolya to a small bluff 2 miles away, where Dmitri Konstantinovich had been buried next to Ivan's father, the Old Man.

Under the heavily overcast sky, the lead-coloured shale of the surrounding cliffs had dulled the gold of the larch trees to a listless ochre. The poplars by the stream, their trunks silvery grey and their next year's buds already formed in purple, had turned their own shade of faint yellow. Even the tabooed cranberries in the spongy moss under our feet seemed drained of colour.

Nothing stirred. It was as if we had arrived on a planet without motion, and were the only moving entities in a time-freeze. Yet the reindeer skull on its post, the box of bones on a pillar, the crushed cigarettes, the torn banknotes, and the broken bottles were recent traces, even victims, of purposeful human action.

Dmitri Konstantinovich looked out at us with the startled, frozen look* of a native in a photographic studio in town, which makes even the most animated person into a dummy. I gazed at the photo for a long time, remembering life back into it.

Dmitri Konstantinovich, are you there? Dmitri Konstantinovich, how come the photo doesn't show those permanent creases at the corners of your eyes from laughing so much? And look at that massive reindeer skin hanging from the tree: are you still fat and heavy as you ride it around the next world?

I laid my offering on his grave. Dmitri Konstantinovich, will you release us to go home?

We set out for the hour's walk back. On the way, we met Kesha and Lyuda setting out on horseback, returning to their camp after the funeral. Kesha was pleased that we had visited his father's grave. Lyuda was pregnant with their third child. Sally was concerned for her out in the camp, but she seemed calm and confident.

As we approached the village, I heard a faint sound. Even Tolya did not believe me, but I knew. The sound grew louder until the helicopter clattered over our heads and landed at the edge of the village. When we caught up with it, we found the crew surrounded by sick people wanting to reach hospital and college students whose classes had long since started.

Tolya commandeered a truck and our baggage was brought from Anna's house in minutes. The captain was anxious to leave. He imposed a strict weight limit, and we briskly selected a few of the beseeching villagers to come with us. The ladder was pulled up and the door closed. The helicopter pulled on its wheels, creating a circular dish of wind across the grass, then wrenched itself off the ground with that familiar sensation, which is not a jolt, but an almost imperceptible sense of parting between two sticky surfaces.

'If I ever make it into *Who's Who*,' said Sally, 'I'll list my hobby as taking off in helicopters!'

We rose to just above the treetops and started zigzagging down the riverbed between the mountains. Looking past the pilots' shoulders and through the front windscreen, Sally exclaimed in astonishment, 'That's exactly the view I saw in my dream!'

For an hour, blizzards swept across our window as we banked and twisted between massive cliffs. Tolya leaned into the cockpit to guide the pilot past each peak and tributary. Suddenly instead of mountains above us there were hills beneath us, and then the forested plains stretching to the horizon. We emerged into the first sunshine we had seen for nearly a month. Two hours later we landed at the aerodrome against an orange evening sky and drove into the city through the exhausts of a traffic jam.

The helicopter had flown to a village near the mountains to repair a generator, and the captain had made the decision to nose his way up the River Tumara to look for us just as I was placing my offering on Dmitri Konstantinovich's grave. To my Eveny friends, such connections are only to be expected. The muted hesitancy of Eveny partings disguises an extreme sensitivity to separation, and a grave can hold you until your farewell is complete.

Three years later, travelling with my student Seona, I tried the same approach. We had found another *mashina* from camp 3 to the village, where I had arranged for a biplane to pick us up. Once again, we were trapped in the village by continued bad weather. Camp 7 had felt sad without Yura's light touch. My last sight of him by the *mashina* had been casual and unreflective – and I had not yet visited his grave. Again I went to the spot, which lay next to the graves of his father and Dmitri

Konstantinovich, and again the aircraft appeared just as I was returning with Seona to the village. Ice dragged on the wings as we struggled to lift ourselves out of the valley, but cloud had already rushed in to close off the village airstrip and there was no turning back. Within minutes, as the aviators handed round caviar and hunks of bread, we had clambered through the upper clouds and were looking down on the jagged snowy peaks of the Verkhoyansk Range as they stretched in the evening sunlight, ridge behind ridge, towards the invisible Arctic Ocean.

EPILOGUE

Outliving the end of empire

Every time I fly over this land, I try to name the places I have visited and recall the events associated with them: the mountain-top in camp 7 where I built a cairn with Gosha and Yura, the river in camp 8 where Kesha told me about his swan as we fished through the ice, the pass nearby where I huddled on a sledge with Lyuda and little Diana, another pass on the way from camp 7 to camp 3 where our girls cried in the blizzard. When I flew back across the area where I had just left Vladimir Nikolayevich to continue his midwinter hunt, I could trace the tracks of our journey together, and even work out some of the sites of our overnight halts. Half a mile above Vladimir Nikolay-evich and his *uchakhs*, the pilot dipped his wing in greeting.

The threads of human movement are so far apart that if you did not understand the signs, you might imagine this landscape was uninhabited. But to the attentive eye, winter reveals every step taken by human or animal, even a mile below. Sometimes, my aircraft inches past a caravan of sledges as they wind along a frozen river, their movement almost imperceptible except for a slowly extending furrow in the virgin snow. There is usually someone on board who can name the solitary figure on the lead sledge and say where he is going and why. The deeply snowdrifted western slopes of the Verkhoyansk Range, where I have never been overland, are too remote and difficult for do-mestic herds, and there are only the occasional sledge-trails of

the most experienced hunters seeking small groups of wild reindeer and single elks. Once made in early winter, an animal's
prints may remain discernible until May or June. If one could
follow them forever, as in a string maze puzzle, one would catch
up with every single creature.

The marks left by a herd of domestic reindeer are quite different. The secret growth of the richest lichen is betrayed through
the snow by the densest churning up of hooves, while the outer
edges of a lichen bed preserve the thinner marks of sallies and
retreats by leaders and followers, like solar flares lashing out
from the seething surface of the sun. The area around the winter
hut of their herders is different again, as an undifferentiated
trampling by human feet thins out into a fan of repeated short
walks to a woodpile or a frozen river, and finally to the separate
traces of single journeys which may extend for a hundred miles
beyond.

The churned-up areas have become fewer through the 1990s,
as the decline in reindeer numbers has led to reductions and
amalgamations among the brigades. The territory of brigade 6
lay fallow for a while to recover after the herd was destroyed
following an outbreak of brucellosis. Brigade 4 was manned
by incompetent herders after Zinovy's retirement and also lay
within reach of outside poachers. By the winter of 2000–1 it
sank to 380 State-owned and 70 private reindeer, and the Farm
amalgamated them with brigade 2 against the protests of Maxim,
the brigadier who received them. Even brigade 5, that once
proud flagship with its woman brigadier, winner of so many
socialist production competitions, was merged for a while with
a newly founded and short-lived brigade 14, though they
pitched their tents separately and visited each other like guests.

The next time I returned to Ivan's family in August 1999, I
found a similar situation. They had moved their camp at Djus
Erekit to the other side of the river*. Nothing had ever come of
Kesha's earlier suggestion of a merger between their families,
but instead, the Farm had amalgamated Ivan's herd with the

smaller herd of the recently re-established camp 13, run by Karl, the son of old Vasily Pavlovich.

The two groups of herders had started to migrate together, but they did not quite live together. Each cluster of tents had its own frontage to the river, but the bank between them was interrupted by a cliff. In order to reach camp 13's group of tents, one had to stagger along the stony riverbed or climb up a hill behind the cliff and down the other side. To a fieldworker whose business was to notice what everyone was saying and doing, the inconvenience of this barrier was eloquent.

In Ivan's cluster there were eight tents, the highest number I had ever seen. But of the family who had lived there since 1972, the only people left were Ivan himself and Gosha. Ivan's brother and close partner Yura had been murdered; Lidia was in the village struggling, with inadequate contacts, to get her elder girl into university; Granny, too, had gone to live in the village, I think not simply because she was older but because the influx of new people had disturbed the intimacy of her family life; Emmie had given birth to a baby of mysterious origin and had joined her mother, who doted on the child.

There were also new people: Granny's nephew Timofei the vet, who talked to me about Freud's libido and the connections between traditional Eveny beliefs and the theosophical ideas of Madame Blavatsky; a brother of Gosha's, freshly released from a distant penal colony; the dying Zinovy, a refugee from the collapse of camp 4, living here despite his terminal cancer since he was bored in the village; Zinovy's robust-talking wife; and their granddaughter Varya, a teenage girl who cooked for the bachelors, with her own fatherless baby whom she breastfed in between milking the reindeer. My student Seona shared her tent and noticed how kind Ivan was to the mother and baby.

The new joint entity was still named camp 7, and in the struggle over who should be brigadier ('Two big bears can't live in one lair'), Ivan had prevailed. Each half of the camp cooked and ate separately in its own communal canteen tent. Ivan insisted that they were not living as two camps and explained the layout by saying that Karl's group did not need to be on our side of the cliff as they did not milk their female reindeer, an activity which was best done on a plateau behind our tents. There were similar tensions between the leading reindeer and I felt I was witnessing the problems of a merger between two families through their animal surrogates. Large segments of the amalgamated herd were repeatedly separating and disappearing as they moved round an enlarged territory which combined the main route of camp 7 with an extended detour around the rich winter lichens of the former camp 13.

Like many old people, Granny became restless in the village and later went back again to the camp. She had a bad fall from an *uchakh* in the summer of 2000, but lived an active life until her death in December 2002. Soon after she died, Ivan did get married, just as Sally had predicted – to a girl twenty years his junior.

By 2004, Ivan had ceded camp 7 to Karl, and had taken over camp 3 in place of Valery, who had been dismissed by the Farm management after he was accused of eating more than his allocation of the Farm's reindeer. Gosha had moved far away, to become brigadier of camp 12 (the position offered to Arkady fourteen years earlier), and Lidia had joined him. Each herder had taken his personal *uchakhs* with him. The experiment in family autonomy initiated by Granny and the Old Man, and the passionate attachment of their children to the peaks of camp 7, now exist only in the fading memories of these scattered herders and their short-lived mounts.

Kesha continues to run camp 8, which remains strong. Diana and her little brother have started going to school and Lyuda now spends her winters in the village, like other mothers. Diana has forgotten much of the extraordinary behaviour and sayings that made her parents suspect she might be a reincarnation of Kesha's shaman grandmother. They now feel that their third child, a boy, has special insights, and I wonder whether they connect this baby with Dmitri Konstantinovich, who died shortly before the baby was born.

Kostya has retired from running camp 10, worn out by continual separation from his family and the increasing futility of his deeply internalized work ethic. But as he told me all those years ago, reindeer herding is the only activity that interests him and when I visit him in his house, his frustration is obvious. Arkady is married and has children, but he has not found satisfaction in his village jobs as boiler-attendant or accountant and plans to return to brigade 10, which is now run by Zhenya, the elder of old Nikitin's two reindeer-enthusiast grandsons. The other grandson, the once little Sergei who so admired Arkady, works alongside him. Sergei is also a singer and poet, and travels throughout the region to perform at festivals.

Vladimir Nikolayevich developed an eye problem, said to be caused by the dazzling sunshine coming through the ozone hole, and underwent a series of operations in Yakutsk. Later, he was also afflicted by stomach sickness, but continued to travel around the mountains in great discomfort. In 2004 he was diagnosed with liver cancer, and at the time of writing he was already not expected to live long. It was as a small repayment of a great debt that I was able to have the account of our winter journey translated into Russian and flown to his bedside a week before he died.

I can only agree with my friends that Sebyan's greatest misfortune has been that Vladimir Nikolayevich's childhood spent working in a prison camp did not give him enough formal education to be appointed to run the Farm: the director trumped him with a degree in veterinary science. The ever-generous Afonya still respects the director for certain canny strategic decisions he made in the 1960s and 1970s. But it is an equal misfortune that the director's power later became so unchallengeable in the face of momentous changes, while his vision became so puny. He did not release his grip on the community's finances until the day of his own death in July 2002. Old enough to have retired two decades earlier and enfeebled by alcohol and a stroke, he continued to the last to groom his successors and sabotage the schooling of their rivals' children, while his people continued to die at his feet. He was succeeded by an honest but unadventurous man, and by 2004 pressure was growing in the community to appoint Kesha.

Afonya remains head of the administration. He has never succumbed to the spirit of alcohol and continues to uphold what I see as civil administration morality. The civil administration is the only reliable source of cash, since it draws on the budgets of federal and regional government*. It is the Government that pays pensions, and when workers in bankrupt Farms are unpaid, their families live on the pensions of their old people. Sometimes a government measure succeeds for a very small cost. In the Sakha Republic, the wages of tent workers were transferred from the Farm to the regional budget in the late 1990s. Suddenly, these women became the only people to receive regular payment for working in the taiga, and several more of them moved out to the camps to soften the herders' lonely lives.

Afonya's concern is to keep his community provisioned, even in areas that are not his formal responsibility. As winter approached in 1999, there was no fuel oil for the boiler-houses that heated the village's public buildings. Just as I was leaving for the city, Afonya tore a chequered page out of an exercise book* and scribbled an appeal, its tone at the same time official and impassioned, which he enjoined me to pass on direct to the President of the Sakha Republic himself.

When I first arrived in 1988 the Soviet Union was at the tail end of a very stable system. From the mid-1960s to the early 1980s, any tensions and faults were held indefinitely in abeyance, and this period of stability was so uneventful that it was later nicknamed 'the stagnation' (zastoy). In the reindeer-herding communities there had been no upheavals since the 1960s when the nomads had been settled in villages, and the progressive decline of the family and increase in herd size had taken place without any significant structural changes. The director of the State Farm in Sebyan and his associates were deeply stagnant people, who matched those stagnant times.

But stability was not resilience: the changes around 1990 were like the snapping of the ice in spring on a river that has already picked up speed beneath the surface. New forms of cooperatives, collectives, production associations, and commercial enterprises emerged throughout the native North of Russia. These are still referred to colloquially as State Farms, and indeed, structurally and procedurally they are often not as different as they sound. Sebyan took the least radical of all possible alternatives and became a GUP (Gosudarstvennoye Unitarnoye Predpriyatiye, State Unitary Enterprise). By the late 1990s, calculators were appearing in the Farm office alongside the abacus, and computer print-outs alongside the pounding of the typewriter ribbon. But the Sebyan GUP was still run by the same people and seemed indistinguishable from a State Farm – except that it failed to provide the centralized support services that had made the State Farm viable.

The reindeer people of Siberia have been compelled to adapt many times over: to their challenging landscape from time immemorial, to the arbitrary violence of Cossack fur-bandits and the casual greed of tsarist officials, to the paternalistic and systematically violent onslaught of the Soviet State, and now to the vacuum left by the withdrawal of that State and the indifference of the so-called market economy. But Ivan, Granny, Lidia and Gosha; Kesha and Lyuda; Kostya, Vladimir Nikolayevich, Vitya, Tolya, and Afonya are not passive victims. They are intelligent, flexible people, politically alert, whose inner spiritual life and reserves of irony allow them to survive and look out for each other, even while they see their world for what it is.

The emotional journey of the reindeer peoples of Russia has been hard, and the feeling of loss that I sense in so many people does not come from naive nostalgia, whether theirs or mine. The Soviet ideal of progress was based on a rejection of the past:

graves were to be forgotten, children were to separated from their primitive, deported, or murdered parents, dead shamans were not to be reborn. This catalogue of sacrifices is now seen to have been in vain: the Soviet project was too ambitious – or perhaps even fraudulent. Now, in addition, they learn that their bodies have been irradiated, their Farm directors have lost direction, their local economies are bankrupt, the helicopters on which they have been led to depend have been grounded forever, and their children are killing themselves.

Russia's population of domestic reindeer has plummeted from 2.2 million in 1990 to 1.1 million today. Is this the beginning of the end of yet another branch of human civilization*, as one Russian reindeer specialist has put it? Eleven thousand years after the great wild herds retreated from central and southwestern Europe and 2,000 years after they were first saddled in Inner Asia, is the Age of Reindeer nearly over?

Certainly, the relation between reindeer and humans in Russia is undergoing a realignment. The decline in domestic reindeer is balanced by a corresponding rise in the number of wild reindeer, which are increasing to fill their vacated ecological niche. From a mere 200,000 in 1961, wild reindeer in Russia have increased till they also number 1.1 million, said to be the first time in 150 years that the figures for domestic and wild reindeer in Russia have been the same. Other practices that formerly limited wild reindeer have also been curtailed. Until around 1990, the huge herd on Taimyr Peninsula was being culled at the rate of 50,000 a year to feed the region's industrial towns, but this has ceased with the disappearance of the aviation to gather up such large quantities of meat.

The increase in wild herds is monitored by biologists and wildlife departments. I have seen young observers from the local Ministry of Natural Resources sitting on the plastic sofa of the aviators' hut playing cards or comparing notes by species and river in their exercise books as they wait to be called to their spotter plane. These boys, plotting the tracks and movements of wild animals in late winter when the revealing snow is still on the ground, have taken over one of the many roles of the flying shaman. Their work has expanded from the shaman's task of guiding hunters to their prey to include modern conservation and management, as well as giving advance warning when wild reindeer threaten to invade and abduct domestic herds.

During the 1990s there also arose a new category which biologists' figures were not designed to accommodate. The decline of morale and attentiveness among some herders did more than expose their reindeer to wolves. It allowed portions of their herds to break off and become feral. Domestic strains in this partly wild state retain a memory of human contact and can still be brought back under human control, unlike truly wild reindeer. But for this very reason, they are less alert than their wild cousins and more vulnerable to predators.

Without the standardizing influence of Soviet policy, the destinies of different herding regions are increasingly diverging*, like the individual destinies of Sebyan's brigades 7, 8 and 10. The species *Rangifer tarandus* continues to retreat from the south, and the heartland of the reindeer now lies firmly in the far North. In the southern forest regions where the species was first domesticated, reindeer herding is steadily dying out, and of the unique Tofalar breed of reindeer on the Mongolian border, fewer than 1,000 animals remain. The native communities that live by reindeer herding in southern regions such as Tuva, Khabarovsk, and Sakhalin are already endangered peoples.

The habitat and lifestyle of the forest reindeer herders in the south are generally closer than those of Sebyan to earlier Tungus forms. I have twice stayed with a camp of Evenki who keep small herds and hunt wild reindeer for food and sables for cash, migrating across the border between the Sakha Republic and the Amur district. Though at first sight their dense southern forest seems vast, they form a small enclave in Russianized territory and are boxed in on several sides by large-scale mining of coal and of the gold whose tailings pollute their rivers.

These herders, too, have waited long periods for their wages.
In 2001 they were paid a backlog – in little gold ingots, each
finely stamped with a woolly mammoth and the name of the
Aldan Gold Company. It was up to the herders to sell these
ingots, perhaps to dentists, to raise the cash to buy groceries.
They thought the rate in the local mining town was too low, so
they sent one of their more sophisticated women to Yakutsk.
She asked me to help and offered to sell them to me – what a
souvenir! – but I did not have several thousand dollars or the
time to resolve the formidable legal problems for a foreigner
wishing to own or export gold. My contacts were out of town
and we did not find any other safe outlet. She returned 370
miles to the herders with the gold still in her pocket, and they
remained without provisions. This was a bizarre reversal of the
extraordinary situation that prevailed during the Aldan gold
rush of the 1920s, when the young Soviet State had such difficulty preventing prospectors from selling their finds on the black
market that the State-run stores were forbidden to accept cash.
They used their monopoly position to sell food only for gold
dust, and alcohol only for gold nuggets*. Now, on the contrary,
the State was so short of cash that it was paying its remote forest
employees in gold bars.

These southern Evenki live near a branch of the trans-Siberian
railway, whose mineral traffic brings them more problems than
benefits. In the far North, by contrast, the viability of reindeer
herding depends on access to transport and the market, and
generally deteriorates as one moves from the relatively well-
connected flatlands at the western end of the Russian Arctic to
the high and remote mountain ranges of the east.

In the European part of the Russian North, the number of
domestic reindeer has stayed fairly constant since 1990, and
across the Urals among the Nenets on Yamal Peninsula it has
even risen. The Nenets people have preserved a strong kinship
system as well as trading networks with the outside world. Even
in the Soviet times some 40 per cent of the region's reindeer

remained in family ownership, and with the liberalization of the 1990s they have moved energetically into selling velvet antlers to middlemen from China and Korea*. Further east in the Sakha Republic, the number of domestic reindeer has dropped by more than half, from 343,400 in 1992 to 136,900 in 2002. As a small component of these figures, the decline in Sebyan does not look at all bad: from 20,000 in 1988 to 16,000 by the mid-1990s, more recently remaining steady at around 17,000–18,000.

In Chukotka in the far northeast, the collapse of reindeer herding has been catastrophic. In Soviet times the homeland of the Chukchi people was turned into a frantically overproductive reindeer factory, which yielded a high proportion of all Russia's reindeer meat. By 1980, the sparse vegetation of this one harsh, windswept district was being chewn by 540,206 domestic reindeer and trampled by their 2,160, 824 hooves.

The Chukchi were much more vulnerable than the Nenets of Yamal. They owned a mere 5 per cent of the reindeer they herded*, making them completely dependent on the State Farms that employed them. When the Farms were broken up in the early 1990s and their assets privatized, the herders were still missing six or eight years' back-pay which they would never receive, and had no other assets. Some Farms were turned into joint-stock companies (*aktsionernoe obschestvo*). Often the herders received the reindeer, but the equipment went to the Farms' Russian specialists who took it out of the area altogether, leaving the herders without technical support. The starving and cashless communities were approached by outside traders who offered groceries in exchange for meat, and laced their offers with lorry-loads of vodka. While an unsympathetic local government looked on, the herders succumbed to firewater, and their neglected reindeer succumbed to four-legged and two-legged wolves. By 2000, the Chukchi had become the most destitute reindeer herders in the world. At the 2002 spring festival in the village of Omolon, where the 30,000 reindeer – more than have ever existed in Sebyan – had been reduced to a mere 6,000, the younger people danced away the night in a disco, while the older herders sat and wept. Omolon was dying from a sickness far more hopeless than the despair of Sebyan. The brigade nearest the village, which suffered most from the outside traders' blandishments, sold or drank their inheritance down to the last reindeer.

The massive herds that the old people were lamenting, and which to them already seemed 'traditional', were in fact a modern industrial development of the previous small-scale relationship between hunters and their *uchakhs*. The decline in herd size is consistent with a wider de-industrialization of the Russian North, as the high tide of the Soviet empire retreats after the shock of *perestroika*, like blood being withdrawn from the outer limbs of an organism to the vital organs at the core.

While business culture flourishes at the centre of the new Russia, the industrial towns that spearheaded the 'mastery of the North' at the Arctic edge are dying. The only thriving industry in the North is the extraction of oil and gas for export. Towns like the coal-mining settlement of Sangar, loading point for the barges of the Lena River Fleet, or the Arctic Ocean ports of Chersky and Tiksi, which serve the caravans of icebreakers escort-escorting merchant ships from Murmansk to Vladivostók or Yokohama during the few weeks of thin summer sea-ice, are becoming ghost towns where one can buy an apartment in a half-empty block for the price of a one-way ticket to Moscow. Unable to fall back on subsistence herding like natives, Russian technical workers are leaving – if they can. The white population of the North depended on aviation, which was plentiful and virtually cost-free. Now, any journey costs the rouble equivalent of hundreds of dollars. Some people are trapped without the cash to leave their remote settlements and may never go anywhere for the rest of their lives. The project of the homogenization of space has been defeated by that space's own vastness. In the great

Soviet vision, distance itself was the North's greatest resource; now, it has become the region's greatest burden*.

When Pushkin wrote that his high art would reach even the 'wild Tungus', he was echoing a poem written in Latin 2,000 years earlier by Horace, court poet to the first Roman emperor Augustus, who consolidated the modern idea of empire as control over a far-flung territory of diverse peoples who feed consumption at the centre in exchange for civilization, their own lives bent to an agenda they can barely comprehend.

'I have wrought a monument more lasting than bronze' (*exegi monumentum aere perennius*), proclaimed Horace as he opened the final poem in his major collection, a fanfare which Pushkin adapted as 'I have raised myself a miraculous monument'* (*Ya pamyatnik sebe vozdvig ne rukotvorny*). Where Pushkin expected to be cited by Tungus peoples such as the Eveny, Horace looked for his primitive peoples at the southern tip of Italy. 'I shall be quoted [*dicar*],' he claimed,

> *qua violens obstrepit Aufidus*
> *et qua pauper aquae Daunus agrestium*
> *regnavit populorum*

> *Where the torrential River Aufidus growls*
> *And where the water-starved Daunus*
> *Once ruled over wild tribes*

Poems can indeed last longer than bronze, and certainly longer than empires. The Roman empire vanished 433 years after* Horace published these lines; in the twentieth century a Soviet empire, which Pushkin could not even have imagined less than a century earlier, lasted for only seventy-four years.

Daunus was a legendary king. Perhaps his tribes had disappeared by Horace's time, perhaps they had been civilized by the emperor: we are not told. But Pushkin's Tungus were 'still' (*nyne*) wild, and remained so until Soviet educators civilized them. The Soviet Union was not only the most psychologically intrusive empire the world has ever seen, but also the shortest-lived. Having subverted what people had of their own, it just as suddenly disappeared.

What kind of people will the northern reindeer-herding natives become as the empire abandons them?

The villages established in Soviet times were designed to occupy a specialized niche in a complex political ecosystem, drawing the nomadic tribes into the State's programme of development as suppliers of meat to industrial settlements. Native communities were made logistically and psychologically dependent on veterinary and medical services, schools and hospitals, all integrated through aeroplanes and helicopters into a tight network of control and fulfilment. Suddenly all these facilities, these goals and satisfactions, have been withdrawn. Sebyan can go for a whole summer without a doctor or any physical contact with the outside world, and sell no meat in the autumn. With the Farms bankrupt and their urban customers emigrating, can such settlements continue to exist? Many native villages were closed as uneconomic in the 1960s and amalgamated with others. Why not the remaining ones?

'Sebyan is slowly dying of alcohol poisoning,' Lidia told me over one of our many pots of tea. 'But they won't close it as long as there's still reindeer herding. The market will encourage us to form kin-based associations* [*rodovaya obschina*], without the State Farm.'

Many visions of renaissance since the late 1980s have been based on a return to modernized versions of pre-Soviet forms, based necessarily on reduced dependence and greater self-sufficiency. The more organized visions have created these self-help associations, collectives that struggle in some regions and occasionally flourish (though several attempts to establish them in Sebyan have failed). But then, combining a contradictory scenario in the same flow of thought, Lidia raised a more anarchic possibility which I have also heard from others.

'If there's a big crisis,' she said, 'people will flee to the taiga, not to the city.'

So the Eveny may take to the mountains again like their ancestors, feral humans riding on *uchakhs* to hunt feral reindeer, the descendants of their own abandoned herds. The Communist Revolution and the might of the Soviet Union; the State Farm and the wonder of free helicopter taxis on demand; the twentieth century, bringer of industrialization, war, and large-scale betrayal: all these will have come and gone, and over a large swathe of the earth's surface they will leave nothing but occasional ruins of wooden poles on an uninhabited landscape that changes so slowly that it will take centuries for the Arctic vegetation to roll over them.

'But there are lots of young people who can't survive in the taiga,' I objected. One had only to remember Lidia's own daughters on Hotorchon Pass during a brief summer snowstorm.

'Yes,' she conceded, 'some will go to the city – the ones with education.' She thought for a moment, then insisted, 'But the rest of us could live in the taiga without the village.'

I do not believe that the Age of Reindeer is coming to an end, but that the people who live with reindeer are moving to a new global awareness that opens up possibilities for new kinds of action, even turning deficiencies into opportunities. As the air transport of heavy reindeer carcasses becomes prohibitively expensive, herders seek to increase the value of every available flight. In the far west, within reach of Scandinavia, they try to sell the meat abroad as an exotic delicacy; in other areas, they load the aircraft instead with velvet antlers for the Korean market.

My German former student Otto reports that Nenets and Komi reindeer herders in the European Arctic see the local oil industry not just as an environmental threat, but also as a source of transport and diversified employment. The coal-mining town of Khalmer-Yu, beyond the old prison-camp capital of Vorkuta, has been abandoned by its population of 30,000–40,000 Russians and is now cut off beyond impassable derelict railway tracks. But it is still accessible to Nenets herders, hunter-gatherers of a new kind who harvest bricks, glass, and wood from these ruins and take them away on reindeer sledges to repair their own villages, or perhaps to re-sell in Vorkuta.

Talk in Russia of total collapse*, which became so prominent in the 1990s, is itself a kind of rhetoric that can also serve as a 'cover' (krysha) for people's own purposeful activities of vyzhiv-anie (surviving, or making do) and entrepreneurship, which crosscut and belie the appearance of dependency. Even while their Farm's inadequacy closes off transport links and makes the marketing of their meat impossible, reindeer peoples continue to reach outwards in attempts to form new connections, asso-ciations, and pressure groups.

In May 2002 I visited Johan Mathis Turi, the Sami president of the Association of World Reindeer Herders at his home in the far north of Norway. We had just been taking part in an international conference in Kautokeino on reindeer herders' legal rights*, held to mark the publication of a comprehensive survey of the current state of reindeer herding throughout the world.

'I went on a trip to Topolinoye in 1990,' he told me as we sat in his upstairs living room, gazing at a fast, shallow river that ran away down the valley just in front of the window, seeming by an optical illusion to flow straight out of the house. Topo-linoye was the showcase Eveny village, 5,000 miles east of the Sami, that I had sidestepped in favour of Sebyan during my first visit in 1988. 'I was amazed,' he continued, 'I still thought we Sami were the only reindeer people in the world.'

Though they live in liberal democracies, the migration of their reindeer still brings the Sami of Norway, Sweden, and Finland into conflict with their governments over rights to land, culture, and language, and their professional and political associations are well developed. The journey to Siberia made such an impres-sion on Johan Mathis that in 1993 he persuaded the Sami to

Poster from the World Reindeer Peoples' Festival, Tromsø 1993.

invite the reindeer herders of Russia to a World Reindeer Peoples' Festival in Tromsø, which was attended by representatives from twenty-four different peoples. Inspired by this, the herders of Russia formed their own association, a more specific pressure group than the Association of Northern Native Peoples whose inaugural meetings in Moscow and Chersky I had attended in 1990. At the banquet in Yakutsk to celebrate the establishment of the regional branch of the Association of Reindeer Herders of the Russian Federation in 1995, I drank vodka and Crimean champagne with friends from many communities around the Sakha Republic and danced with elegantly dressed women whom I had last seen in floral frocks or baggy tracksuits boiling meat in a tent. In 1997 the Association of World Reindeer Herders was established at a congress in Nadym and Johan Mathis was elected president.

The functioning of reindeer-herding communities is often

undermined by the extraction of oil, gas, and other minerals, and their associations must work in an uneasy blend of opposition and cooperation with their governments, also drawing on international support to stimulate official interest at national and local levels. The meeting that brought me to Johan Mathis's home was attended by high-level representatives from the Norwegian and Russian governments, as well as by Nenets leaders, the director of the only reindeer Farm in Russia that made a profit (selling meat to the military bases on the Kola Peninsula), and the Chukchi Vladimir Etylin, dedicated Member of Parliament (Duma) in Moscow for the region of Chukotka, who had just promoted a law that would protect communal reindeer pasture* from being sold off to private owners.

In the late winter of 2001, Tolya and I brought the last practising Tungus shaman in all of Siberia to Sebyan. The spirits of the village had agreed to allow this old Evenki man, who lived as a recluse in a forest hundreds of miles away, to perform a séance.

The village hall was packed. Row after row of serious faces filled the darkened seats, while more people stood in the aisle and along the side walls. The audience wore fur coats, fur gloves, fur muffs, fur hats: reindeer, fox, wolf, mountain sheep, marmot, wolverine, otter, squirrel, musk rat, ermine. In any other setting, this would have been a glorious fashion display. But in this hall at −30°F, their dinner of meat and fat and their covering of the outer skins of mammals were all that kept these destitute humans alive.

The backdrop to the stage was a stylized reindeer prancing beside a *chum*, a tepee in the old conical Eveny style which few living people have seen. In the dust at the front of the stage lay the sphere that projects coloured rays of light at discos and song-and-dance cabarets. The audience looked up over this sphere onto the platform, where the light of a single electric bulb showed the shaman's young woman assistant lighting a fire of larch shavings and herbs to fumigate and purify his costume and drum. The hidden art of an authentic shaman had come out of the forest and onto the stage, a setting that for eighty years throughout Siberia had carried the exaggerated manic convulsions of actors performing hostile propaganda. Eleven years ago in this very space, its walls heavy with the accumulated reprimands of leaden voices telling the villagers what to do, think, and feel, I had watched the downfall of a theatrical shaman, naively played by a schoolchild who could have understood little of what he was enacting.

The shaman sat on a rectangular mat made from the brown head-fur of two wild male reindeer, the slits of the eye-holes and antler-holes sealed with the white fur of a male domestic reindeer. Patches of wolverine, bear, and lynx were sewn around the border with thread of wild reindeer sinew. His robe was made from tanned elk hide, processed with a dye obtained from the inner bark of an alder tree, and embroidered with a brass sun as well as iron representations of the shaman's own skeleton and of his animal and bird spirits. Every item was laden with meaning. From the shaman's waist hung tassels embroidered with tufts of fur gathered from a reindeer's throat, the seat of its soul, in such a way that not a single hair was scattered or lost. His headdress was made of two crossed strips of decorated cloth, representing the four corners of the universe, from which a fringe of fur tassels covered his eyes to conceal the ordinary world as he made his voyage of insight into another reality.

The four struts of the cross-piece on the inside of his flat drum likewise represented the four directions. The shaman himself stood at the centre of the universe, which for the duration of this ritual would be located in the village hall of Sebyan. The inside of the drum was ornamented with iron models of his helping spirits: an eider-duck and a bear. A pair of bear's ears would enable him to hear the speech of spirits, and twelve pine cones represented each of the levels of the cosmos to which he was qualified to fly.

The shaman was already singing into the interior of the drum in an unbroken chant and beating the drum with a flat, paddle-shaped drumstick in a steady 1 – 1 – 1 – 1 rhythm, the bells inside the drum and round his costume jingling with a slight delay after each beat. His eyes were half open, but he was already in another state of consciousness. On the biplane to the village, the shaman had been just a frail old man, blinking shyly as many people do here, avoiding eye contact even more than most. Now, an entire community was transfixed, watching every flicker of his half-hidden face. The tone changed as he moved from invocation to full trance, and both chant and beat acquired a sudden extra force as his body stood up and his soul took off. The fur tassels hanging from his waist swirled, while the iron animals and the brass sun and bells jingled with the swaying of his torso.

This ritual was not to locate wild animals, but to heal. Tolya had selected two women for treatment, one of them his own sister Anna, who suffered from crippling arthritis in her hands.

The shaman's tassels swirled as he bent down to his first patient. His spirits continued to sing through him, but the throbbing beat stopped for a moment as he put his hands on her head. He drew out her illness, which now appeared as a smear of blood and gore on the inside of the drumskin which his assistant held out to the audience. Until now they had been silent, but there was a hubbub as the people sitting in the first few rows and standing around the walls surged forward, adults and children together, their various fur coats pressing together like a herd of mixed animals. His assistant wiped away the blood and the shaman resumed his drumming while she whispered to the patient and helped her down from the platform.

Now the shaman was bending down in front of Anna, while his assistant talked to her gently and helped her to her feet. The beat stopped again while he crouched in front of her and brushed her over and over again with his drumstick, repeating the movement with his bare hands, drawing her arthritis down from the wrist and out through the tips of her fingers. Anna blinked, dazed, as he passed his drum right around her body, doing his best to gather up her illness into its concave interior. Tolya turned the drum towards the audience, shining my head-torch onto another splash of blood, and they surged forward again.

The spirits had declared that Anna's illness was the working out of a curse which had also manifested itself through the recent death of her boy, who had been victimized by gangs when he went to study in the city and had mysteriously fallen down a stairwell in a student hostel. She was suffering for an ancestor who had killed a sacred piebald reindeer; her arthritis was a punishment for the sin of his hands that had committed the deed. There would be no easy cure. Anna's eyes brimmed with tears and the assistant comforted her in a low murmur as she stepped down from the stage and joined her surviving son.

The ritual was coming towards the end. The assistant wiped the shaman's face, which was drenched in sweat. She removed his headdress, then his drum. He sank exhausted onto his mat of wild reindeer fur, his soul beginning to return from the twelfth level of the sky. His chanting began to slow down as she wiped his face repeatedly and fed him sips of water.

Instructed by Tolya, others donned the robe and picked up the drum, to draw the spiritual arousal away from the shaman and bring his soul safely down to earth. Kesha was the first to imitate his drumbeat and twirling dance. The pace was the same as before, but the intensity was diminishing as the robe and drum passed from one person to the next. The shaman lay huddled on his mat like a baby, still singing softly in rhythm to the beat. He put on a woollen skullcap and started to wipe his own face, gradually falling silent as he lay under a coat, smoking a cigarette. At length, all drumming ceased*.

Tolya had waited years for an opportunity to show the people of his community what they had lost from the heart of their own culture. While his aim was to raise cultural awareness, I believe that native leaders and activists like Tolya and Vladimir Etylin have taken on a core role of the old shamans, that of protecting their people by cultivating a specialist knowledge of other worlds. They sense, in a way that the old shaman cannot, how the hidden reality behind the surface of daily appearances has changed. Instead of the spirits of animals, land, and sky, modern native leaders must understand regional and federal government agencies and international organizations. Like the upper and lower worlds of shamanic cosmology, the purposes of these bureaucratic worlds are inscrutable to ordinary people, yet they have the power to nurture or destroy them. A drilling licence granted over a community's head to a multinational company can sever the reindeers' migration routes and smother their pasture under devastating oil spills; a nature reserve created with the uncomprehending encouragement of an international wildlife organization can destroy a community and starve the families who are forbidden to catch food in their own home; a nomads' territory mistakenly marked as uninhabited and unused on the plan of a project that does not consult the people, can be lost forever.

This is the capriciousness of Bayanay carried into new realms. Where shamans would traditionally 'fly' on reindeer or drums made from their skins to locate wild animals for harvesting and to fend off hostile spirits, these new native leaders fly in aeroplanes between ministries, parliaments, and expert committees, petitioning and bargaining for their communities, making sure a native voice is heard, harvesting new laws about land rights and natural resource management, and remaining ever vigilant for new kinds of threat. Like shamans, they cannot work alone with their own limited strength but need helper spirits, who include sympathetic scholars, lawyers, and activists in Russian cities and abroad.

In an attempt to help the destitute Chukchi camp at Omolon, Vladimir Etylin has flown around the local aviation authorities, the office of a new, sympathetic governor in Anádyr, and the Sakha Ministry of Agriculture in Yakutsk. By teaming up with

Tolya, he has come up with an extraordinary plan: to re-stock this camp by buying 1,000 reindeer from the Eveny of Sebyan and Topolinoye and transporting them to Chukotka to reconstitute an entire herd.

Small groups of reindeer were moved between districts in Soviet times, usually stud males to improve the stock. Even on this scale, there were problems of integration and adaptation: breeds had to be chosen to suit the forest or taiga environment, and incoming animals had to fight with local reindeer for supremacy. When Sebyan bought some Tofalar males from the original area of reindeer domestication in south Siberia, they died of the northern cold; when they sold their own reindeer to a Farm 250 miles away in Zhigansk, the animals ran home.

The reindeers' homing instinct can be thwarted by airlifting them. In 2001, 150 males were transported several hundred miles across Kamchatka in helicopters, avoiding a year's run overland and heavy losses on the way. But the operation envisaged in Chukotka will be unprecedented*. Rather than a few males to tone up existing stock, an entire herd of breeding males, females, *uchakhs*, and calves will be delivered to a brigade that has lost every one of its previous animals. Chukchi herders will fly in advance to the Eveny villages to work alongside local herders and become acquainted with the animals, and the Eveny will return with them to Chukotka to settle the reindeer into their new pasture.

A smaller airlift of this kind was carried out 3,000 miles west of Chukotka in April 2003, when 500 reindeer were driven to an airfield in batches and carried in five flights of a large Antonov-24 plane over the 800 miles between Yamburg and Baykit. Further flights were made in March 2004, timed earlier the second year to minimize distress to the pregnant females.

The Sakha–Chukotka airlift is planned on a scale that has never before been conceived, even under the technical might of the Soviet empire. A gigantic Mi-26 military helicopter will collect 150 animals at a time right from where they graze and

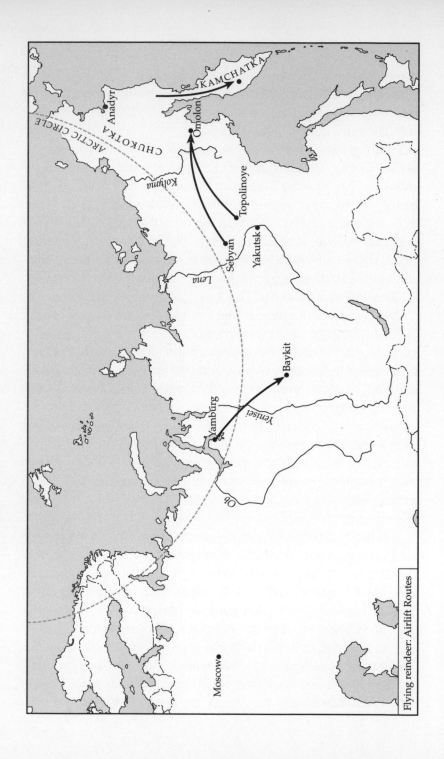

Flying reindeer: Airlift Routes

carry them on a 1,200-mile swoop over the massive peaks of the Chersky and Kolyma Ranges, decanting them directly into their new pasture.

Will the reindeer try to circle anticlockwise inside the cavernous body of the helicopter, as they do in a pen or when contained by a ring of gesticulating women and children? Lacking room to throw their lassos, how will the herders tie them down to distribute their weight evenly and safely over the interior? Will the reindeer grunt in the echo-chamber of the aircraft, or will they endure in silence, like Zinovy's four reindeer which half filled our little Mi-8 helicopter, soothed by the vibrations of the engine and the calming presence of their own herders? I hope I shall be there to find out.

This assistance from the Eveny to the Chukchi will be the biggest airborne herd of reindeer in history, a huge modern manifestation of the flying reindeer that have helped humans for millennia. Specific manifestations of belief can change. When the Old Man, Granny, Yura, Dmitri Konstantinovich, and Tolya's mother died, they were not buried on platforms high up in trees as they would have been 200 years ago, or under stones carved with reindeer sprouting huge winged antlers, like people in western Mongolia two millennia earlier; when Ivan, Kesha, and Tolya strip to the waist to chop wood on a hot day, they are not revealed as tattooed with reindeer on their shoulders, like the mummified people of Pazyryk.

Yet the journeys made beyond the grave on a sacrificed *uchakh* show that neither Christianity nor Communism has succeeded in completely wiping out the association between reindeer, flight, and human salvation. The discovery of how to ride a reindeer must have been a literal fulfilment of a long-held fantasy, allowing the ancestors of today's Siberian peoples to partake in the speed with which reindeer run for huge distances over plains, mountains, ice, water, bog, and scree, with an ease of movement that had otherwise been available only to the shaman in trance.

In taking off from the earth, the reindeer at the old mid-summer festival did not merely support the Eveny physically on a saddle: they also carried their souls. Since Tolya's early conversations about the festival with Eveny elders, in a language with 1,500 specialized words for expressing human relations with reindeer, almost every one of those old people has ridden their *uchakh* on to the next world. If I had understood those firelight conversations at the time, I might have pressed Tolya to ask them what they 'really' believed, what they 'really' felt at the moment when they were said to be flying. But that would have been giving in to my newcomer's impatience. My quest to enter the inner world of the Eveny could not be fulfilled by such direct, crude questions, but only by sharing their daily work, witnessing their life stories, and reflecting on their experiences of spirits and dreams. The life of another people is a mystery one can never plumb to the full; but my reward for living with the Eveny has been some wonderful friendships, and a glimpse into the enduring relationship between a community of humans and a species put on earth to nourish them with its flesh, insulate them with its fur, and exalt them with its soul.

I have come home from afar,
I have not beheld you for so long.
With all my heart I love you,
My homeland!

How fine you are in spring,
Sebyan, Sebyan,
Your lake surrounded by peaks,
Your pure water, Sebyan!

The autumn leaves fall,
My voice echoes far.
My song is about you, my homeland,
Birthplace of my ancestors!

If the reindeer do not come,
If the herd turns away,
If the reindeer do not come,
There will be no more Eveny!

Song by Motya the Music Woman, daughter of old Sofron in camp 1

NOTES

PROLOGUE

Soul-flight to the sun

3. In the Verkhoyansk Mountains of northeast Siberia, Eveny nomads are on the move:
The name is Even, but throughout this book I have called them Eveny (which is actually the Russian plural) since the word 'Even' would be unmanageable in English (imagine trying to begin sentences with 'Even women . . .' or 'Even men . . .'). Basic historical and cultural overviews of the Eveny in English are given by Levin and Potapov (1964: 670–84) and Arutyunov (1988), and at greater length in Russian by Turayev et al. (1997). A comprehensive bibliography of works on the Eveny (almost all in Russian) has been compiled by Alekseyeva (in press). When I saw a draft it already amounted to over 1,000 entries (many of them admittedly slight), or one for every seventeen Eveny people alive today.

During the Soviet period, 26 indigenous minority peoples were officially recognized in the Russian North, the cultures of many of them based on reindeer herding, and in recent years this number has increased to over 30. The 1989 census counted 181,517 persons; a semi-official and problematic census in 2002 counted 212,209 (IWGIA, 2004: 40–48). Concise overviews are given in Levin and Potapov (1964, focusing on 'traditional' cultures and containing little information about the twentieth century, apart from brief glorifications of Soviet rule), and in a more up-to-date way by Vakhtin (1994) and Vitebsky (1996). A people-by-people summary and bibliography is given in Funk and Sillanpää (1999). For grand histories of native Siberia, see Forsyth (1992) and Slezkine (1994). For contemporary politics, see various publications by IWGIA (International Work Group for Indigenous Affairs), especially Køhler and Wessendorf (2002).

The northern indigenous peoples belong to several language families. Historically, the Eveny and Evenki, of the Tungus family, have used reindeer to spread over the largest distances. Populations of the Tungus family in the 1989 census were:

Eveny (also known in the older literature as Lamut)	17,055
Evenki (also known in the older literature as Tungus)	29,901
Nanai (also known in the older literature as Gol'd)	11,883
Ul'cha	3,173
Udegey	1,902
Negidal	587
Oroch	883
Orok	179

The controversial 2002 census gives 35,377 Evenki and 19,242 Eveny. All figures are open to qualification. Identity can shift during a lifetime according to government benefits, or to whether it is shameful or a point of pride to be native. On reaching the age of 16, each Soviet (later Russian) citizen had to declare an ethnic identity which is written into their identity document: Eveny, Russian, Jew, Tatar, Georgian . . . Education, migration and mixed marriages may lead one to declare one identity on one's sixteenth birthday and others at censuses over the following decades.

'Siberia' is popularly used both in Russia and elsewhere for the whole of Russia east of the Urals, though specialists use 'the Russian Far East' for the regions bordering on the Pacific Ocean. In Russia the 'North' is likewise a loose term, and includes much of this region plus the parts of European Russia close to the Arctic Ocean. Technically, the boundary of the 'Arctic' is the Arctic Circle, but in terms of human life this term is often extended to a large area further south which is also sometimes called the sub-Arctic. For reasons of geography, climate, and hardship, the 'North' in Siberia is considered in the Far East to extend as far south as the border with China and Japan.

Research by Russian scholars on northern minority peoples, and on their reindeer, has a long, rich history and in these notes I can acknowledge only a few key sources. As the debate on the origin of reindeer domestication will show, the questions asked by these researchers are very different from those of their western counterparts. The only western scholars in modern times to carry out fieldwork in Siberia before the mid 1980s (not among reindeer peoples) were Humphrey in the 1960s–1970s (1998 [1983]) and Balzer in the 1970s (1999). Their achievements were remarkable, though they were allowed only short, tightly controlled visits. My own fieldwork, described in this book, began in 1988, and I was probably the first western researcher in Siberia since the Russian Revolution either to be free-range or to study reindeer people. From the early 1990s, a succession of doctoral students based in western Europe and North America began long-term anthropological fieldwork among Siberian reindeer-herding peoples. Academic monographs already published include Anderson (2000a), Bloch (2004), Gray (2004a), Kerttula (2000), Rethmann (2001), and Ssorin-Chaikov (2003). Several more PhD dissertations are completed but as yet unpublished, e.g. Donahoe (2003), Habeck (2003), King (2000), Kwon (1993), Stammler (2004), Ventsel (2005); still others are in progress.

6. endless succession of short migrations:
I have called each short stage of the herders' annual movement a 'migration'. The English word 'migration' can cover both repetitive seasonal movements that bring one back to the same place (as in the annual migrations of reindeer) and movements of populations into new territories (as in the historical migrations of the Eveny and Evenki). The meaning should be clear from the context, though the distinction is not absolute: animals and humans may explore new possibilities even while travelling the 'same' route (see chapter 13). It will be clear from this book that nomads do not wander erratically, but move in carefully planned cycles.

6. European or American ideas about Santa Claus:
St Nicholas was a fourth-century bishop of Myra (now Demre in Turkey), who used to give presents to poor children. Under the name of Santa Claus, he was taken by Dutch immigrants to New York. He was first associated with reindeer in the poem ' 'Twas the night before Christmas', published there in 1823. The eight reindeer in this poem were later supplemented by a ninth, Rudolf, who was invented in 1939 by an advertising copywriter.

6. ancient 'reindeer stones' dating from the Bronze Age:
Savinov (1994) summarizes the literature and gives a typology. Deer stones exist in Tuva, Gorny Altay, Zabaikal'ya, and eastern Kazakhstan. The greatest concentration is in northern and western Mongolia, where over 500 surviving examples have been found, often concentrated in groups of up to ten stones (Savinov, 1994: 29). For deer stones in Mongolia, see Volkov (1981). Both the Academy of Sciences in Novosibirsk and the Smithsonian Institution in Washington DC have current deer-stone research programmes.

8. one of the horses sacrificed in a grave ... clearly dressed up to imitate a reindeer:
This interpretation is supported, among other authors, by Skalon (1956: 95). The mask is described by Rudenko (1970: 179) and illustrated in his plate 119.

8. the wife of the Khan ... rode into battle on 'a reindeer with branching antlers':
Skalon (1956: 95), who also interprets the 'reindeer' in this way. Skalon's source is an unspecified Mongolian chronicle.

9. tattoos on their bodies. After death they were ... mummified:
An accessible account of the Pazyryk tombs, with numerous illustrations, is given by Rudenko (1970: the numbering of figures does not correspond to the Russian original). For recent discussions of the Pazyryk graves and mummies, see Polosmak (2000); Polosmak and Molodin (2000). All three works are in English translation and include bibliographies. A technical account of the mummies is given (in Russian) in Derevyanko and Molodin (2000).

It seems quite plausible that the function of these tattoos is similar to that of the images on the deer stones, and that an inscribed upright stone is a representation of a tattooed person, expressing their continuing existence after death. Stone and skin would be two versions of the same idea: the human being wrapped around on the surface by sacred animals which symbolize or assist the soul within, and link the person's existence in this life and the afterlife.

Polosmak (2000) describes and illustrates the three tattooed bodies and confirms the location of deer on the shoulders (pp. 95, 100, and illustrations). She, too, sees the deer as a guide to the next world

(p. 100), but her suggestion (following Gryaznov) that it is tattooed on the shoulder because the shoulder is displayed as one flings off one's cloak misses the point of flight and seems banal.

Allowing for the griffin's heads on some of the deer's bodies and the griffins or birds at the tips of some of the antlers, are these *reindeer*? The depiction of the slender, elegant bodies and finely branched antlers usually distinguishes these deer clearly enough from the stocky bodies and more paddle-like antlers of elk, which also appear in Pazyryk art. Russian archaeologists sometimes speak of *olen'*, deer (of any species) and sometimes of *severnyy olen'* (literally 'northern deer'), meaning specifically reindeer (*Rangifer tarandus*). Rudenko (1970: 143, 249) considers that it is not always possible to tell whether the images are of reindeer or red deer (maral). Articles made from both animals are found in the tombs, while 'almost all the saddle cushions were stuffed with reindeer hair' (p. 249). The key seems to lie in the outgoing impulse of antler growth: the motif of flowing antlers is so pervasive that they even appear on a tiger (p. 271 and fig. 137b).

9. the water that flowed into the graves:
Pazyryk is south of the main permafrost zone. The extraordinary process is explained by Rudenko (1970: 7–12), who concludes that the ice was formed in two phases, one at the time of construction and another by the much later act of tomb-robbing, and was then insulated by the structure of the graves (p. 10). Molodin (2000) warns that the preservation of organic remains in such graves is now threatened by global warming.

11. this ritual was one of Tolya's discoveries:
Described by him in Alekseyev (1993: 27–31). The sketch on his p. 27 is
by Tolya, based on a rougher drawing by his informant Vasily Pavlovich.
Caroline Humphrey has reminded me, in line with the argument in
Humphrey and Onon (1996), that there may be an opposition between
the role of the shaman (who was absent from this rite) and of the elders
who conducted it.

**13. suckled by a white reindeer . . . high in the tree that links earth
and sky:**
Ksenofontov (1930: 179–83); miming the act of killing, in Vasil'evich
(1969: 249–54), summarized and analysed (in French) by Hamayon
(1990: 468); the shaman combing the forest, in Prokof'eva (1981),
Hamayon (1990: 461–4). Hamayon's book is the most thorough
analysis of Siberian shamanism in any western language. English translations of diverse Russian sources are given in Michael (1963); Balzer
(1990); Diószegi and Hoppál (1978). For reasons that will become clear
in this book, such practices are almost entirely extinct today.

PART I
THE PARTNERSHIP OF REINDEER AND HUMANS

1 The prehistoric reindeer revolution

17. 'the Age of Reindeer':
A phrase coined by French prehistorians in 1877. Some sources suggest
that humans may have been hunting reindeer for half a million years.
For wide-ranging discussions of prehistoric and modern reindeer hunters, see Jackson and Thacker (1997); Spiess (1979); and Burch (1972).
All have extensive bibliographies and reveal many uncertainties and
controversies among specialists. For a specific modern ethnography
of reindeer hunters in North America, see Gubser (1965); in Siberia,
Popov (1966). The word *caribou* is from the Micmac language.

17. a form of domestication that is unlike that of any other animal:
The reindeer biologist Nicholas Tyler reminds me that ethologists in

western languages generally speak of 'semi-domesticated' reindeer, since the animals range over their natural pasture throughout the year and herders usually have little control over their movements on a day-to-day basis (though they maintain considerable control on a monthly or seasonal basis). Russian authors distinguish 'domesticated' (*domashniy*) from 'wild' (*dikiy*) reindeer. For the sake of simplicity, I use the term 'domesticated' throughout this book. Among the Eveny, there is an obvious distinction between the general herd kept for breeding and meat, and the smaller inner circle of reindeer that are selected for naming, riding, milking, or sacrifice (see chapters 11–13). The former are indeed no more than 'semi-domesticated'. Strangely, the more intimate taming may be a historically earlier development than the keeping of large herds of semi-domestic reindeer.

For a theoretical discussion of terms like 'domestication' and 'taming' in English, see Ingold (1980). The papers by authors in Anderson and Nuttall (2004) show how intensely forms of indigenous and State management are applied to domesticated and wild reindeer throughout the Arctic. The journal *Rangifer*, published in Tromsø, is devoted to scientific articles on reindeer, mostly in biology.

20. the earliest droppings so far analysed:
Van der Knaap (1986).

20. Julius Caesar:
In his memoirs, *De bello gallico* ('The Gallic War' or 'The war in Gaul'), Book 6, sections 21–8 (in Latin: translations should retain this section numbering). The mention of an ox-like physique suggests an elk, but the high, straight antlers suggest a reindeer.

20. the shore of Lake Ladoga . . . to Ufa in Bashkortostan:
Syroechkovskiy (1995: 27). The dwindling reindeer areas around the Sayan Mountains and over the border in Mongolia are now decisively cut off by the trans-Siberian railway, which was built in the late nineteenth century and opened up a corridor of Russian settlement, agriculture, and industry.

21. Scandinavian Lapland:
The literature is enormous, and there is much in English. The following

combine expertise with readability: Vorren and Manker (1962); Paine (1994); Beach (1993).

21. from a peak of 2.4 million in 1970 to 1.2 million in 2002:
My figures for wild and domestic reindeer in Russia are a compromise between differing sources. Jernsletten and Klokov (2002: 58–9), in the most comprehensive recent survey of the state of reindeer herding around the world, write that in 2000 Russia contained 1,357,700 wild and about 1,246,000 domestic reindeer (Klokov is a master of Russian reindeer statistics: see e.g. Klokov, 2000; 2001). In the same year, Baskin (2000: 23), also a well-informed Russian expert, stated that Russia contained around 1.6 million domestic reindeer.

Figures for domestic reindeer are built up regionally from local reports, each with its own bias. They are generally exaggerated in times and places where there is a subsidy to be gained per head of reindeer (as in recent decades), or a punishment for losses (which in Stalin's time could even be death). Conversely, they are downplayed where there is a wish to conceal large numbers, such as the many privately owned reindeer of the Nenets on Yamal Peninsula when this was against Soviet policy. Once, when I naively questioned whether a former State Farm official would be able to help me interpret some old statistics, he replied, 'No problem – I remember falsifying them!'

In the 1960s and 1970s, reported figures were fairly reliable if slightly exaggerated, and a sudden fall in numbers in a particular location can often be correlated with a known epidemic or destructive weather event. From around 1986, figures became less reliable as auditing became weaker. Figures from the early 1990s may still have been massaged upwards, either to cover up local failure or to capture any remaining subsidies, but the decline in numbers in most areas of the Russian North was so precipitous that nothing could conceal it (see the Epilogue to this book).

23. instinct for aggregating in herds:
Paine (1994: 201–9) gives a short summary of the behavioural characteristics of the species; see also Spiess (1979: 36–40). For a detailed account in Russian, see Baskin (1970).

23. Even where they invented skis:
Levin and Potapov (1961: 79–105) give details of ski types across native

Siberia. Though people in Sebyan have skis, the use of reindeer for transport is so thoroughgoing that I have never seen them used there.

24. roll a trumpet out of birch bark to imitate the sound of a male:
Illustrated in Levin and Potapov (1964: 629).

25. wild reindeer almost impossible to domesticate today:
Some sources given in Pomishin (1990: 42–3). There are anecdotal reports of the occasional taming of an individual wild reindeer, but this is a long way from drawing a group of them into an enduring relationship of reproduction under human management. Pomishin points out that individuals of some other species of ungulates such as elk, Manchurian deer (*izyubr*), and roe deer can be domesticated (or tamed?) quite readily.

26. she discovered that the milk was delicious:
For a survey of reindeer milking throughout Siberia, see Fondahl (1989).

28. a Reindeer Revolution:
Some of the many important Russian texts on the origins of reindeer domestication include Vasil'evich and Levin (1951), Skalon (1956), Vasil'evich (1964), Vainshtein (1970–71), and Pomishin (1990). It seems likely that reindeer were used as pack animals (involving a specially designed pack-saddle) before they were ridden.

Though the issue is not familiar to a western audience who know little of North Asia, the Russian historian Tolstov (1961: 112, cited in Vainshtein, 1980: 121) called this one of the major questions of world history. Most westerners would at least agree with Vainshtein himself when he writes of the domestication of herd animals (p. 130), 'The origin of pastoralism, of which [rein]deer-herding is a type, is one of the greatest problems in the study of man's cultural history.' Vainshtein's study of south Siberian pastoralism also covers horses, cattle, and other animals. His translated chapter on the origins of reindeer herding, the most thorough available in English, omits some sections of the Russian original. For a brief discussion in English of the arguments, see Khazanov (1984: 41–4, 111–14).

28. isolated and unrelated languages:
Chukchi is probably related to Kamchadal (Itel'men), see Fortescue
(1998: 37–44), but other languages such as Yukaghir have no known
relatives. The languages of northeast Siberia are of particular interest
for world prehistory since they lie in the path of major waves of
migration into North America. For a short review see Krauss (1988);
for a detailed scholarly discussion, see Fortescue (1998).

**28. whether or not herders milk their reindeer, use dogs to round
them up, or use stirrups to ride them:**
Vainshtein (1980: 130) summarizes the distinctions made on this basis
by many Russian scholars between five basic types of reindeer-herding
system.

28. numerous drawings on rocks and cliffs:
Sources in archaeological reports (some mentioned in Vainshtein, 1980:
120, 134) can be hard to find, apart from Devlet (1965). The interpretation of illustrations, like their dating, is not always straightforward.

30. the distinctive riding position that is still used in the Sayan today:
Vainshtein (1980: 128) reports from modern times, 'This method of
saddling, whereby the bulk of the weight rested almost on the middle
of the animal's back, caused many deer to die, since the spine could
not withstand the strain.'

30. where Marco Polo reports them as riding:
In the thirteenth century the Arab traveller Rashid ad-Din and the
Chinese text called *Yuan'-shi* describe the use of reindeer around this
region as pack animals, without mentioning riding (cited in Vainshtein,
1970: 8, and 1980: 132). However, Marco Polo mentions the riding of
reindeer (Benedetto's edition in English, 1931: 90, fuller edition in
Italian, 1932: 91). The text suggests that Marco had not seen the area
at first hand. Benedetto (1932: 438) places this region somewhere
between the Mongol Khan's capital at Karakorum and Lake Baikal (so
it could well be in the Sayan area).

**31. merging with local populations to form the modern Nenets and
Nganasan peoples:**
In recent times the Samoyed peoples have come to be called Nenets,

Nganasan, Enets, and Sel'kup. It is well known that continuities and changes in race, language and culture do not necessarily run in parallel, but 'A comparative evolutional genetic analysis of the mitochondrial gene pool showed that the Pazyrykians were closest to the contemporary Samoyeds of West Siberia' (Polosmak and Molodin 2000: 84). Vasil'yevich and Levin (1951: 73) argue that the reindeer techniques of the Sel'kup are of the Sayan type.

31. An alternative theory:
Argued by Vasil'yevich and Levin (1951), who link differences in saddle design and riding technique to separate Turkic and Mongol influences. The claims of a single or a dual origin of domestication is the focus of a long-running and highly technical debate among Russian scholars. Vainshtein (1980: 137–44) does not accept the theory of dual origin. He argues (1) that reindeer were used as pack animals before they were ridden; (2) that the underlying structure of pack saddles in both regions is similar, while that of riding saddles is different; (3) that therefore domestication originated in only one region (which for him is the Sayan).

32. The reindeer sledge . . . was developed relatively late . . . sledges pulled by dogs:
The imitation of dog-sledging was argued by Vasil'yevich and Levin (1951), and is now widely accepted. Dog-harnesses have been found along the lower Ob River dating from the first millennium BC, but dog-sledging in the far north is likely to be much older than this. For reindeer sledges from 1,400 years ago on Yamal Peninsula: Fedorova (1998); Fedorova and Fitzhugh (1998).

33. King Alfred the Great of England received a visit from a Viking chief:
The Viking was Ottar, also spelled Ohthere. Alfred inserted Ottar's account into his Anglo-Saxon translation from the Latin of Orosius' *Historia adversus paganos* (A history against the pagans).

33. Santa's theme park:
Near Rovaniemi, on the Arctic Circle, where Santa also runs a post office (in competition with his clones in Greenland and elsewhere). One current Internet travel advertisement promises an 'unspoilt vista

of snow-clad pine trees, grazing reindeer and cosy log cabins where flickering log fires ensure the warmest of welcomes'. Some tour prices include use of thermal underwear.

33. Russian adventurers moved into Siberia:
The classic account of the Russian colonization of Siberia is Armstrong (1965). Other histories, focusing on the native peoples, are Forsyth (1992), and Slezkine (1994). Native peoples on the European side of the Urals had much older trading relations with Norway, supplying not meat, but reindeer furs.

34. a transformation had already begun in the native peoples' relationship with their reindeer:
Discussed by Krupnik (1993: ch. 5). Krupnik powerfully demonstrates the change that swept across every reindeer region of Eurasia in the 1600s–1800s, building up numbers of domesticated reindeer and squeezing out wild herds, with an associated shift from hunting to pastoralism. To take just one example, the Nenets on the European side of the Urals probably had 10,000–20,000 domesticated reindeer at the end of the 1600s. Within a century, the number of domestic reindeer in this area alone had grown ten times to an estimated 160,000 (pp. 162–3). Krupnik also coins the phrase 'reindeer revolution' (p. 164), but uses it not as I do for the first, ancient domestication but for this reindeer population explosion in the eighteenth century. He argues that the transition to intensive reindeer herding was not caused simply by Russian colonization, but was also related to internal socioeconomic developments, combined with a cooling of the climate. Between the eighteenth and nineteenth centuries, he writes (p. 182), 'the entire structure of human relationships with the environment was radically transformed . . . One can only be amazed at the flexibility of this adaptive strategy, its capacity for rapid adjustment, and its malleability in unique sociohistorical or natural conditions.'

I hope that the continuing flexibility and adaptability of my reindeer herding friends will be apparent throughout this book.

34. confiscated and placed collectively in large herds:
In general, reindeer people in the North and their animals were gathered up into Collective Farms (*kolkhoz*), in which animals were owned in common and members shared profits and liabilities. In the

early 1960s most Collective Farms in the North were restructured as State Farms (*sovkhoz*), in which the animals belonged to the State and the herders were employees, on a system of wages and bonuses. When I first arrived in 1988 Sebyan was a State Farm, but most people believed the Collective Farm had offered a better incentive to work. A classic account of a Siberian Collective Farm (in the far south, specializing in cattle, sheep, and other southern livestock) in its heyday is Humphrey (1998 [1983]). Recent anthropological studies of reindeer regions in the North (see the first note, above) are set in State Farms during their collapse and restructuring in the 1990s.

35. the doomed, sickly labour of the Gulag prison camps:
There is an extensive literature on this (though not from the perspective of northern natives). The most recent study in English is by Applebaum (2003).

35. the great development of oil and gas:
With proven reserves of 60 billion barrels, Russia is the world's second largest exporter of oil after Saudi Arabia, and the world's number one exporter of natural gas with reserves of 1,680 trillion cubic feet. Oil and gas contribute well over half of all Russia's export revenue, and high world prices in the 2000s have been a significant factor in the improvement of the country's economy since the 1990s. Much of this resource base lies under the pastures of reindeer herders.

36. the Sora, an aboriginal tribe on the margins of Hindu civilization:
Vitebsky (1990a; 1993).

36. first encountered Siberian shamans in the eighteenth century:
See Flaherty (1992) for the fascination this exerted on the European mind, a fascination that continues today in New Age forms.

36. the first murmurings of *perestroika*:
Literally 'reconstruction', the process initiated in 1986 by Mikhail Gorbachev, who realized that the Soviet Union (USSR) was no longer economically or politically viable in its current form. I took up a post as Terence Armstrong's successor at the Scott Polar Research Institute in Cambridge in January 1986, a few weeks before Gorbachev's famous Murmansk speech promised to open up the Soviet Arctic.

Perhaps no one living today will ever again have the sensation of entering a world that is so huge, so complete, and so closed, as the few western researchers who worked in the Soviet Union before 1990. Ernest Gellner called his first weeks in Moscow some of the loneliest in his life (1988: viii); the same could be said of the first two out of my three winter months in Leningrad. The city looked European, but contained none of the cafés that define European city life. There was nothing to do but sit in one's room (or a silent library in which one knew nobody) or walk for sixteen hours a day around streets that offered nowhere to rest or eat.

38. the Sakha or Yakut people:
A people speaking a Turkic language and numbering over 300,000, occupying the Sakha Republic which until recently was known as Yakutia. Yakut is the Russian name by which they have been known in the outside world for four centuries, while Sakha is their own name for themselves and has become increasingly widely (and officially) used since 1990. The traditional occupation of the Sakha, which still has strong cultural and political significance, is the herding of distinctive hardy strains of horses and cattle. For accounts in English, see Vitebsky (1990b); Balzer (1995); Jordan and Jordan-Bychkov (2001); Argounova-Low (2004).

2 Civilizing the nomads

40. –96°F:
Just below –71°C. The meteorological stations in the two small towns of Verkhoyansk and Oymyakon compete to record the lowest temperature in the northern hemisphere, and both make claims around this figure. The low temperature is related to the relative lack of wind: temperatures on the Arctic Ocean coast to the north are not as low, but the fierce winds there make them harder to endure. The world's lowest recorded temperature, at the Russian scientific station of Vostok in Antarctica, is –128.56°F (–89.2°C).

43. This fixity is just what the Soviet authorities intended:
'Nomadism as a way of life' and 'industrial nomadism' are my translations of *bytovoe kochevanie* and *proizvodstvennoe kochevanie* respectively

(see Vitebsky, 1992: 232–3; Vitebsky and Wolfe, 2001: 83–5). Gray (2004a: ch. 4) writes of 'indigenous people as moveable parts' in the new socialist system, perhaps an ironic reference to their new kind of (greatly reduced) mobility.

44. This increase was made up mainly of breeding females:
This strategy, which gives the most rapid increase in herd size, was also followed by wealthy herd owners in pre-Soviet times. Poor families would have had few animals left for breeding after they had met their essential need for a basic minimum of castrated males for transport (cf. Krupnik, 1993: 103–4).

44–5. 'the able-bodied population' . . . 'unutilized labour resources':
Phrases from Lashov (1973: 94).

45–6. Pushkin . . . *the still wild Tungus*
From the poem headed in Latin with the phrase from the Roman poet Horace *Exegi monumentum*, 'I have wrought a monument' (see also note to page 381, below). I translate the adverb *nyne* as qualifying *dikiy*, rather than as qualifying *nazovet*, which would give a weak meaning. For a discussion of the 'wild Tungus', see Anderson (2000b) (though he quotes from a translation that misses *nyne* altogether).

46. young Communist vigilantes who noticed that I was reading a book on shamanism:
This harassment was an expression of the ideological uncertainty of the time. *Izbranniki dukhov* (Chosen by the spirits) by Vladimir Basilov (1984), a member of the Communist Party, was published by a party publishing house in a print run of 100,000 (so I cannot have been the only person reading it on the metro). So hungry was the Soviet public for material on shamanism that the edition quickly sold out and when the author lost his own copy he was unable to obtain another. Rather than go through the difficult procedure of requesting permission to use a xerox machine (still all but forbidden), he asked me to xerox my own copy in England and bring it to him in Moscow. My plan to translate the book and publish it in England fell through, but some chapters were later translated in Balzer (1990).

47. one official textbook explained:
Lashov (1973: 94).

47. 'mastery' (*osvoyeniye*) of the North:
See Horensma (1991); McCannon (1998).

48. On the Chukotka Peninsula facing Alaska ... The Nenets of the Yamal Peninsula:
For refences to the Chukchi and Nenets peoples, see note to the Epilogue, below.

48. massive oil and gas development that has made the lives ... wretched:
As soon as it became possible to publish such things, the native Khanty writer Aipin (1989) wrote that his ancestral fishing grounds had been polluted by oil 'two inches thick ... killing all life along the way ... [near] the rich oil fields of Samotlor, the flicker of whose flares have been for decades dancing on all this human misery'. The problem is now a regular topic in the literature and websites of numerous environmental and other organizations. For just one recent example, see Wilson (2003).

48. almost total absence of mining or mineral exploitation:
Oil and gas accumulate in low-lying areas, but mountains may contain toxic workings of valuable metals such as gold and uranium. The territory of Sebyan contains recently abandoned lead and tin mines at Endybal (see page 201).

50. real power was held by the Farm's director:
Humphrey's analysis (1998) of a Collective Farm in south Siberia shows that this kind of situation was already well established in the 1960s.

51. *perestroika*:
See note above to page 36.

51. forming a new district that would be attuned to Eveny needs:
This district had been formed in the 1930s and abolished in the 1960s (Vitebsky, 1992).

55. a total of 26,000:
It was more than ten years before I came to realize that there had probably never been more than 20,000 reindeer (the figure I have quoted earlier in this book). The Farm management in Sebyan kept two sets of figures, one revealing the true situation, another for sending on to the outside world. This practice was probably widespread throughout the Soviet North.

56. a brave and punishing struggle to improve his people's life:
I have presented oral or written Russian translations of parts of this book to some of the main characters, and expect to consult them even further when preparing a Russian edition. When this chapter was read out to Tolya, he responded, 'You don't have to make me out to be such a hero. You should add some salt – I'm a difficult person, and anyone who knows me will know that, too!' I did not change the wording, but hope that this book conveys my admiration for a loyal friend with whom a close relationship can sometimes be stormy.

58. the languages they used when talking to each other:
The boundary between Eveny and Sakha is not always sharp, and many people who claim one ethnic identity also have ancestors of the other. Eveny, along with its cousin Evenki the northernmost member of the Tungus-Manchu language family, was the mother tongue of the oldest people I knew in Sebyan, though with younger generations they spoke Sakha, the northernmost member of the Turkic family. With the expansion of Sakha traders over the past 300 years, Sakha has become a lingua franca across northeast Siberia, operating at a level between local languages and Russian. There are other Eveny villages, such as Topolinoye, where Sakha is not spoken and Eveny competes directly with Russian. For a range of northern language situations, see Vakhtin (1993; 2001).

59. the autumn tasks of sawing antlers and cutting down larch trees to build a corral:
These are explained in chapter 6.

59. he took me on some wonderful journeys through the mountains:
Described in the Interlude and chapter 14.

PART II A TALE OF TWO HERDS

3 The massacre of Granny's 2,000 reindeer, camp 7

66. include them in a television documentary:
Siberia – After the shaman, in the series 'Nomads', made for UK's Channel 4 television, shot in spring and autumn 1990, first broadcast in 1991 and subsequently shown in many countries. Graham Johnston (director-cameraman) won first prize for this film at the 1992 film festival of the European Foundation for the Environment in Paris, while Lindsay Dodd (sound) was nominated for a BAFTA award in 1993.

68. hardier and meatier:
Nicholas Tyler asks whether this was the result of selective breeding or a phenotypic response to a high level of nutrition over several generations. I cannot tell: it may even be a point of politics and rhetoric rather than of biology.

69. the antlers for medicinal use in Korea:
Young sprouting (velvet) antlers look very phallic and are widely believed to enhance male sexual potency. They are turned into medicinal preparations in pharmaceutical factories (Yudin, 1993). The label on a bottle of tablets, authorized on 2 April 1986 by Yu.G. Bobrov, Head of the Department for the Introduction of New Medical Substances and Medical Technology of the Ministry of Health of the USSR, reads somewhat tamely, 'Rantarin raises the capacity for physical and mental work . . . Rantarin is employed as a tonifying remedy . . . Side-effects may include nausea.' The medicine is not to be used with a range of heart, artery and kidney conditions. A bottle of drops reads simply, 'Institute of Biological Problems of the Cryolithic Zone of the Siberian Division of the Academy of Sciences. Extractum fluidum EPSORINUM, extract of reindeer antler. Keep in a dark place at room temperature.' Ideas about the end-use of dry, mature antlers are vaguer. I was told in Sebyan that the Chinese grind them up and feed

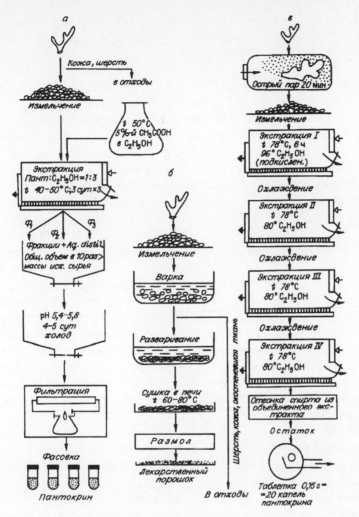

Chemistry of an aphrodisiac: how to turn a reindeer antler into a pill.

them to oysters to form pearls. Hiroki Takakura (personal communication) was told in Sakkyryr that the Chinese add the powder to cement for building.

72. Bells tinkled behind us:
Bells are hung round the necks of *uchakhs* and herd leaders to make them easier to locate and to warn wolves that the animals are under protection from nearby humans. They may be made from condensed milk cans with spoons as clappers, or occasionally from Buddhist temple bells of heavy Mongolian brass embossed with Tibetan prayers. In the mountain air with no competing engine noises, the sound carries for miles.

4 Granny's herd restored: late summer site, 1–2 August

80. to keep up with the reindeer as they flowed upstream:
In north Norway and on either side of the Urals, they would be making a comparable spring movement northwards out of the forest towards the open tundra on the coast, with migration routes running in parallel or crossing each other over the course of several hundred miles. In pre-Soviet times, wealthy families commanded more labour and migrated further than poor families, thereby enhancing the condition of their herd with better nutrition (cf. Krupnik, 1993: 95–6).

83. uncle Terrapin:
This was how they seemed to pronounce the Russian name Trofim.

91. the intensity of feeding had diminished:
Nicholas Tyler qualifies this by pointing out that the peak season for feeding continues from July well into August. Increased daily rates of movement in June and early July are partly due to the need to escape from biting insects. Reindeer cease growing in winter, however much food is available, and so need less pasture.

92. competing to be the first to find the next tasty leaf:
A similar scene is described by Baskin (1970: 59–60).

92. this bunching instinct provides the foundation of the herders' technique:
Reindeer have the same response to any potential predator, whether wolverine, lynx, or human.

93. a list of 1,500 words:
Dutkin (undated). There are several dialects of Eveny, and he does not say where each word comes from. I have given a range of terms from his list, but not all of these seem to be recognized in Sebyan. No single person or community uses all these words.

5 Migrating into autumn, 3–8 August

113. you should work it out for yourself:
Perhaps a moral dimension of what David Anderson calls Evenki pedagogy: 'An apprentice . . . will be expected to find useful things to do all by themselves, and then have their faults mercilessly exposed with insults . . . The men NEVER explain anything but will quickly do up a harness or wind a lasso even twice [as a demonstration] before getting frustrated' (Anderson, 2000a: 33).

117. The women were less interested in reading:
This may seem contradictory, since women tend to be more interested in education (chapter 7). But this interest in education is for the sake of advancement (Vitebsky and Wolfe, 2001). Reading for its own sake, especially on far-flung topics of the imagination, is commoner among men.

122. We don't say goodbye – it's taboo:
Perhaps more properly translated as a *breach* of taboo. *Grekh* is the Russian translation of the New Testament Greek *hamartia* (sin), literally meaning a 'missing of the mark' or a 'mistake' which violates the order of things.

6 Kostya's mushroom crisis, camp 10

135. the State-owned herd of 2,000:
This figure was probably accurate, at least for camp 10. Each brigade

did have this number of animals when functioning fully: the Farm inflated its overall figure to 26,000 by covering up the fact that two or three brigades tended to be in abeyance at any one time.

142. our Manchu cousins took the Chinese empire:
The Manchu dynasty lasted from 1644 to 1911. Linguistically, at least, the Manchu and Eveny are related.

INTERLUDE: SOLITUDE AND SILENCE

Vladimir Nikolayevich's winter hunt

151. Each of these soft or fluffy furs:
Almost every traditional Soviet ethnography of a northern people will have a section on traditional fur clothing. For accounts in English, see Chaussonnet (1988) and the oddly titled *Spirit of Siberia* (Oakes and Riewe, 1998), which is largely about boots. The most sophisticated discussion of mimicry between Siberian humans and animals is Willerslev (2002).

154. the threshold of −40°F
This is, coincidentally, the point at which the Fahrenheit and Centigrade (or Celsius) scales intersect.

158. the hut was full of thick fog:
Formed from condensed water vapour from our breath and from damp wood in the stove. As the fire warmed the hut, the air's capacity to hold water vapour increased, so the droplets progressively re-vaporized and the fog dispersed. At this temperature there would be around 25 per cent less oxygen in the air, weakening the burning power of the candle. In addition, since a candle burns by vaporizing fat and igniting the vapour, a greater input of heat would be required to bring the fat to the vapour state (thanks to Gareth Rees for physics information).

164. the Eveny writer Platon Lamutsky:
A pen-name from 'Lamut', an old term for the Eveny; real name Platon

Afanas'evich Stepanov (1920–86). A native of Sebyan, he wrote the first Eveny novel, *Sir icchite*, 'The spirit of the land'.

177. a better method of harnessing their reindeer to sledges:
This innovation is known as the Samoyed style of harnessing, said to have spread widely across the Russian North from west to east since the 1970s. Reindeer are harnessed in pairs. In the traditional Eveny method, a strap is passed over the head and round the inside leg of each reindeer. This puts a strain on the outer side of the animal's neck. The innovation was to link this strap more securely to a broader one which fits round the reindeer's midriff and spreads the strain more evenly (cf. Levin and Potapov, 1961: 49, figs 2d and 3d).

PART III BEADS FOR THE NATIVES

7 Frightened children and the disdain of women

184. recalled riding thousands of miles:
Paradoxically, the advent of aviation has greatly reduced mobility, as young people lose the skills to travel overland. David Anderson calls these skills 'knowing the land'. When an old Evenki lady told him about her childhood sledge journeys to destinations thousands of miles apart in all directions, his young, educated interpreter smiled with embarrassment and apologized in Russian: 'You have to forgive the old woman – she must be getting senile – it is impossible for her to have seen those places.' The old lady was probably right and the interpreter wrong (Anderson, 2000: 131).

189. rounded up by helicopter . . . and taken away from their parents:
The shouting teacher is shown in my film *Siberia – after the shaman*. The third film in Heimo Lappalainen's beautiful trilogy *Taiga nomads* dwells on children's movement from an Evenki camp to a boarding school. While my film explains everything, Lappalainen explains nothing, leaving the viewer to work things out in true taiga style.

189. giving his address as Pacific Fleet Headquarters in Vladivostók:

Tolya was the only Asian on his ship of Europeans and his commanding officer called him Dersu Uzala, after the exotic Nanai hero of Arsen'ev's book (1922; made into a film of the same name by Akira Kurosawa).

191. swapping boarding-school atrocity stories:

The situation has many parallels with the experiences of Australian Aborigines and Native Americans (Duran and Duran, 1995) around the same period.

By 1988, the Nenets writer Anna Nerkagi was able to write openly, 'I spent ten years in a boarding school. I left my parents' camp as a simple little girl and returned as a growing adult. I remember my teachers well, and am grateful to them. But I hardly remember my mother' (Nerkagi, 1988). In an earlier short story, an educated young native woman clearly representing Nerkagi is recalled in an emergency to the family tent:

> Her father was approaching with hurried, awkward steps.
>
> 'Father!' she suddenly realised, and was immediately struck with fear and confusion. Could this completely unknown and pathetic old man really be her father? Did she have to hug him, pick up the pieces of the feelings she once had . . . ? She grabbed her briefcase and placed it before her, as if trying to protect herself . . . the old man reeked of smoke, tobacco, and an unwashed body . . . How could she abandon it all: college, the theatre, movies, dances, discussions with friends about art . . . put on a leather robe, live by the light of a kerosene lamp and . . . get old?!

(Nerkagi, 1982: 48–9, adapted with small changes from translation in Slezkine (1994: 370–71))

Gray (2004a: ch. 4) and other anthropologists touch on native boarding schools, but the fullest study is by Bloch (2004), who identifies a complex blend of nostalgia, resistance, accommodation, and ambivalence, and argues that the Evenki of the Tura region have coopted the boarding school into a programme of native identity-formation. This did not happen in the region around Sebyan.

193. the continuum from wild to civilized:
Vitebsky and Wolfe (2001). The only substantial publication so far about women's experience in a Siberian reindeer-herding community is Rethmann (2001), but the outcome of Virginie Vaté's long-running research on Chukchi women is awaited.

8 Men fulfilled and men in despair, camp 8

207. one-third of all deaths are through accident, murder, or suicide:
Pika (1993); Pika and Bogoyavlensky (1995). Aleksandr Pika was a Russian anthropologist and champion of native causes, who himself died with several other anthropologists and native friends in a boating accident in the sea off Chukotka (see memorial issue of *Arctic Anthropology*, 1997). The high death and alcohol rates in native Siberia cannot be blamed entirely on Soviet rule, since they are not very different in native Arctic communities in Scandinavia, Greenland, and North America. The cause seems related to the impact of southern colonialism and loss of control over their lives in a much broader sense (cf. Anderson, 2004: 6, in relation to wildlife management regimes and their restrictive effect on natives: 'state systems have promoted a monoculture from one side of the Arctic to the other').

209. their bodies washed up against moraines of smashed trees and ice-rubble:
The policeman Little Bird himself disappeared without trace in 2002 while fishing with two friends on the same river near Sangar.

9 Landscape with Gulag: brushed by White Man's Madness

212. their life expectancy was eighteen years lower:
See note to page 207 on accident, murder, and suicide.

212. among the rows of suits and medals:
An account of this meeting is given in IWGIA (1990), which on p. 5 mistakenly places the meeting in 1989 instead of 1990. The association

later became RAIPON, the Russian Association of Indigenous Peoples of the North.

213. to some Russian settlers, it was home:
See Vakhtin et al. (2004); Thompson (2003a, 2003b).

216. 70 per cent of the local Eveny are dying of cancer:
No doubt an exaggeration, but in any case there is no way that accurate figures could have existed then or even now. Britain, France, China, and the USA have also been accused of irradiating native populations in remote regions with nuclear tests.

226. a monumental description of the dialect of Sakkyryr:
Sotavalta and Halén, 1978.

230. One researcher who has studied the KGB archives:
Vieda Skultans, author of Skultans, 1998.

10 Killing the shaman and internalizing betrayal

232. dropped out of an aircraft and challenged to fly:
However, many people were relieved when shamans disappeared. As well as healing, they were widely feared for their power to harm.

234. the Greek and Trojan kings of Homer's Bronze Age:
As in the *Iliad*, with retinues of non-members who rose or fell with their patrons. The part of the manipulative gods on Mount Olympus was played by bosses in the regional party office and the Moscow Kremlin. 'The son of a bandit!' is true to the idiom of the 1920s: the phrase 'enemy of the people' became widespread only in the 1930s.

237. your 55th birthday!:
I have received similar citations on my birthdays, signed by my Eveny friends. I, too, am susceptible: to be included in rituals fits the anthropologist's fantasy of belonging.

237. these former party believers:
I assumed the woodcutter was a party member. Years later, I learned

that he had never been 'received' (*prinyat*) into the party, on the grounds that he had been a 'hooligan' in his youth and was therefore unreliable. This makes the psychology of the situation even more subtle: the woodcutter was perhaps additionally susceptible because he was forever seeking approval and redemption. His nickname was *Agent* ('KGB officer'), and it is said that he died from overwork.

238. people succumbed one by one to a zombie plague:
When Tolya was chairman of the Village Council in the late 1980s, he persuaded the women to vote for prohibition on the grounds that their men would be sober. Gray (2004a) explains why the country's centralized structure made it more difficult for individual communities to ban alcohol than in North America. The Native American therapist was Eduardo Duran, co-author of Duran and Duran (1995), whom I visited with my wife Sally.

243. Soviet rule (*sovetskaya bylast*):
The Sakha pronunciation of the Russian word *vlast'*, meaning 'rule', 'power', or 'authority'. Sakha incorporates political and bureaucratic terms from Russian. There is no 'v' in Sakha. Being a Turkic language, it will not begin a syllable with a double consonant, but will break up the two consonants with a vowel (which will harmonize with any other vowel in the word, here a back vowel).

244. If I criticized each herder individually, I could go on forever!:
Here was the culture of indirectness taken to a new level. The herders being criticized were not in the audience; they had not even been invited to their own public reprimand, because it had been a secret until it actually occurred. Silence at meetings is often taken by white people to mean assent. This is convenient for Russian officials as well as for multinational oil companies who are now obliged under Russian law to hold public 'consultation meetings' before they risk polluting local reindeer pastures. But the director must have known something that white people do not: that his own people's silence can mean dissent when it is too dangerous, pointless, or contemptible to speak out.

250. the Sausage King of Alaska:
Roasting is not an indigenous technique, and would probably be

impossible inside a tent. John and Andrew continue to organise links between Alaska and northeastern Siberia to this day. In 2004 I sat in on a briefing they gave in Anchorage to US drug enforcement officers about to cross the Bering Strait to meet their counterparts in Chukotka.

253. a people on the verge of extinction . . . that samurai who went wild in the jungle!:
For the rhetoric of extinction, see Vitebsky (2002). Samurai: the Russian media had recently reported that a Japanese soldier had been found on a Pacific island, unaware that the war had ended fifty years earlier.

255. From a song improvised by old Vasily Pavlovich:
Original Eveny text with Russian translation and discussion in Alekseyev (1993: 55, with photo of the singer on p. 46).

PART IV SPIRITS OF THE LAND

11 Animal souls and human destiny

259. The spirit of a phenomenon represents its essence:
Vitebsky (1995a: 12–14).

260. the spear-like sharpness of the antlers:
Antlers can symbolize not only flight, but also virility. Hamayon (1990: 316–18, 504–5) shows how a shaman may use them aggressively to defend clients from the spirits that hunt them just as humans hunt game.

261. almost no shamans of this old type exist anywhere today:
The upsurge of interest in neo-shamanism in Siberian cities during the 1990s (Vitebsky, 1995b; Balzer, 1993; Humphrey, 2002) is a very different phenomenon and has touches of New Age and globalization (in May 2004, the Russian press reported that someone claiming to be the 'Supreme Shaman of Siberia' wished to exorcise the Duma (parliament) building in Moscow from evil spirits).

263. school had never prepared him for this question:
For an analysis of how generations may continue the same practice while changing their understanding of those practices' meaning, see Krupnik and Vakhtin (1997).

265. A male hunter may dream of sexual intercourse:
This association, like the secret language mentioned below, is widespread in hunting cultures. Neither my informants nor the anthropological literature can give much indication of how this works for female hunters.

270. As Vasily Pavlovich sang into our microphone:
Eveny text, Russian translation, and discussion in Alekseyev (1993: 57).

273. the Farm represented pure predation:
Not all Farms were like this. Some had good directors who struggled to make their enterprises prosper. The early 1990s were the hardest period. Sebyan was exceptionally remote, and exceptionally shielded from outside scrutiny.

273. the Nenets drove their animals to the slaughter house:
The main slaughtering season is much later here than in Sebyan (October). Snowmobiles are used because the herds are near a town and the entire region is more accessible (in European Russia and near a railway) and better provisioned.

278. a reindeer that had been specially consecrated:
Vainshtein (1980: 123) states that the Tofalar reindeer herders in the Sayan region (where the domestication of reindeer is likely to have begun, see chapter 1) use the word *kuddai* for a riding reindeer, and gives the etymology of this as *kuu + dai*, 'grey horse'. He interprets this (p. 136) as evidence that the riding of reindeer was copied from the riding of horses. This matches my feeling that the relationship with one's *kujjai* is a variant of the relationship with the reindeer one rides.

283. believed to lie in a sunken plane:
The story of Lidia's brother-in-law will be told in chapter 12.

On 12 August 1937 Levanevsky's crew of six men set out from Moscow in an experimental Antonov-6 plane. They were to fly to the

North Pole, then south to refuel in Fairbanks, Alaska (a journey that should have lasted no more than thirty hours in all), before continuing on to New York. They had already passed the pole when bad weather and mechanical failure were followed by loss of radio contact. A year of searches by Soviet and American rescue planes failed to find any trace of the flight. Though this was way off their course, it was later claimed that a memorial board had been seen on the shore of Sebyan Lake, with a partially obscured inscription referring to Levanevsky ('Here perished . . .') and dated 13 August 1937. This board (if it ever existed) has now disappeared. A team of Polish divers, some years before I arrived in Sebyan, failed to find any traces of the plane in the lake (thanks to Mike Hewitt for additional information).

12 Dreams of love and death

290. a veiled declaration of a feeling that could not be told:
Rethmann (2001: ch. 7) shows how Koryak women sew reindeer-fur clothes for men as a delicate form of courtship.

13 Sacrificing at a nomad's grave

311. Sora spirits were elaborately named and distinguished:
Vitebsky (1993).

315. the animals' instinct towards cyclical repetition:
Regular, long migration routes were most frequent among the wealthiest reindeer owners in the pre-Soviet period, while poor people often had to go where they could, making do with short, frequently changing routes. Clearly, their reindeer would do less well. For a wide-ranging collection of papers on reindeer movement, see Beach and Stammler (forthcoming).

328. that damp seaside cottage in Yorkshire:
Though the temperature was above freezing, the cottage was damp and poorly heated, and the Siberians felt the cold. The conference in Rovaniemi, Finland, was on reindeer herding, that in Budapest on shamanism.

328. The older belief was that you would be reborn:
In the North, reincarnation beliefs are particularly persistent today among Inuit/Eskimo peoples: see chapters in Mills and Slobodin (1994). For a reinterpretation of this by young Yupik in Siberia, see Krupnik and Vakhtin (1997).

14 Bringing my family

335. Don't let them change you, Katya!:
The intimate form in Russian for Ekaterina (Catherine).

335. *Boletus* mushrooms . . . Siberian ginseng:
The mushrooms are dried for use in winter soups, and ginseng is steeped in vodka to make a medicinal tincture. Both are Russian customs, not much followed by native peoples.

339. how confined the world of women has become:
Vitebsky and Wolfe (2001); Rethmann (2001).

340. to think that he might depend on me in other ways:
While I tried to work out meanings in cultural and social terms, Sally would explain the same situation in the idiom of the psychoanalysts Freud, Klein, and Winnicott. Her training as a therapist combined with her lack of local languages to make her specially alert to non-verbal clues to people's feelings.

344. Hiding bullets, weapons, or tools made of metal under one's pillow:
Several symbolic strands of the destructive and protective power of iron meet here. Guns are associated with Russian colonization and the violence of killing; while throughout Siberia the iron reindeer antlers of a shaman's headdress and the models of spirits which are sewn onto the shaman's robe are made by a blacksmith, who has a hereditary magical power parallel to that of the shaman – the power to shape iron into aggressive or protective objects. The blacksmith is said to be the one person who is immune to the shaman's power.

346. Bayanay received the vodka:
This seemed to be both a joke and a serious explanation – perhaps inevitably so, as it was simply the way one speaks of such things.

349. they didn't understand the names of months:
Through talking to old people, Tolya has reconstructed a traditional Eveny lunar calendar, in which the months were marked off on the human body, moving cyclically from the crown of the head in midwinter, down the left arm to the genitals in midsummer and back again up the right arm (Alekseyev, 1993: 7–15).

15 How to summon a helicopter

351. How could we have pitched it in such a stupid place?:
Perhaps an example of Anderson's idea of being left to learn from one's own mistakes (2000a: 33–7). I noticed the difference from tribal India, where my Sora friends were openly interventionist in their attention to my safety and comfort.

352. the helicopter avoided our mountains ... but a bar of chocolate would persuade them:
Years later, when I was setting up the location for a documentary film (*Arctic aviators*, National Geographic, 2001), I came to realize how meticulous these swashbuckling men are about safety, their sorties additionally controlled by meteorologists who fine-tune their assessment throughout the day on the basis of messages that filter in from observation stations and village radio operators.

355. a hostile voice said over the radio:
Negotiations were taking place in the new grey area between the public and private sectors, not in Russian but in two native languages I did not understand well, and I was hearing only one side of the conversations. But it was a reminder that as well as friends, I also had enemies.

355. There's no one to marry in the village:
Dependence since the 1960s on aeroplanes (which fly mostly to the towns and cities) has restricted the movement of young people

between Eveny villages and cut off previous flows of marriage partners.

363. startled, frozen look:
Proud, passionate individuals among the Sora also wore a blank expression in town: photographs taken there and in the jungle could hardly be recognized as being of the same person. I sense something similar in many old photos of indigenous people in remote places around the world.

EPILOGUE

Outliving the end of empire

370. They had moved . . . to the other side of the river:
I suspected that one reason for having moved across the river was that the layout of the old site of Djus Erekit would not easily have accommodated this separated existence, but would have forced everyone into an intimacy that they did not feel. Maybe, too, the old site was too full of camp 7's family memories.

374. The civil administration . . . draws on the budgets of federal and regional government:
In the Sakha Republic today, subsidies are more reliable than during the 1990s and are paid through enterprises such as Farms or kin collectives.

374. Afonya tore a chequered page out of an exercise book:
The Farm had given up on the problem, and Afonya took it upon himself – thereby, of course, laying himself open to criticism if he failed.

376. the beginning of the end of yet another branch of human civilization:
Title of a presentation by Vitaly Ignat'evich Zadorin at a workshop on 'The human role in reindeer/caribou systems' in Rovaniemi, Finland, February 1999.

When I read Tolya a Russian translation of this chapter, he commented at this point, 'The Eveny are a footprint in the snow, and when the snow melts they will disappear.'

377. the destinies of different herding regions are increasingly diverging:
John Tichotsky's joke is to adapt the opening sentence of Tolstoy's *Anna Karenina* to read, 'All happy reindeer Farms are alike, but each unhappy Farm is unhappy in its own way.'

For accounts of the Nenets, see Golovnev and Osherenko (1999); Stammler (2004).

378. sell food only for gold dust, and alcohol only for gold nuggets:
Sources in Tichotsky, 2000: 94.

379. selling velvet antlers to middlemen from China and Korea:
It seems the herders in Sakha have less scope to do this because of a State monopoly there.

379. They owned a mere 5 per cent of the reindeer they herded:
A figure confirmed in conversations with Patty Gray, who adds that in many cases the reindeer went not to the herders but to Russians, who simply slaughtered them for cash. In the early 1990s, some brigades broke away from the Farms, but in 1998 their herds were snatched back by the administration (Gray, 2000; 2001), so that today not a single herd in Chukotka is owned by herders. For the industrialization of reindeer herding in Chukotka see Gray, 2004b, and for the wider picture Gray, 2004a.

I have not been to Omolon, and my information comes mainly from Vladimir Etylin, who was born there. I alone am to blame if I have misrepresented anything he has told me over years of conversations. Patty Gray has given me the following figures for Omolon State Farm, obtained from the local Department of Agriculture:

Year	Headcount
1976	30,300
1980	34,502
1985	33,898

Year	Headcount	
1995	8,986	(but by now several brigades had split from the Farm with their reindeer, so they are presumably not counted)
1996	8,034	
1997	6,435	
1998	4,411	
1999	–	(the number is blacked out in the data; by now, any private reindeer have presumably been taken back)
2000	2,407	
2001	2,576	
2002	1,656	

381. distance itself was the North's greatest resource; now, it has become the region's greatest burden:
The main message of Vitebsky (2000). Hill and Gaddy (2003) present a similar idea, though their understanding is narrowly economistic. Thompson (2003a, 2003b) gives a more nuanced picture which goes beyond economics to encompass senses of identity and belonging in the North.

381. I have raised myself a miraculous monument:
Literally, a monument 'not wrought by human hand'. The word *nerukotvorny* is a calque from the Greek *acheiropoietos*. Pushkin hints at the Spas *Nerukotvorny* of Orthodox iconography, the image of Christ 'the Saviour *not wrought by human hand*', rather than the bronze of Roman imperial statuary referred to by Horace (thanks to Simon Franklin for advice on this point). Horace's poem is from his Odes (*Carmina*), Book 3, number 30, which rounded off his collected books 1–3, published together in 23 BC.

381. The Roman empire vanished 433 years after:
The end of the empire is conventionally dated to AD 410. The process was, of course, one of a long transformation into successor political formations.

382. kin-based associations:
The so-called 'clan association' (*rodovaya obschina*): see Fondahl et al.

(2001); Sirina (1999). I have been told that no *obschinas* in the Sakha Republic (Yakutia) function without subsidies from the regional government.

384. Talk in Russia of total collapse:
See Vitebsky (2002).

384. an international conference ... on reindeer herders' legal rights:
I am grateful to the conference organizers for inviting me to attend the meeting and give the concluding summary. The report is Jernsletten and Klokov (2002).

386. a law that would protect communal reindeer pasture:
The situation on land law is changing rapidly, and the tone of optimism or pessimism often reflects the moment when a piece is written. Novikova (2004) contains a recent collection of papers, including the text of the law, presented by Etylin himself (pp. 221–9). In English, see Fondahl (1998), Habeck (2002), Donahoe (2003: ch. 8). All of these could be made obsolete by a law now being discussed in Moscow which may privatize all land throughout Russia, including forests and reindeer pastures. In Sakha, the expectation of the likely privatization of all remaining State-owned reindeer is leading families to send their young people back to the taiga, in anticipation that this will make reindeer herding into the main, necessary basis of every family's economy. But many herders now fear that the proposed privatization of land will lead to its sale to big capitalists, undermining this move towards viability and removing the very foundation of their existence. They argue that pasture should be vested in herders through immemorial use. The paternalism of the Soviet Union towards northern minorities also gave them a measure of protection, and they are not equipped to compete on equal terms in the open market.

389. At length, all drumming ceased:
A rumour was later spread that the shaman died as soon as he got home, killed by the spirits of Sebyan. This says more about human factionalism in the village than about its spirits. At the time of writing, the old shaman is alive and well.

391. the operation envisaged in Chukotka will be unprecedented:
Kamchatka airlift: thanks to Katharina Gernet for information based
on her own involvement in the project. Airlift between Yamburg and
Baykit: from the RAIPON website.

BIBLIOGRAPHY

Note: the name Eveny is generally given here in the more correct form, Even (see note on page 399).

Aipin, Ye. (1989) Not by oil alone, *Moscow News* 2: 9–10, reprinted in *IWGIA (International Work Group for Indigenous Affairs) Newsletter* 57: 136–43

Alekseyev, A. A. (1993) *Zabytyy mir predkov* [Forgotten world of the ancestors], Yakutsk: Sitim

Alekseyeva, V. N. (in press) *Bibliograficheskiy ukazatel' po yazyku, literature, fol'kloru, etnografii i istorii evenov* [Bibliographical index of the language, literature, folklore, ethnography and history of the Even, Yakutsk: Institut Problem Malochislennykh Narodov Severa

Alfred the Great (890 AD (approx.)) *The Old English Orosius, by Paulus Orosius and Alfred, King of England*, London and New York: Oxford University Press for the Early English Text Society, 1980 (other editions also exist)

Anderson, D. (2000a) *Identity and ecology in Arctic Siberia: the number one reindeer brigade*, Oxford: Oxford University Press

Anderson, D. (2000b) Tracking the 'Wild Tungus' in Taimyr, in Schweitzer, P., Biesele, M. and Hitchcock, R. (eds), *Hunters and gathers in the modern world*, New York and Oxford: Berghahn, pp. 223–43

Anderson, D. (2004) Introduction in Anderson, D. and Nuttall, M.

Anderson, D. and Nuttall, M. (eds) (2004) *Cultivating Arctic landscapes: knowing and managing animals in the circumpolar North*, New York and Oxford: Berghahn

Applebaum, A. (2003) *Gulag: a history of the Soviet camps*, London: Allen Lane

Arctic Anthropology (1997) [Memorial issue for Arctic anthropologists and their native companions who died in a boating accident], 34(1)

Argounova-Low, T. (2004) Diamonds: contested symbol in the
Republic of Sakha (Yakutia), in Kasten, E. (ed.), *Properties of culture
– culture as property: pathways to reform in post-Soviet Russia*, Berlin:
Dietrich Reimer Verlag, pp. 257–65

Armstrong, T. (1965) *Russian settlement in the North*, Cambridge:
Cambridge University Press

Arsen'yev, V. K. (1922) *Dersu Uzala*, Moscow: Gosudarstvennoye
izdatel'stvo geograficheskoy literatury; tr. *Dersu the trapper*, 1939,
London: Secker and Warburg (other editions also exist)

Arutyunov, S. A. (1988) Even: reindeer herders of eastern Siberia, in
Fitzhugh and Crowell (1998), pp. 35–8

Balzer, M. (ed.) (1990) *Shamanism: Soviet studies of traditional religion in
Siberia and Central Asia*, New York: M. E. Sharpe

—— (1993) Two urban shamans: unmasking leadership in
fin-de-Soviet Siberia, in Marcus, G. (ed.), *Perilous states:
conversations on culture, politics and nation*, Chicago: University of
Chicago Press, pp. 131–64

—— (1995) A state within a state: the Sakha Republic (Yakutia), in
Kotkin, S. and Wolff, D. (eds), *Rediscovering Russia in Asia:
Siberia and the Russian Far East*, New York: M. E. Sharpe,
pp. 139–59

—— (1999) *The tenacity of ethnicity: a Siberian saga in global perspective*,
Princeton: Princeton University Press

Basilov, V. N. (1984) *Izbranniki dukhov* [Chosen by the spirits],
Moscow: Politizdat

Baskin, L. L. (2000) Reindeer husbandry/hunting in Russia in the
past, present and the future, *Polar Research* 19(1) [Proceedings of
the human role in reindeer/caribou systems workshop]: pp. 22–9

Baskin, L. M. (1970) *Severnyy Olen': Ekologiya i povedeniye* [The
reindeer: ecology and behaviour], Moscow: Nauka

Beach, H. (1993) *A year in Lapland: guest of the reindeer herders*,
Washington: Smithsonian Institution

Beach, H. and Stammler, F. (eds) (forthcoming) Special edition of the
journal *Nomadic Peoples* devoted to 'People and reindeer on the
move'

Bloch, A. (2004) *Red ties and residential schools: indigenous Siberians in a
post-Soviet state*, Philadelphia: Pennsylvania University Press

Burch, E. S. (1972) The caribou/wild reindeer as a human resource,
American Antiquity 37(3): 339–67

Caesar (Caius Julius Caesar) (52 BC) *De bello gallico* [The Gallic War] (in Latin: numerous modern translations)

Chaussonnet, V. (1988) Needles and animals: women's magic, in Fitzhugh and Crowell (1988), pp. 209–26

Derevyanko, A. P. and Molodin, V. I. (eds) (2000) *Fenomen altayskikh mumiy* [The phenomenon of the Altai mummies], Novosibirsk: Institut Arkhaeologii i Etnografii SO RAN

Devlet, M. A. (1965) Bol'shaya boyarskaya pisanitsa [The Great Boyarskiy inscriptions], *Sovetskaya Arkheologiya* 3: 124–42

Diószegi, V. and Hoppál, M. (eds) (1978) *Shamanism in Siberia*, Budapest: Akadémiai Kiadó

Donahoe, B. (2003) A line in the Sayans: history and divergent perceptions of property among the Tozhu and Tofa of south Siberia, PhD thesis, Department of Anthropology, Indiana University

Duran, E. and Duran, B. (1995) *Native American postcolonial psychology*, Albany: State University of New York Press

Dutkin, Kh. I. (undated) *Terminy olenevodstva v evenskom yazyke* [Reindeer-herding terminology in the Even language], Yakutsk: Yakutskiy Filial SO AN SSSR (booklet, 22 pages)

Fedorova, N. (ed.) (1998) *Ushedshiye v kholmy: kul'tura naseleniya poberezhiy severo-zapadnogo Yamala v zheleznom veke* [Gone to the hills: the culture of the coastal population of northwestern Yamal in the iron age], Ekaterinburg: Izdatel'stvo Ekaterinburg

Fedorova, N. and Fitzhugh, W. (1998) Ancient Legacy of Yamal, in Krupnik, I. and Narinskaya, N. (eds), *Zhivoy Yamal/Living Yamal*, Moscow: Sovetskiy sport

Fitzhugh, W. and Crowell, A. (1988) *Crossroads of continents: cultures of Siberia and Alaska*, Washington: Smithsonian Institution

Flaherty, G. (1992) *Shamanism and the eighteenth century*, Princeton: Princeton University Press

Fondahl, G. (1989) Reindeer dairying in the Soviet Union, *Polar Record* 25(155): 285–94

—— (1998) *Gaining ground? Evenkis, land, and reform in southeastern Siberia*, Boston: Allyn and Bacon

Fondahl, G., Lazebnik, O., Poelzer, G. and Robbek, V. (2001) Native 'land claims', Russian style, *The Canadian Geographer/Le géographe canadien* 45(4): 545–61

Forsyth, J. (1992) *A history of the peoples of Siberia*, Cambridge: Cambridge University Press

Fortescue, M. (1998) *Language relations across the Bering Strait: reappraising the archaeological and linguistic evidence*, London: Cassell

Funk, D. and Sillanpää, L. (eds) (1999) *The small indigenous nations of Northern Russia: a guide for researchers*, Social Science Research Unit, Publication No. 29, Vaasa: Åbo Akademi University [in English and Russian]

Gellner, E. (1988) *State and society in Soviet thought*, Oxford: Basil Blackwell

Golovnev, A. B. and Osherenko, G. (1999) *Siberian survival: the Nenets and their story*, Ithaca, NY: Cornell University Press

Gray, P. (2000) Chukotkan reindeer husbandry in the post-Soviet transition, *Polar Research* 19(1) [Proceedings of the human role in reindeer/caribou systems workshop]: 31–8

—— (2001) The obshchina in Chukotka: land, property and local autonomy, *Working paper No. 29*, Halle: Max Planck Institute for Social Anthropology

—— (2004a) *The predicament of Chukotka's indigenous movement: post-Soviet activism in the Russian North*, Cambridge: Cambridge University Press

—— (2004b) Chukotkan reindeer husbandry in the twentieth century: in the image of the Soviet economy, in Anderson and Nuttall (eds) (2004), pp. 136–53

Gryaznov, M. P. (1984) O monumental'nom isskustve na zare skifo-sibirskoy kul'tury stepnoy Azii [Monumental art at the dawn of the Scythian-Siberian culture of the Asian steppe], *Arkheologicheskiy sbornik Gosudarstvennogo Ermitazha*, 25: 76–82

Gubser, N. (1965) *The Nunamiut Eskimos: hunters of caribou*, New Haven: Yale University Press

Habeck, J. E. O. (2002) How to turn a reindeer pasture into an oil well, and vice versa: transfer of land, compensation and reclamation, in Kasten, E. (ed.), *People and the land: pathways to reform in post-Soviet Siberia*, Berlin: Dietrich Reimer Verlag

—— (2003) What it means to be a herdsman: the practice and image of reindeer husbandry among the Komi of Northern Russia, PhD thesis, University of Cambridge

Hamayon, R. (1990) *La chasse à l'âme: esquisse d'une théorie du chamanisme sibérien* [Hunting the soul: outline of a theory of Siberian shamanism], Nanterre: Société d'ethnologie

Hill, F. and Gaddy, C. (2003) *The Siberian curse: how communist planners left Russia out in the cold*, Washington: Brookings Institution

Horace (Quintus Horatius Flaccus) (23 BC) *Carmina* [Odes] (in Latin: numerous modern translations)

Horensma, P. (1991) *The Soviet Arctic*, London and New York: Routledge

Humphrey, C. (1998) *Marx went away, but Karl stayed behind*, Ann Arbor: University of Michigan Press (revised edition of her *Karl Marx Collective: economy, society and religion in a Siberian Collective Farm*, Cambridge: Cambridge University Press, 1983)

—— (2002) Shamans in the city, ch. 10 in Humphrey, C., *The unmaking of Soviet life: everyday economies after socialism*, Ithaca, NY and London: Cornell University Press (first published separately in *Anthropology Today* 15(3), 1999: 3–10)

Humphrey, C. and Onon, U. (1996) *Shamans and elders: experience, knowledge and power among the Daur Mongols*, Oxford: Clarendon Press

Ingold, T. (1980) *Hunters, pastoralists and ranchers: reindeer economies and their transformations*, Cambridge: Cambridge University Press

IWGIA (1990) Indigenous peoples of the Soviet North, *IWGIA Document No. 67*, Copenhagen: IWGIA (International Work Group for Indigenous Affairs)

—— (2004) *The indigenous world 2004*, Copenhagen: IWGIA (International Work Group for Indigenous Affairs)

Jackson, L. J. and Thacker, P. T. (eds) (1997) *Caribou and reindeer hunters of the northern hemisphere*, Aldershot: Avebury

Jernsletten, J. and Klokov, K. (2002) *Sustainable reindeer husbandry*, Tromsø: Centre for Saami Studies, for the Arctic Council

Jordan, B. B. and Jordan-Bychkov, T. G. (2001) *Siberian village: land and life in the Sakha Republic*, Minneapolis: University of Minnesota Press

Kerttula, A. (2000) *Antler on the sea: the Yupik and Chukchi of the Russian Far East*, Ithaca, NY and London: Cornell University Press

Khazanov, A. (1984) *Nomads and the outside world*, Cambridge: Cambridge University Press

King, A. (2000) *Trying to be Koryak: Soviet constructions of indigeneity in Kamchatka, Russia*, PhD thesis, University of Virginia

Klokov, K. (2000) Nenets reindeer herders on the lower Yenisei River: traditional economy under current conditions and responses to economic change, *Polar Research* 19(1) [Proceedings of the human role in reindeer/caribou systems workshop]: 39–41

Klokov, K. B. (2001) *Olenevodstvo i olenevodcheskiye narody Severa Rossii: Chast' I, Respublika Sakha (Yakutiya); Chast' II, Sever Sredney Sibiri* [Reindeer herding and reindeer people of northern Russia: Part I, Sakha Republic (Yakutiya); Part II, north-central Siberia], St Petersburg: St Petersburg State University and five other organizations

Køhler, T. and Wessendorf, K. (eds) (2002) *Towards a new millennium: ten years of the indigenous movement in Russia*, Copenhagen: IWGIA (International Work Group for Indigenous Affairs)

Krauss, M. (1988) Many tongues – ancient tales, in Fitzhugh and Crowell (1988), pp. 144–50

Krupnik, I. (1993) *Arctic adaptations: native whalers and reindeer herders of Northern Eurasia*, Hanover: University Press of New England

Krupnik, I. and Vakhtin, N. (1997) Indigenous knowledge in modern culture: Siberian Yupik ecological legacy in transition, *Arctic Anthropology* 43(1): 236–52

Ksenofontov, G. V. (1930) *Legendy i rasskazy o shamanakh u yakutov, buryat i tungusov* [Legends and stories about shamans among the Yakut, Buryat, and Tungus], Moscow: Bezbozhnik

Kwon, H. (1993) Maps and actions: nomadic and sedentary space in a Siberian collective farm, PhD thesis, University of Cambridge

Lashov, B. (1973) *Nekotoryye voprosy razvitiya natsional'nykh rayonov Kraynego Severa* [Some questions of the development of ethnic regions of the Far North], Yakutsk: Yakutskoye Knizhnoye Izdatel'stvo

Levin, M. G. and Potapov, L. P. (1961) *Istoriko-etnograficheskiy atlas Sibiri* [Atlas of historical ethnography of Siberia], Moscow and Leningrad: Institut Etnografii AN SSSR

Levin, M. and Potapov, L. (1964) *The peoples of Siberia*, Chicago: University of Chicago (Russian original 1956)

McCannon, J. (1998) *Red Arctic: polar exploration and the myth of the North in the Soviet Union*, Oxford: Oxford University Press

Michael, H. (ed.) (1963) *Studies in Siberian shamanism*, Toronto: University of Toronto Press

Mills, A. and Slobodin, R. (eds) (1994) *Amerindian rebirth: reincarnation belief among North American Indians and Inuit*, Toronto: University of Toronto Press

Molodin, V. I. (2000) The burial complex of the Pazyryk culture: Verkh Khaldin I and the problem of modern climate warming, *Archaeology, Ethnology and Anthropology of Eurasia*, 2000(1): 101–8

Nerkagi, A. (1982) Aniko and the Nogo clan, in Burkova, I. (ed.), *Blizok Krayniy Sever: sbornik proizvedeniy molodykh pisateley narodnostey Severa i Dal'nego Vostoka*, [The Far North is close at hand: a collection of work by young writers from the peoples of the North and the Far East], Moscow: Sovremennik, pp. 11–106

—— (1988) Letter in the magazine *Severnyye Prostory* 1988(3): 25–6

Novikova, N. (ed.) (2004) *Olen' vsegda prav: issledovaniya po yuridicheskoy antropologii* [The reindeer is always right: studies in legal anthropology], Moscow: Strategiya

Oakes, J. and Riewe, R. (1998) *Spirit of Siberia: traditional native life, clothing and footwear*, Washington DC: Smithsonian Institution Press

Paine, R. (1994) *Herds of the tundra: a portrait of Saami reindeer pastoralism*, Washington: Smithsonian Institution

Pika, A. (1993) The spatial-temporal dynamic of violent death among the native peoples of Northern Russia, *Arctic Anthropology* 30(2): 61–76

Pika, A. and Bogoyavlensky, D. (1995) Yamal Peninsula: oil and gas development and problems of demography and health among indigenous populations, *Arctic Anthropology* 32(2): 61–74

Polo, M. (approx. 1298) *The travels* (also known as 'Il milione' and 'The marvels of the world' – in Old French: many translations). References are to edition by L. F. Benedetto: English version entitled *The travels of Marco Polo*, London: Routledge, 1931; Italian version entitled *Il libro di Messer Marco Polo*, Milan and Rome: Fratelli Treves, 1932

Polosmak, N. V. (2000) Tattoos in the Pazyryk world, *Archaeology, Ethnology and Anthropology of Eurasia*, 2000(4): 95–102

Polosmak, N. V. and Molodin, V. I. (2000) Grave sites of the Pazyryk culture on the Ukok plateau, *Archaeology, Ethnology and Anthropology of Eurasia*, 2000(4): 66–87

Pomishin, S. B. (1990) *Proiskhozhdeniye olenevodstva i domestikatsiya severnogo olenya* [The origin of reindeer herding and the domestication of the reindeer], Moscow: Nauka

Popov, A. (1966) *The Ngansan: the material culture of the Tavgi*

Samoyeds, The Hague: Mouton and Bloomington: Indiana University Uralic and Altaic Series, vol. 56 (Russian original 1948)

Prokof'yeva, Ye. D. (1981) Materialy po shamanstvu sel'kupov [Materials on the shamanism of the Sel'kup], in Anon. (ed.), *Problemy istorii obschestvennogo soznaniya aborigenov Sibiri* [Problems in the history of social consciousness among the aborigines of Siberia], Leningrad: Nauka, pp. 42–68

Pushkin, A. S. (1836) 'Ya pamyatnik sebe vozdvig ne rukotvorny' [I have raised myself a miraculous monument] (in Russian: numerous translations in collections of his poems)

Rethmann, P. (2001) *Tundra passages: history and gender in the Russian Far East*, University Park, Pennsylvania: Pennsylvania State University Press

Rudenko, S. (1970) *Frozen tombs of Siberia: the Pazyryk burials of iron-age horsemen*, London: J. M. Dent (Russian original 1953)

Savinov, D. G. (1994) *Olenyye kamni v kul'ture kochevnikov Yevrazii* [Deer stones in the culture of the nomads of Eurasia], St Petersburg: Izdatel'stvo Sankt-Peterburgskogo Universiteta

Sirina, A. (1999) *Rodovyye obschchiny malochislennykh narodov Severa v Respublike Sakha (Yakutiya): shag k samoopredeleniyu?* [Kin-based communes of the minority peoples of the North in the Republic of Sakha (Yakutia): a step towards self-determination?], Moscow: Institut Etnologiya i Antropologiya, Studies in Applied and Urgent Ethnology, No. 126

Skalon, V. N. (1956) Olennyye kamni Mongolii i problema proiskhozhdeniya olenevodstva [Reindeer stones of Mongolia and the problem of the origin of reindeer herding], *Sovetskaya Arkheologiya*, vol. 25, pp. 87–105

Skultans, V. (1998) *The testimony of lives: narrative and memory in post-Soviet Latvia*, London and New York: Routledge

Slezkine, Y. (1994) *Arctic mirrors: Russia and the small peoples of the North*, Ithaca, NY and London: Cornell University Press

Sotavalta, A. and Halén, H. (1978) *Westlamutische Materialien* [West-Lamut materials], Helsinki: Suomalais-Ugrilainen Seuran Toimituksia, vol. 168

Spiess, A. E. (1979) *Reindeer and caribou hunters: an archaeological study*, New York: Academic Press

Ssorin-Chaikov, N. (2003) *The social life of the state in subarctic Siberia*, Stanford: Stanford University Press

Stammler, F. (2004) When reindeer nomads meet the market: culture, property and globalisation at the End of the Land [i.e. Yamal Peninsula], PhD thesis, Halle (Germany): Max Planck Institute for Social Anthropology

Syroechkovskiy, E. (1995) *Wild reindeer*, New Hampshire: Science Publishers (Russian original 1986)

Thompson, N. S. (2003a) Migration and resettlement in Chukotka: a research note, *Eurasian Geography & Economics* (formerly *Post-Soviet Geography*), 45(1): 73–81

—— (2003b) Administrative resettlement and the pursuit of economy: the case of Chukotka, *Polar Geography*, 26(4): 270–88

Tichotsky, J. (2000) *Russia's diamond colony: the Republic of Sakha*, Amsterdam: Harwood Academic

Tolstov, S. P. (1961) Nekotoryye problemy vsemirnoy istorii v svete dannykh sovremennoy istoricheskoy etnografii [Some problems of world history in the light of the data of contemporary historical ethnography], *Voprosy istorii*, 11

Tugolukov, V. (1985) *Tungusy: evenki i eveny sredney i zapadnoy Sibiri* [The Tungus: Evenki and Even of central and western Siberia], Moscow: Nauka

Turayev, V. A. et al. (1997) *Istoriya i kul'tura evenov* [History and culture of the Even], St Petersburg: Nauka

Vainshtein, S. I. (1970–71) Problema proiskhozhdeniya olenevodstva v Yevrazii [The problem of the origin of reindeer herding in Eurasia], Part I, *Sovetskaya Etnografiya* 1970(6): 3–14; Part II, *Sovetskaya Etnografiya* 1971(5): 37–52

—— (1980) *Nomads of South Siberia: the pastoral economies of Tuva*, Cambridge: Cambridge University Press (Russian original 1972)

Vakhtin, N. B. (1993) Towards a typology of language situations in the far North, *Anthropology and Archeology of Eurasia: A Journal of Translations*, 32(1): 66–92

—— (1994) Native peoples of the Russian far North, in Anon. (ed.), *Polar peoples: self-determination and development*, London: Minority Rights Group, pp. 29–80 (first published as a separate booklet in 1992)

—— (2001) *Yazyki narodov severa v XX veke: ocherki yazykovogo sdviga* [Languages of the peoples of the North in the 20th century: studies in linguistic displacement], St Petersburg: Evropeyskiy Universitet Sankt-Peterburga

Vakhtin, N. B., Golovko, E. and Schweitzer, P. (2004) *Russkiye starozhily Sibiri: sotsial'nyye i simvolicheskiye aspekty samosoznaniya* [Russian Old Settlers of Siberia: social and symbolic aspects of consciousness], Moscow: Novoe Izdatel'stvo

van der Knaap, W. O. (1986) On the presence of reindeer (*Rangifer tarandus* L.) on Edgeøya, Spitsbergen, in the period 3800–5000 B.P., *Circumpolar Journal* (2): 3–9

Vasil'yevich, G. M. (1964) *Tipy olenevodstva u tungusoyazychnykh narodov* [Types of reindeer herding among the Tungus-speaking peoples], Moscow: Nauka

—— (1969) *Evenki: Istoriko-etnograficheskie ocherki (xviii–nachalo xx veka)* [The Evenki: historical-ethnographic essays (18th–early 20th centuries)], Leningrad: Nauka

Vasil'yevich, G. M. and Levin, M. G. (1951) Tipy olenevodstva i ikh proiskhozhdeniya [Types of reindeer herding and their origins], *Sovietskaya Etnografiya* (1): 63–87

Ventsel, A. (2005) *Reindeer, rodina and reciprocity: kinship and property relations in a Siberian village*, PhD thesis, Halle (Germany): Max Planck Institute for Social Anthropology

Vitebsky, P. (1990a) Piers Vitebsky, interview in Dinnage, R. (ed.), *The Ruffian on the Stair: Reflections on Death*, London: Viking, pp. 38–52 (reprinted Harmondsworth: Penguin, 1992)

—— (1990b) Yakut, in Smith, G. (ed.), *The nationalities in the Soviet Union*, London: Longmans, pp. 302–17

—— (1992) Landscape and self-determination among the Eveny: the political environment of Siberian reindeer herders today, in Croll, E. and Parkin, D. (eds), *Bush base, forest farm: culture, environment and development*, London and New York: Routledge, pp. 223–46

—— (1993) *Dialogues with the dead: the discussion of mortality among the Sora of eastern India*, Cambridge and New York: Cambridge University Press (reprinted Delhi: Foundation Books, 1993)

—— (1995a) *The Shaman*, London: Macmillan, and Boston: Little Brown (reprinted London: Duncan Baird, 2001; reprinted as *Shamanism*, Norman: University of Oklahoma Press, 2001)

—— (1995b) From cosmology to environmentalism: shamanism as local knowledge in a global setting, in Fardon, R. (ed.), *Counterwork: managing diverse knowledges*, London and New York: Routledge, pp. 182–203

—— (1996) The Northern Minorities, in Smith, G. (ed.), *The*

Nationalities question in the post-Soviet states, London: Longman, pp. 94–112

—— (2000) *Coping with distance: social, economic and environmental change in the Sakha Republic (Yakutia), Northeast Siberia*, Cambridge: Scott Polar Research Institute for Royal Geographical Society and Gilchrist Educational Trust

—— (2002) Withdrawing from the land: social and spiritual crisis in the indigenous Russian Arctic, in Hann, C. (ed.), *Postsocialism: ideals, ideologies and practices in Eurasia*, London and New York: Routledge, pp. 180–95

Vitebsky, P. and Wolfe, S. (2001) The separation of the sexes among Siberian reindeer herders, in Tremayne, S. and Low, A. (eds), *'Sacred custodians' of the earth? Women, spirituality and the environment*, New York and Oxford: Berghahn, pp. 81–94

Volkov, V. V. (1981) *Oleniye kamni Mongolii* [Reindeer stones of Mongolia], Ulaan Baatar (Ulan Bator): Nauchnyy mir

Vorren, Ø. and Manker, E. (1962) *Lapp life and customs: a survey*, Oxford: Oxford University Press

Willerslev, R. (2002) In-between self and other: hunting, personhood and perception among the Upper Kolyma Yukaghirs of north-eastern Siberia, PhD thesis, University of Cambridge

Wilson, E. (2003) Freedom and loss in a human landscape: multinational oil exploitation and the survival of reindeer herding in northeastern Sakhalin, the Russian Far East, *Sibirica: Journal of Siberian Studies* 3(1): 21–47

Witsen, N. (1672) *Noord en Oost Tartaryen* [North and East Tartaria], Amsterdam: M. Schalekamp

Yudin, A. M. (1993) *Panty i antlery: roga kak lekarstvennoye syr'ye* [Velvet and dry antlers: horn as medicinal material], Novosibirsk: Nauka

INDEX